Elsevier *Global Energy Policy and Economics* Series

China's Oil Industry & Market

Haijiang Henry Wang

Titles of related interest

Books

Petroleum Exploration and Exploitation in Norway
Sedimentary Basins and Petroleum Geology of the Middle East
Applied Geothermics for Petroleum Engineers
Computing Risk for Oil Prospects: Principles and Programs
Hydrocarbon Exploration & Production
Hydrotreatment and Hydrocracking of Oil Fractions
Quantification and Prediction of Hydrocarbon Resources
The Practice of Reservoir Engineering

Journals *(free sample copy sent on request)*

Energy
Energy Economics
Energy Policy
Fuel
J. Petroleum Science and Engineering
Marine and Petroleum Geology
Resource and Energy Economics
Resources Policy
Utilities Policy

For further details please contact your nearest Elsevier Regional Sales Office – see back of this book.

Elsevier *Global Energy Policy and Economics* Series

China's Oil Industry & Market

Haijiang Henry Wang

1999

ELSEVIER

Amsterdam – Lausanne – New York – Oxford – Shannon – Singapore – Tokyo

ELSEVIER SCIENCE Ltd
The Boulevard, Langford Lane
Kidlington, Oxford OX5 1GB, UK

First edition 1999.
British Library Cataloguing in Publication Data
A catalogue record from the British Library has been applied for.

China's oil industry & market / by Haijiang Henry Wang.
p. cm. — (Elsevier global energy policy and economics series)
Includes bibliographical references and index.
ISBN 0-08-043005-8 (hc.)
Petroleum industry & trade — China. I. Wang, Haijiang Henry.
II. Title: China's oil industry & market. III. Series.
HD9576.C52C5235 1999
338.2'728'0951–dc21 99-21565
CIP

ISBN: 0 08 043005 8

The paper used in this publication meets the requirements of ANSI/NISO Z39.48-1992 (Permanence of Paper).

Printed in The Netherlands.

To my wife, mother, and daughter

Foreword

The authority of this book on China's oil demand and government policies and practices rests on two essential foundations: first and foremost on the author's informed background of China's oil situation and prospects and his access to Chinese energy literature and data; secondly, on that insight afforded him and therefore his readers from his fluency in Chinese — an asset not always possessed by Western analysts, who have to rely on someone else's "interpretation".

The author analyzes the Chinese oil market and the rising pressure on Beijing to reform policies which constrain China's ability to meet soaring demand and to pay for crucial imports in a period of growing political and economic uncertainties: who is to succeed Deng with what tensions? Which policies will be retained, which will be reformed? These are crucial unknowns.

Dr. Wang acknowledges the importance that China meets its growing domestic oil demand, if at all possible, through national production. The strategic significance to China's regional neighbors (and to the US as well) of China's efforts to obtain foreign supplies are familiar issues to other importers elsewhere. But the sheer weight of China's population and their burgeoning requirements as industrialization spreads into most regions dwarf the needs of others and will place unprecedented strains on international oil trades.

What role will the state allow Chinese companies and investors to play? What role will foreign companies and investors play in the light of the huge sums needed to build infrastructure and to pay for domestic exploration? How will China welcome these players and how durable will government policies prove to be as leaders contend for political control? Dr. Wang stresses the fact that the outcome is hard to define yet the time required to tackle the nation's energy needs is not limitless. Moreover, Dr. Wang reminds the reader of the perennial difficulty in meeting widely disparate economic and energy needs in different regions of the vast country.

This book deserves careful reading and thought both within and outside of China. It deals with exceptionally important issues.

Melvin A. Conant
Washington, DC

Preface

The economic awakening of the People's Republic of China (hereafter referred to as China) began with Deng Xiaoping's accession in 1979, but it is only now beginning to have an impact on the world oil market and industry. Like most developing countries, China chose to liberalize other segments of its economy first, while keeping oil under strict government control. And unlike many developing or industrialized countries, China had the geological resources to remain self-sufficient in oil despite strong growth in demand. But now that China has reached the limit of its present production capabilities, the government appears to understand that it cannot develop an efficient oil sector without bringing in foreign investment and new technology.

In 1993, China became a net oil importer. Not coincidentally, in 1994, it also began to initiate important changes in its oil policies, ranging from inviting foreign firms to explore the Tarim Basin to raising many of its internal oil prices. China's petroleum industry is in a period of profound transition. Through its policy of opening up to the international market, China is acquiring capital and more sophisticated technology. This book focuses on China's oil industry and market: its history, current development, and prospects. It addresses a number of critical questions: How fast will oil demand and supply grow in the "brave new world" of a modernizing China? How fast will refining capacity expand in China? Will the refining capacity in China over the next decade be adequate to meet the country's thirst for fuels? Where will China's refiners get their crude supply? How much will have to be obtained from outside the region? How will China's oil situation affect the dynamic Asian oil scene and Singapore margins? How will that affect world markets?

Most past studies of oil demand and the Asia-Pacific oil market have reluctantly excluded China due to a lack of information. Now, the China factor has become impossible to ignore when examining oil on a regional or even global basis. Improved availability of statistics/data have made conditions better than ever for focused study. *China's Oil Industry and Market* is an in-depth appraisal of China's oil outlook. Oil producing and refining companies with high fixed costs require a long-term outlook that develops a view of the oil market over the next 10 years before making costly investment decisions. The Chinese government also needs policy strategies that will lead to long-term political stability and economic growth. Consequently, forecasting the long-term supply and demand balance of oil, while still vulnerable to the pitfalls of forecasting, is not an exercise in futility. It engages us in a thoughtful process to organize what we know and think about China into a coherent view of the future. This book is aimed at governments; at oil producing, refining, and trading firms; and at associated service companies, shipping companies, and banks.

The Framework of this Book

This book provides a specific outlook on the next 10 years that is based in part on a quantitative, econometric analysis of the supply and demand of oil, and in part on a broad conceptual framework for thinking about the political, financial and regulatory factors shaping the production and consumption of oil in China.

This book sees China's oil industry and market evolution through the lens of economic liberalization and political conservatism. Our emphasis builds from a few "facts of life" in today's — and tomorrow's — China. The book is organized as it is illustrated in Fig. P.1. First of all, politics and policy are the foundation of overall economic growth, as well as the development of the oil industry. Second, an analysis of the political structure and the process of liberating the economy from its regulatory shackles is important, since these two factors are the predominant determinants of the pace of petroleum product demand, refinery expansion, and petroleum product supplies. As we saw in 1994, product demand in China is particularly sensitive to government policy. Third, refining capacity expansion and crude exploration are a function of investment and oil industry policy, which are linked to foreign policy and macroeconomic strategy. Fourth, the Chinese oil industry will, by its very growth, create opportunities and challenges with which the Chinese government must deal.

Finally, we believe that the crude and petroleum product "balances" that we present in this book reveal the source and magnitude of issues that China will face in the coming decade. For the purposes of this book, the *crude balance* is a comparison of domestic crude production (actual output) and consumption (refinery throughput plus crude direct burn). The petroleum *product balance* is based on a simple comparison of product demand and refinery "producibility" for gasoline/naphtha, middle distillates, and fuel oil. The changes in China's crude balance will influence and be influenced by the world oil market, while its product balance will have a strong impact on Asian refining and import/export markets.

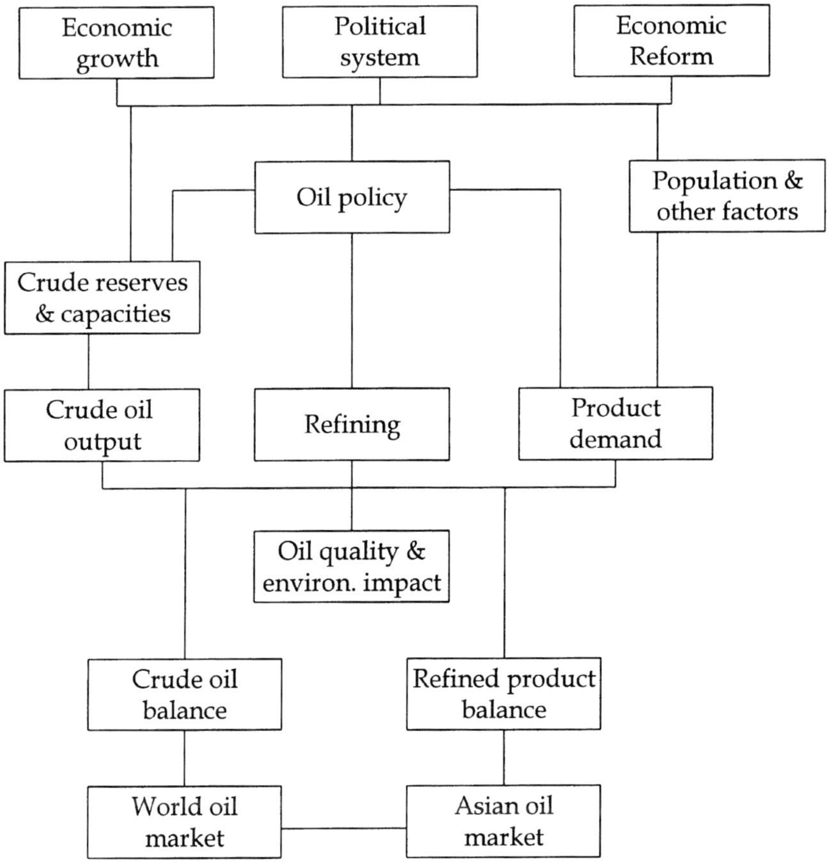

Fig.P.1 Framework of this book

What's New in this Book?

China Savvy: We present a realistic picture of overall Chinese politics, current energy policy, oil industry regulation, price reregulation, import licenses and quotas, the product distribution system, infrastructure, and the impact of economic reform on the oil business.

Forecasts of Demand: The book contains product-specific demand forecasts (gasoline, diesel oil, kerosene, jet fuel, naphtha, and fuel oil), forecasts by province for gasoline and diesel oil, and a total petroleum product demand forecast based on principal product demand projections. Methodologies used for demand forecasts include econometrics, country comparative analysis, and historical comparative analysis.

Upstream Outlook: We include a crude production overview and outlook by oilfield. Two methods are applied for crude production projection: (1) a "top down" approach, which relies on Chinese-sourced proven reserves data and adjusted R/P ratios; and (2) a "bottom up" approach, with details on each oilfield.

Particular Attention to the Tarim Basin: Prospects and problems for Tarim Basin oil exploration are presented and explored.

Crude Quality: The book provides an analysis of China's crude quality, based on field by field crude quality information from the China National Petroleum Corporation.

Refining: We show total Chinese capacity by refining process, and list new refinery projects we believe are likely to be completed, with information (where available) on location, capacity, investors, initial capital, and scheduled completion dates. Product supply forecasts are built on the basis of expected new refining capacity.

Trade Balances: The book presents crude and product balances, including net crude imports, and product imports/exports over the next decade. It also lists current sources of imports and destinations of exports for crude and products.

Broad Market Impact: The book details China's probable impact on Asian oil markets: crude deficits, refining capacity, and prices.

Detailed Information: The book provides Chinese oil data and information, some of which have never been published outside of China.

Acknowledgments

This book is a rewritten and updated version of a multi-client study conducted by author in Energy Security Analysis, Inc. (ESAI) in 1994. I am grateful to Dr. Edward Krapels, Sarah A. Emerson, Monica M. Ferreri, Raoul Leblanc, George Beranek, Robert C. McNally, Steve Fine, and Paul B. Phelps at ESAI for their helpful comments and assistance.

I would like to thank the invaluable information and assistance received during the writing of this book from Zhang Shu (China Statistic Bureau), Tian Dadi (China Petroleum Information Institute, CNPC), and Sun Yongsheng (China Petrochemical Information Institute, Sinopec).

Finally, while many of the data presented in this book are based on official Chinese sources, the author is solely responsible for both the data and the analyses that follow.

Haijiang Henry Wang
Maryland

ACRONYMS AND INITIALS

ASEAN	–	Association of Southeast Asian Nations
ANS	–	Alaska North Slope
BFET	–	Bureau of Foreign Economic and Trade
CNODC	–	China National Oil and Gas Development Corporation
CNOOC	–	China National Offshore Oil Corporation
CNPC	–	China National Petroleum Corporation
COEs	–	Collective Ownership Enterprises
CONHE	–	China Offshore Nanhai East Corporation
CPC	–	The Chinese Petroleum Corporation
CPEs	–	Centrally planned economy
CRS	–	Contract responsibility system
E&D	–	Exploration and development
EIT	–	Enterprise income tax
ESZ	–	Economic special zone
ETC	–	State Economic and Trade Commission
FETM	–	Foreign Economic and Trade Ministry
FSU	–	The former Soviet Union
GATT	–	General Agreement on Tariffs and Trade
GDP	–	Gross domestic product
GNP	–	Gross national product
IEA	–	International Energy Agency
MFET	–	Foreign Economic and Trade Ministry
MFN	–	Most-favored-nation status
MGMR	–	Ministry of Geology and Mineral Resources
MOF	–	The Ministry of Finance
MOPI	–	The Ministry of Petroleum Industry
MTBE	–	Methyl Tertiary Butyl Ether
NIC	–	Newly industrialized countries
OECD	–	Organization for Economic Co-operation and Development
OPEC	–	Organization of Petroleum Exporting Countries
PBC	–	The People's Bank of China
PIT	–	Personal income tax
PLA	–	People's Liberation Army
PPCS	–	The Petroleum and Petrochemical Company of Shanghai
SEZ	–	Special economic zone
Sinochem	–	China National Chemcials Import and Export Corporation
Sinopec	–	China Petrochemical Corporation
SOEs	–	State-owned Enterprises
SPC	–	State Planning Commission
SPDC	–	State Planning and Development Commission
TCE	–	Ton of coal equivalent
ULCC	–	Ultra large cargo carriers
Unipec	–	China International United Petroleum and Chemicals Corporation

VAT	–	Value added tax
VLCC	–	Very large cargo carriers
WTI	–	West Texas Intermediate
WTO	–	World Trade Organization

CONVERSION FACTORS

Barrels per metric ton

Crude Oil:
Crude (excluding Shengli) in China = 7.3
Crude in Shengli Oilfield = 7.35
Crude imported from Yemen, Papua New Guinea, and Russia into China = 7.35
Crude Imported from Other Countries into China = 7.3

Products:
Asphalt = 6.06
Diesel oil = 7.46
Hydrocarbon solvents = 7.00
Kerosene = 7.46
Jet fuel = 7.93
Light oil for chemical industry = 8.50
Liquefied petroleum gas (LPG) = 11.60
Lubricants = 7.00
Motor gasoline = 8.50
Naphtha = 8.50
Paraffin wax = 7.87
Petroleum coke = 5.50
Residual fuel oil = 6.66

Contents

Table of Tables

Table of Figures

Chapter 1
Economic and Political Outlook

The future of global oil demand growth centers on the Asia/Pacific region, which in turn hinges on developments in China. Strong, sustained economic growth could ultimately turn China into the region's preeminent economic and political power, transforming China's political and strategic role in global affairs.

This change will have significant consequences for global energy and oil markets, where China is already a giant. China is the third largest producer of energy in the world. In 1997, the country's primary commercial energy supply and consumption amounted to 1.33 and 1.44 billion tons of standard coal equivalents (TCE).[1] China's oil resources are enormous. In 1997, China ranked sixth in crude production, fourth in refinery throughput, and fifth in total petroleum product consumption. With 1.2 billion people, however, China ranks much lower in per capita energy production and consumption. But China's size and rate of growth promise to have a dramatic impact on energy — and oil — production and consumption trends in the region.

This chapter addresses two of the most important assumptions underlying any analysis for the oil industry and market: economic growth and political stability. It evaluates economic and political trends and assesses their direction and scopes over the next 10 years. It also describes and assesses certain reforms and transformations underway in China's oil industry. Most importantly, it draws general conclusions about how economic and political events may change China's oil industry.

1.1 Economic Performance: Overview and Outlook

China has achieved sustained and high rates of economic growth over the past 15 years. This growth has been distributed unequally among

[1] Tons of standard coal equivalent (TCE) is an energy unit in Chinese practice, defined as 7 million kilocalories. The applied conversion factors are roughly 0.7 for coal, 1.4 for crude oil, 0.35 for electric power, and 1.2 for natural gas.

regions and across economic sectors, however. The rate and distribution of economic growth are key variables that will determine how economic change will alter oil consumption and production.

According to the World Bank,[2] China's GNP in 1996 totaled US$ 906 billion (excluding Hong Kong), which surpassed the GNP in Canada and Spain, and was close to the GNP in Italy and the UK, making China the seventh biggest GNP country in the world, following the US, Japan, Germany, France, Italy, and the UK. The same source from the World Bank also revealed that China ranked third in annual growth rate of GNP per capita from 1990 to 1996, showing 11.0 percent, only lower than Equatorial Guinea's 15.9 percent and Kuwait's 15.7 percent.

China achieved economic growth of about 9 percent per year through the 1980s. Since 1991, the Chinese economy has grown at a double-digit rate ranging from 10 percent to 14 percent. No other major economic power has been able to sustain such growth. Economic growth rates in individual provinces over the last decade are mixed because the regions are so varied. The economic regions in China can be divided into three groups: (1) the rapidly growing coastal regions in the southern and eastern provinces; (2) the poorer regions in western China; (3) the internal provinces between them.

The economic surge stemmed from Deng Xiaoping's strategy of "growth-at-all-cost". The predominant view was that China, as a late-coming developing country, must consider modernization as its fundamental goal and must adopt the policy of high-speed development to catch up with the advanced countries in the West. The fast-growth policy favors several major provinces in the southern and eastern coast because the economic conditions there are considerably better than those in the inland provinces. In the future, as in the past, a high growth rate can be maintained only if the coastal areas continue to develop much faster than other, poorer regions. Coastal provinces will be the "development engine" leading China's economic take-off. There are several factors behind the rapid economic growth of the coastal regions. First, coastal areas received preferential treatment in conducting economic reforms and were the first regions to open themselves to the outside when those economic reforms started in 1978. For instance, the first 14 cities and four special economic zones which were set up by the central government to serve as models of economic reform and openness are located mostly in coastal areas. Special economic zones have privileges such as the right to offer foreign companies tax exemptions of up to five years and reduced taxes thereafter.

---

[2] See *World Development Report, 1998,* the World Bank.

Second, these coastal regions are geographically closer to international markets and have historical links with the outside world. They have easy access to foreign technology, investment funds, and management skill. Third, before the 1960s, fearing aggression from the US and Taiwan, China located only a small part of the country's state-owned industries in the southern coastal regions. As a result, these regions have relatively fewer old, inefficient and indebted state-owned firms to reform. Finally, the coastal regions have traditional collective and township (nonstate-owned) enterprises in rural areas and towns, which are easily developed with economic liberalization.[3] This market-oriented growth was most pronounced in the collective and private enterprises, which emerged as the most dynamic sector of the Chinese economy.

Market-oriented growth in the coastal provinces dominated the Chinese economy in both the 1980s and 1990s, while inland regions were undergoing the least change. In 1996, GDP grew by 19.8 percent in Guangdong, 16.3 percent in Fujian, 12.6 percent in Jiangsu, 14.3 percent in Shandong, and 11.4 percent in Zhejiang, while national GDP growth averaged 9.9 percent. Despite rapid economic development, oil use in these coastal provinces remained low relative to the national average. In 1996, for example, annual gasoline consumption per capita was 8.9 gallons in Guangdong, 5.1 in Fujian, 9.6 in Jiangsu, 8.8 in Shandong, and 8.2 in Zhejiang in 1996, but 9.5 in the country. 1996 annual diesel oil consumption per capita was also lower in coastal provinces than the national average — 6.5 gallons per capita per year in Fujian, 9.9 gallons in Jiangsu, 8.9 gallons in Shandong, and 7.0 gallons in Zhejiang in 1996, compared with 10.5 gallons per capita per year for the nation — except in Guangdong Province (25.5 gallons). In the future, therefore, oil demand and oil intensity will continue to grow faster in the east and south coast provinces.

Growth has also been uneven across sectors of the economy. During the past 10 years, the annual rate of real economic growth in the industrial sector (including transportation) averaged about 12.2 percent. Agricultural growth, in contrast, was much lower in the early 1990s (less than 3.2 percent) than in the 1980s (6.2 percent). Growth in the service sector also fell from more than 10.8 percent in the 1980s to 2.5 percent in the early 1990s. In China, the industrial sector uses oil more intensively than any other sector, consuming 28 percent of total diesel oil and 49 percent of total fuel oil in 1995, while accounting for about 34 percent of total GDP. In contrast, the agricultural sector consumed 23 percent of total diesel oil and about 1 percent of kerosene in 1995, but contributed about

[3] See William A. Byrd and Liu Qingsong (eds.), *China's Rural Industry: Structure, Development, and Reform*, New York: Oxford University Press.

27 percent of total GDP. The commercial and residential sector has increased its share in total GDP from 28 percent in 1985 to 31 percent in 1995. The expectation is that the share of GDP contributed by the industrial and service sectors will continue to increase, while agriculture's share of GDP will decline. As the Chinese economy continues to develop, however, the growing oil intensity of the industrial sector will be offset to some degree by the lower rate of oil demand in the expanding service sector.

1.1.1 Historic ratios of economic, oil consumption and production growth

The energy sector plays an essential role for economic development and systemic economic transformation from a strictly planned regime to a market-oriented economy.

Table 1.1 presents Chinese economic, energy and oil growth over the past 20 years. China's economic growth has been uneven, exhibiting a "stop–go" cycle: a peak in economic growth always followed by consequent economic sluggishness leading to a deflationary trough. But despite these swings, the Chinese economy, on average, has still sustained high economic growth for the past two decades.

China's GDP more than doubled between 1993 and 1997, from 3.5 trillion yuan to 7.5 trillion yuan in normal terms, growing at an average rate of 11.3 percent in real terms. During this period, energy use grew from 1.11 billion tons TCE to 1.44 billion TCE (5.7 percent per annum), while total petroleum product consumption grew even a little faster, from 2.8 million b/d to 3.5 million b/d (5.8 percent per annum). However, the link between energy consumption and GDP is not stable over time and can fluctuate in the short run, as shown by computing energy intensities (that is, the amount of energy needed to produce one unit of GDP).

The elasticities of growth in total energy consumption and production relative to growth in GDP were quite low in China during the past decade compared with other Asian industrializing countries, just 0.5 compared with 0.9 in South Korea, Thailand, and Indonesia, and 1.0 in India. The specific elasticities were 0.3 for crude oil production and 0.5 for petroleum product consumption respectively since 1993. China's low oil/GDP growth elasticity during recent years represents a sharp break from the trend of previous years, before the economic reform, when some oil industry success in the Daqing oilfield created overly optimistic expectations about China's oil reserves and ability to retrieve physical barrels. Nevertheless, oil consumption has increased by 5.8 percent per year during the years from 1993 to 1997, and crude production and refining have lagged behind demand. As a result, China's

Table 1.1 Growth of GDP, energy production, energy consumption, crude oil production, and petroleum product consumption in China

	Growth rates					Elasticities			
	GDP	Energy		Oil		Energy		Oil	
		Production	Consumption	Production	Consumption	Production	Consumption	Production	Consumption
1971	7.6	13.9	17.8	28.6	28.4	1.8	2.3	3.8	3.7
1972	3.5	7.1	8.1	15.9	16	2.0	2.3	4.5	4.0
1973	8.3	5.9	4.9	17.4	13.5	0.7	0.6	2.1	1.6
1974	1.1	4.0	2.6	21.0	14.5	3.7	2.4	19.1	13.2
1975	8.3	17.1	13.2	18.8	15.1	2.1	1.6	2.3	1.8
1976	–5.4	3.3	5.3	13.1	14.9	–0.6	–1.0	–2.4	–2.8
1977	7.9	12.0	9.5	7.4	7.6	1.5	1.2	0.9	1.0
1978	11.7	11.3	9.1	11.1	9.7	1.0	0.8	0.9	0.8
1979	7.6	2.9	2.5	2.0	–1.7	0.4	0.3	0.3	–0.2
1980	7.8	–1.3	2.9	–0.3	–2.0	–0.2	0.4	0.0	–0.3
1981	5.2	–0.8	–1.4	–4.5	–5.2	–0.2	–0.3	–0.9	–1.0
1982	9.1	5.6	5.4	0.9	–0.2	0.6	0.6	0.1	0.0
1983	10.9	6.7	5.4	3.7	1.1	0.6	0.5	0.3	0.1
1984	15.2	9.2	7.4	8.2	3.3	0.6	0.5	0.5	0.2
1985	13.5	9.9	8.1	9.0	6.0	0.7	0.6	0.7	0.4
1986	8.8	3.0	5.4	4.6	8.7	0.3	0.6	0.5	1.0
1987	11.6	3.6	7.2	2.6	4.4	0.3	0.6	0.2	0.4
1988	11.3	5.0	7.3	1.9	3.4	0.4	0.7	0.2	0.3
1989	4.1	6.1	4.2	0.8	5.1	1.5	1.0	0.2	1.3
1990	3.8	2.2	1.8	0.5	–0.2	0.6	0.5	0.1	0.0
1991	9.2	0.9	5.1	1.9	6.5	0.1	0.6	0.2	0.7
1992	14.2	2.3	5.2	0.3	5.5	0.2	0.4	0.0	0.4
1993	13.5	3.5	6.2	2.5	7.6	0.3	0.5	0.2	0.6
1994	12.6	6.9	5.8	1.0	–0.7	0.5	0.5	0.1	–0.1
1995	10.5	8.7	6.9	1.3	6.5	0.8	0.7	0.1	0.6
1996	9.7	2.8	5.9	6.0	6.8	0.3	0.6	0.6	0.7
1997	10.1	0.7	3.6	3.2	8.7	0.1	0.4	0.3	0.9
71–80 avg	5.8	7.6	7.6	13.5	11.6	1.2	1.1	3.1	2.3
81–90 avg	9.4	6.1	5.1	2.8	2.7	0.6	0.5	0.2	0.3
91–97 avg	11.4	3.7	5.5	2.3	5.8	0.3	0.5	0.2	0.5

Notes:
(1) Commercial energy includes coal, natural gas, oil and hydro power.
(2) Total oil consumption includes total petroleum product consumption and crude direct burn.

Sources: *China Statistical Yearbook, 1997*, and *A Statistical Survey of China, 1998*, China Statistical Publishing House, State Statistical Bureau, and author's calculation.

ability to export crude has been falling, and its need for imports has been growing.[4]

1.1.2 Pace of macroeconomic reform and its impact on the oil industry and market

China has had a centrally planned economy for more than four decades, but before the late 1970s, it produced poor economic results. In 1979, the central government revamped its economic strategy to improve efficiency and promote the growth of the economy. The new strategy called for limited decentralization and the introduction of market elements into the planned economy. Since the early 1980s, the Chinese economy has been moving toward greater decentralization and an increased reliance on market forces. In 1992, China adopted the concept of a "socialist market economy" and gave a green light to virtually all market mechanisms and techniques used in a genuine capitalist system, giving its population a great deal of freedom to set up private businesses.[5]

The economic reforms focused on three strategies: (1) rearranging the economic balance between state-owned enterprises and the government, and between the local government and the central government; (2) attracting foreign investment; and (3) private sector development (not privatization). These strategies have produced three significant changes:

1. Nonstate enterprises, including joint ventures with overseas investors, collectively owned,[6] and private investors, are increasingly surpassing their state-owned counterparts in economic output.
2. Decision-making at the provincial/local level is becoming much more independent from the central government.
3. The Chinese economy is increasingly integrated into the world economy.

The performance of macroeconomic reforms can be measured in many ways, including output, growth, efficiency, stability, security, inequality,

---

[4] It was reported that, initiated by the National People's Congress, China National Petroleum Corporation is drafting the country's first petroleum law, and was to submit it to the State Council in 1995. This petroleum law is expected to protect the limited oil resources and to use them efficiently.

[5] In 1992, the Chinese Communist Party's 14th National Congress advocated introduction of a market economy and acknowledged that the market provides the most effective mechanism for distributing resources. See *People's Daily* (Overseas Edition), October 14, 1992, p. 1.

[6] Collectively owned enterprises (COEs) are usually owned by the local community or local government of the townships and villages in which they are located. COEs obtain a small proportion of their supplies from the state and distribute a small percentage of their sales through the state.

and freedom (Bornstein, 1989). In a theoretical model developed by Gregory and Stuart (1992), economic growth develops and evolves as a function of a series of influencing forces, including the economic system, environmental factors, and policies pursued by the government. This model emphasizes four general attributes (probably overlapping) as determinants of the economic system's transition: (1) decision-making structure, (2) mechanisms for information and coordination, (3) property rights, and (4) incentives.[7] These four attributes can also be used to characterize the economy during the process of economic transition in China. Decision-making is increasingly decentralized and coordinated by market mechanisms. Central planning is moving from command-oriented to a more flexible orientation. While ownership is primarily collective and private, material incentives are increasingly used to motivate participants.

Table 1.2 The economy at three stages of economic reform in terms of four criteria

Attributes of economic reform	Stage of economic reform		
	Before	During	After
Decision making	Primarily centralized	Mix of centralized and decentralized system, varying in sector and region	Primarily decentralized
Mechanisms for information and coordination	Primarily planned	Increasing market mechanisms with "guidance" planning	Primarily market with indicative planning
Property rights	Primarily state ownership with collective ownership	Sate, collective and private ownership	Primarily collective and private ownership
Incentives	Primarily moral	Material and moral	Primarily material

Source: Author's adoption of Gregory and Stuart's model fitting the process of economic reform in China.

Reform of the oil sector is among the most demanding and challenging tasks faced by a transitional economy. The importance of the oil sector to the overall economic growth, macroeconomic stability, fiscal balance, international trade, and to enterprise reforms should be given high priority in formulating and implementing reform strategies.

In 1993, the central government took decisive action, drafting a 50-article manifesto calling for sweeping reforms in banking, taxation, trade, and public ownership of enterprises, which were to be implemented

[7] For more detail, see Paul R. Gregory and Robert C. Stuart, *Comparative Economic Systems*, Fourth Edition, Houghton Mifflin Company, 1992, pp. 16–22.

within two years. In early 1998, the central government called again for reform of state-owned enterprises and the banking system. It was thought that the tax and banking reforms were likely to move relatively quickly, given the urgency of the government's desire to strengthen state macroeconomic control.

1.1.2.1 Corporate restructuring

In China, economic reform primarily involves reducing the dominance of state-owned enterprises (SOEs), many of which are relatively inefficient and high-cost operations (of the total state industrial enterprise losses, oilfield enterprises account for some 14 percent).[8] Two different reforms have taken place to reshape SOEs and to reassign responsibilities: (1) the creation of a contract responsibility system (CRS), whereby the SOEs are held accountable under performance contracts concluded with the government; and (2) the creation of stock-sharing firms.

CRS means that management operates free from direct state control, but must provide an agreed quantity of output at fixed prices to the government. Beyond these quotas, however, enterprises can sell their products at market prices. The essence of contract responsibility is to expand the self-management powers of the enterprise and to create an incentive mechanism based on profit sharing between government and enterprises. This gives SOEs with CRS greater decision-making powers and freer play within the market, as well as exposing them to progressively more market discipline. At the present time, most oil companies — upstream, downstream, and trading — are taking this form.

There are two significant disadvantages of CRS. First, the issue of financial accountability is unsettled; most CRS enterprises remain responsible for profits, but not for losses. Second, CRS enterprises may be more short-term profit-oriented at the management level; the long-term success of an enterprise is not in the interest of firm managers. The World Bank (1992) indicated that mechanisms are needed to force contractors to achieve long-term revenue-maximizing objectives, such as using lease terms and renewal conditions to minimize shortsighted behavior, subjecting the process of awarding leases to maximum public scrutiny, and following transparent and elementary rules to maximize fairness.

Joint-stock SOEs have been advocated in China as a way to maintain state interest in state-owned firms, while obtaining much needed capital, technology, and expertise. According to statistics, China has more than 8400 SOEs taking on pilot reform programs, in which only 1.4 percent or about 120 enterprises have been permitted to sell their shares to

[8] World Bank Report No. 12265-CHA, February 17, 1994, p. 60.

the public. Stock in these companies may be bought by both individual and institutional investors. Theoretically, joint-stock ownership creates longer-term incentives than contract responsibility, but two problems may make their success in China problematic: (1) asset valuation will be difficult; and (2) a constrained capital market will confine the feasibility and operation of joint-stock firms.[9]

Turning state enterprises into limited-liability companies and in some cases joint-stock companies reflects the government's anxiety to speed the "corporatization" of the state sector, aimed at making enterprises more accountable for their profits and losses, and reducing the responsibility of the state for faltering industries. However, the central government insists that public ownership remain the "mainstay" of the economy. Then, indicating that it realized that it has to give up a little in one area in order to gain in the other, the central government stated that public ownership may not be necessarily dominant in *certain regions and industries*. This "loophole" may eventually lead to widespread private sector development — especially given the continued rapid expansion of private business in the economy.

Within the oil industry, before corporate restructuring, the state guaranteed oil enterprises' supply of materials or a market for their output. They bought and sold at controlled prices. Their workers had guaranteed employment. Furthermore, the oil industry had few worries about repayment of debt; most of the operating and investment capital was allocated — at no interest — by the central government.

After corporate restructuring, most oil companies have some form of CRS. Oil enterprises receive only guidance from a central plan, and the central government has moved away from direct allocation of capital to emphasis on bank loans. Oil enterprises can decide themselves what raw materials, semi-finished products, and components to use; what, how, and in what quantities to produce; whether to buy materials at home or abroad; from whom to buy and at what price; what kind of and how much credit they raise; and how many and what kind of workers they employ and what wages they pay. On the side of sales and after sales, the system of profit remittance was changed to direct taxation (see below). Oil enterprises can decide by themselves whether to sell on the domestic or the foreign market and at what price (observing the regulations of pricing set by central government and restraints on export licenses); and how much of their profits should be used for the development of the enterprise, for the stimulation of their workers, and for improving their workers' living standard.

[9] As of this writing, a few stocks from previous SOEs are traded on stock exchanges, which are located in Shenzhen and Shanghai.

In March 1998, newly elected prime minister, Zhu Rongji, proposed a bold sweeping change to give 100,000 industrial SOEs three years to shape up. Most are to be ended with government guaranteed bank loans, and to be cut loose from the state. In contrast, nonstate enterprises have become respectable new economic saviors and a vital source of jobs.

1.1.2.2 Tax reform

According to the tax system in place prior to the end of 1993, there were four categories of businesses: SOEs, collectively owned enterprises (COEs), private firms, and foreign-owned/joint venture firms. All of these entities had to pay business taxes and income taxes. For the state-owned and collectively owned firms, the taxes were based on revenue contracts signed between them and local or central governments.[10] All private and foreign enterprises were also subject to taxation. Local governments were responsible for collecting tax revenues from all firms except for large state-owned ventures, and were supposed to remit part of the revenue to the central government though a separate revenue contract system. The central government thus garnered tax revenues from two sources — central SOEs and local governments — with the contribution of the latter growing ever larger in proportion to the former.

Yet in the years leading up to 1993, it became increasingly apparent that China's tax collection system was working to the disadvantage of the central government. The tax structure protected some enterprises by guaranteeing profits through controlled prices, and it routed all funds through local channels. Regional and local governments, closer to economic activity, began to retain a larger share of the tax revenue pie. Early attempts by the central government to enlarge its split of the tax revenue met with stiff resistance. Thus, businesses and local authorities prospered in relative terms, while the central government became increasingly impoverished. One statistic in particular convinced the central government that it needed to revamp the tax system substantially if it hoped to retain its dominant role: from 1978 to 1993, central government revenue dropped from 35 percent of GDP to 17 percent.

---

[10] The SOEs sign contracts with either local governments or the central government, depending on whether they are local SOEs or central SOEs. The collective-owned enterprises sign revenue contracts with the local government. Governments and firms first negotiate on a fixed contract profit basis (a minimum amount of profit to be realized), usually the average of the previous three to five years' profits. Then they negotiate the tax rate on extra profits the firm earns. Profits over the contracted quota are either fully retained by the enterprises, or taxed at a lower rate. SOEs or COEs failing to meet their minimum profit objective must, in principle, pay their tax liability from their self-owned fund. Contracts typically last for three to five years.

In 1993, the central government called for major reforms in taxation. In particular, many officials thought that tax reform would push other systematic economic reforms, with far-reaching implications for China's political and economic future. Nearly a year later, the significant overhaul of the tax system made rapid progress. Reforms aimed at simplifying collection procedures and boosting the central government's share of revenues. It was characterized by two differences from the previous taxation regime: (1) the contract revenue system was being replaced by distinct local and central government taxes, collected directly by local and central government tax authorities; and (2) the dizzying panoply of taxes under the old system were merged into three basic taxes — a value-added tax (VAT), an enterprise income tax (EIT), and a personal income tax (PIT).

The VAT, the principal corporate tax, has been expanded to include most of the previous product taxes and some of the business taxes. It has been applied universally (for the first time ever) on all goods for distribution and resale at a uniform flat rate of 17 percent. The VAT covers both the industrial and commercial sectors and is the single largest revenue source for governments. Central and local governments share the VAT, with 75 percent accruing to the central treasury and the remaining 25 percent flowing into local government coffers.

The EIT was unified to 33 percent for all enterprises. This represents an improvement for large and medium state firms, who formerly had to pay 55 percent. Preferential tax policies for foreign enterprises and joint ventures have survived the reforms and remain intact. Previous revenue contracts between enterprise and government were abolished. The collection mechanism also changed only partially; the central government administers the EIT for central government-owned SOEs, and local governments do so for local government-owned SOEs and COEs.

The PIT changed with the introduction of a uniform system that applies to foreigners as well as Chinese citizens. Nine tax rates were imposed, ranging from 5 percent of base income for poor workers to 45 percent for higher income residents. Provincial and local governments administer the PITs.

The net effect of these changes is that the central government has received more tax revenues both in absolute terms and in comparison to local governments. Competition between the center and provinces in tax collection — the two were previously locked in a fierce struggle for a share of revenue — may ease during the years to come.

Will the new taxation system affect foreign investors? The central government is aware of the country's need for foreign investment and the understandable wariness of many investors concerning tax schemes. To help quell these fears, at least for the medium term, the government has promised that if it increases tax rates for foreign firms at any time, foreign investors will receive a rebate. Thus, preferential taxation for foreign enterprises and joint ventures would not be affected initially; farther down the road, however, the advantage might disappear as the private/foreign rate could be brought into line with the corporate rate. A key exception is preferential taxation that foreign investors have acquired in China's special economic zones, which are still unaffected.

As a result, in 1994, the central government's tax revenues grew rapidly, 7.5 percent higher than expected, and the central government's share of the tax almost doubled reaching 65 percent in national total.

Tax restructuring also affects the oil industry and the market. Table 1.3 presents the major changes in taxes and prices before and after the tax reform and price reregulation. For example, the new VAT is being applied to crude production, refining, and wholesale. Oilfields are subject to other resource taxes and fees, while refineries are imposed product sales taxes — 0.10 yuan per liter (US 1.1 cents) on diesel oil and 0.20 yuan per liter (US 2.3 cents) on gasoline. Because oilfields and refineries are owned by the central government, EITs and after-tax revenues are handed over to the central government. But provincial sales companies are owned by provincial government and their EITs go to provincial governments, with 75 percent going to the central government and 25 percent to provincial governments.

Before 1996, Chinese motor vehicle owners were subjected to at least 20 taxes and fees. After a two-year tax experiment in Hainan province in Southern China, it has been proposed that these multiple taxes and fees be combined into a single fuel-consumption tax. While no specific timetable has been set for its implementation, this proposed tax consolidation would likely stabilize tax rates, making auto ownership increasingly attractive. This move reflects two goals of the Chinese government: to promote the domestic automotive industry by simplifying the motor vehicle taxation structure, while at the same time, maintaining high taxes on gasoline.

1.1.2.3 Financial and currency reform

Financial reform centers on converting the People's Bank of China (PBC) into a Western-style central bank, commercializing the specialized state banks, and eliminating the dual exchange rate system. The new central bank ceased its lending responsibility and transferred all its commercial business, focusing on liability management (base money) rather than asset management (credit management). Other state-run specialty banks were converted into commercial banks while three other banks — the Import-Export Bank, the State Development Bank of China (SDBC), and the Agricultural Development Bank — took over the financial functions

Table 1.3 Tax and price changes before and after tax reform and price re-regulation

	Oilfield		Refinery		Wholesale		Retail prices
	Prices	Taxes	Prices	Taxes	Prices	Taxes	
Before tax reform and price reregulation	70% controlled 30% market price	Product tax 5–12% Natural resource tax 4–6 yuan/ton	20% controlled 80% market price	Product tax 15–40%	20% controlled 80% market price	Business tax 10%	Uncontrolled
After tax reform and price reregulation	Controlled	VAT 17% Operating tax 3–5% Income tax 33% Natural resource tax 8–12 yuan/ton Mineral resource compensation fee 1%	Controlled	VAT 17% Income tax 40% Sales tax: gasoline 278 yuan/ton, diesel oil 118 yuan/ton	Uncontrolled 4–6% lower than retail prices	VAT 17% Income tax 40%	Controlled based on 35 cities benchmark prices

Notes:
(1) Before tax reform, oilfield, refinery and wholesale should also pay income tax (varying rates), urban maintenance and construction tax (7%), and other taxes.
(2) After tax reform, oilfield, refinery and wholesale are also subject to urban maintenance and construction tax (7%), education appended fee (3%), and stamp tax.
(3) 75% of VAT should be handed over to central government, and 25% to provincial government.

of the State Planning Commission and assumed responsibility for virtually all of the government's project and policy lending functions. These three development banks are not profit oriented.

This redesign of the development banks is rooted in two specific reforms of the Chinese economy: investment system reform and credit reform. Investment reform refers to the government's effort to replace the boom–bust investment cycles characteristic of a planned economic system with the more rational system of market regulation. Operationally, investment reform depends on credit reform, which in turn is one of the fundamental government initiatives aimed at creating a market-based banking system. SDBC will assume most of the nonmarket aspects of the banking system in China (with the exception of policy lending to state enterprises), thereby freeing up the rest of the system to work more on market principles. Its primary emphasis will be in three main categories of investment: infrastructure projects, mainstay new industries, and raw material industries (including oil).

The dual exchange rate for the yuan, which included both a state controlled and a swap market exchange rate, was abolished effective January 1, 1994. The single exchange rate floats in the swap market at a rate between 8 and 9 yuan per US$, up from 5.78 yuan per US$ at the end of 1993.

To legislate a wave of new reforms does not mean that it can be quickly or successfully turned into a workable system. Since the central government lumbers through the reform process with a mixture of gradualism and experimentalism, the process will almost certainly be long and bumpy. It would be unrealistic to expect China to establish a workable market system in a short period of time; China does not expect to establish a "preliminary" market system until the end of the century.

The impacts of financial reform on the oil industry are twofold. First, if the oil industry remains under central government control, its financial loans will probably come from both the SDBC and other commercial banks. Its state-owned position will make it easy for the oil industry to receive government loans, rather than commercial loans. Second, oilfields and refineries are no longer required to remit a portion of their foreign exchange earnings to the central government, leaving some leeway for their management.

Under the pressure of the Asian financial crisis in 1998, China tightened the reins on its currency. It may thank the fact that renminbi is not fully convertible. With all crushed currencies in Asia, risk in Chinese economy increased. Chinese exporters were squeezed by the advantages Southeast Asian competitors gained from plunging currencies. In the oil market, diesel oil and fuel oil imported from the rest of Asia, particularly from South Korea, became cheaper than those made at home.

Devaluation of renminbi might keep domestic refined products from being displaced by cheap imports, while it would be of little help to China's limited oil exports. China faced a dilemma: either to devalue renminbi to support its economy, which could trigger another round of global financial turmoil, or to maintain the value of renminbi but hurting its exports and domestic economy.

1.1.3 The prospects for economic growth

Although the Chinese government faced fiscal pressures and inflationary implications associated with subsidizing the loss-making SOEs, the robust economic growth was still in line with the government's efforts to prevent overheating and to rationalize investment in SOEs.

The year 1994 was a tough year for the Chinese economy. In March, the central government set three targets for 1994: (1) a 10 percent inflation rate; (2) internal investment in fixed assets below 1.3 trillion yuan, primarily in state-owned projects; and (3) a 9 percent economic growth rate. China did not achieve any of these targets. Despite government reregulation of some key commodity prices (including oil), the inflation rate skyrocketed to about 20 percent, the highest level since China started its economic reform. Investment in fixed assets surpassed its goal by 10 percent, about 40 percent higher than 1993, which further boosted government budget deficits. Accordingly, the 1994 GDP growth rate reached 12.6 percent. The root of these problems is unprofitable state enterprises, which absorb most government loans and credit, yet are not responsible for their losses. The lack of competition in markets dominated by state-owned industrial firms meant that monetary control and tight credit could not restrict inflation in 1994.

Chinese oil markets were dramatically affected by macroeconomic policy. The future of oil consumption depends on the pace of overall economic growth, where it will occur and in which sectors, and on how economic and political reforms affect the climate for investment in crude production and refining.

1.1.3.1 The engines behind economic growth

Will China maintain high, sustained economic growth over the next 10 years? We believe it will, despite many challenges — the threat of hyperinflation, political crises, etc. — for three principal reasons:

1. Expanding activity by nonstate enterprises;
2. Expansion in the industrial and service sectors; and
3. Continued openness to foreign investment.

The nonstate sector includes collectively owned businesses, private ventures, foreign-owned enterprises, and joint ventures between Chinese and foreign investors. This sector has played a major role in achieving double-digit growth. Farming is mostly quasi-private, and retailing and other services are increasingly concentrated in private hands as well. According to Chinese statistics, the share of private and foreign-owned businesses in total industrial output increased from 3 percent in 1985 to about 33 percent in 1997, and the share of COEs also increased, from 32 percent in 1985 to 41 percent in 1997.[11] In 1997, the total value of the industrial production from nonstate-owned enterprises increased by nearly 12 percent, while the value of products produced by state-owned industries grew by only 1.5 percent. Nonstate enterprise will continue its robust growth and claim a rising proportion of the economy. In terms of the oil sector, however, these enterprises may actually slow the growth of oil demand somewhat, since most nonstate enterprises involve light industrial activities, which consume less energy than state-owned heavy industry.

Experience from other newly industrialized Asian countries shows that the industrial share of total economy production may rise as high as 50 to 60 percent before the country enters the more mature phases of industrialization, at which time the service industry begins to grow and ultimately represents as much as 60 percent of the total economy. So far, the engine behind China's rising GDP is the industrial sector (manufacturing, exploration, construction, energy, etc.), which in the past few years has grown at an average rate of 17 percent, 50 percent higher than overall GDP growth rates. China is in the process of expanding its service sector as well; growth was about 9 percent annually over the past decade. In the next 20 years, China's industrial sector is expected to expand and increase its share in the economy, but it may take more than 10 years for the service sector's share to rise to 50 percent (17 percent in 1985 and 26 percent in 1997). The high growth in the industrial sector will be supported by a high capital accumulation rate. In the past, China has spent 34 to 35 percent of state revenues on industrial expansion, one of the highest rates in the world. If China maintains such high investment accumulation rates during the next decade, its economy would remain on the fast track. China will indeed follow such a path.

The process of transforming a centrally planned economy to a market-oriented economy has gained momentum, fueled by past successes, and it is commonly accepted that the economic reform is irrevocable in

---

[11] In contrast, the state share in industrial output declined from 69 percent in 1985 to about 27 percent in 1997. See *A Statistical Survey of China, 1998*, China's Statistical Publishing House, 1998, p. 10.

China. In addition, the sizable Chinese market has created a tempting environment for foreign investors. For example, actual foreign direct investment (part of committed investment) reached $51 billion in 1997, compared to only 7 billion in 1990. Considerable progress in attracting large sums of foreign investment is expected to continue to stimulate China's economy.

1.1.3.2 Trends and prospects

China was able to make its per capita GDP rise during the 1980s and 1990s, even with a population growth rate of about 1.3 percent per year over the course of the period (see Fig. 1.1). However, in the late 1980s, GDP per capita stagnated at the level of $400 for several years, although this slowdown results partially from lack of statistics and partially from the distorted official foreign exchange rate. There were certainly encouraging signs of per capita income growth after 1990, as China's GDP per capita rose to $750 in 1996, an annual increase of about 10 percent (Fig. 1.2).[12]

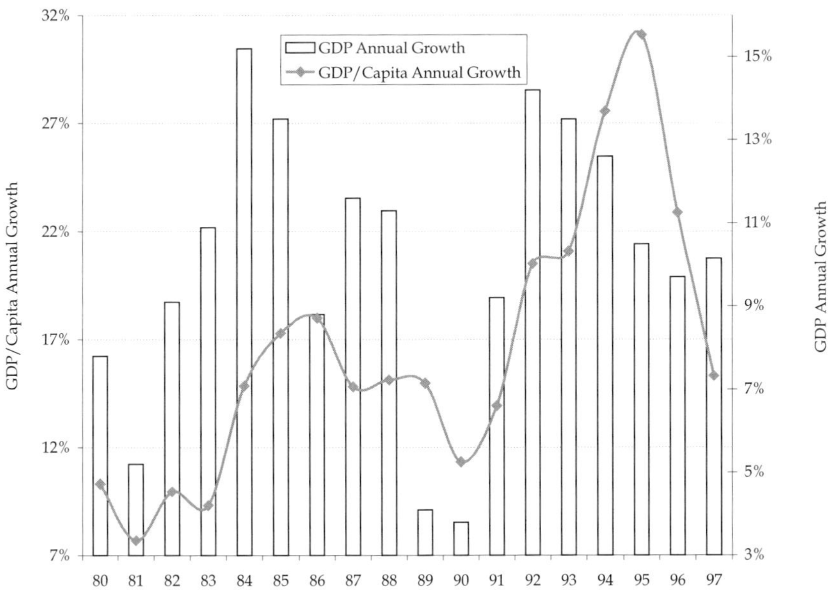

Fig. 1.1 GDP growth, and GDP per capita growth
Source: *Statistical Survey of China, 1998*, China Statistical Publishing House.

[12] See *World Development Report, 1998*, the World Bank.

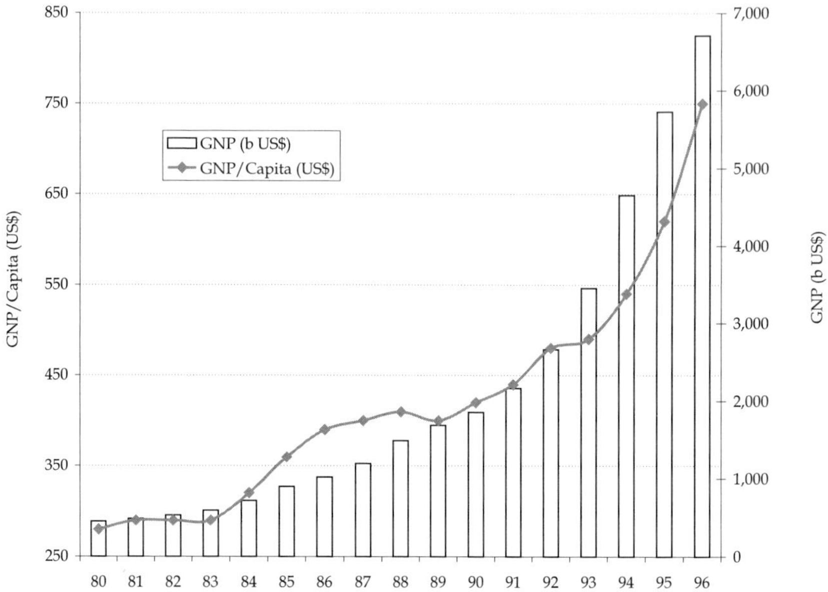

Fig. 1.2 GDP and GDP per capita
Source: World Development Report, 1998, the World Bank.

Turning briefly to the question of population growth, China is a unique country that has instituted an extremely aggressive birth control policy that aims to limit the size of Chinese families to only one child, and thus stabilize China's population at 1.28 billion and 1.35 billion people by the year 2000 and 2005, respectively. This ambitious goal is likely to be accomplished if the central government policy remains in effect over the next decade. There is no good reason to doubt this.

China has shown an acute awareness of the negative effects of too rapid economic growth on its development plans as shown by the reforms of 1993 and 1994. Our analysis foresees a significant risk that the next decade will have episodes of periodic high inflation, provoking periods of tighter controls, which in turn will create periods of lower growth. For these reasons, we will advocate taking a somewhat sober view when choosing an economic growth path for the sake of forecasting China's oil demand growth.

Taking the experience of the past 20 years into account, we predict that China's economic growth in the late 1990s and the 20th century will fluctuate between 6 and 12 percent per annum, averaging about 9 percent per year. Based on these perceptions, total GDP in China is expected to reach US$ 1.0 trillion in 2000 and US$ 1.2 trillion in 2005 (both in 1996 dollars, assuming a 3 percent deflation rate for the US dollar over the coming years). Translating these GDP values into dollars per capita, this means that GDP per capita will average about US$ 930 by 2000 and US$ 1200 by 2005 (both in current value). Our demand model for petroleum products will integrate these data and other assumptions. However, this is still a comparatively low level of GDP per capita by 2005, equivalent to the level of Taiwan in 1977, South Korea in 1978, and Thailand in 1989. It should also be noted that the scale of China's aggregated economy of US$ 1.2 trillion (1996 value) by 2005 remains only less than 25 percent of the Japanese economy and about 16 percent of the US economy in 1996.[13]

1.1.3.3 Some caveats

While China is making this economic transition, it is faced with the necessity of persevering with tough measures to reorganize its economy, creating new jobs for 12 million workers who are laid off from SOEs, reducing its budget deficits, and keeping low inflation. China's success at sustaining high-speed growth, without recurrence of the past economic "go–stop" cycles, depends primarily on the central government's ability to rein in and to stimulate the economy when appropriate. Although we are optimistic, some formidable hurdles remain:

- *Loss-making SOEs* China has 305,000 state enterprises. About two-thirds of them are debt ridden. China's policy is to turn loss-making state enterprises into independent, market-driven enterprises. The government channeled 75 percent of bank credit into state enterprises in the past. When the bad loans keep pilling up in the state-owned banks, it threatens the health of the entire economy. Such state enterprise reform requires an overhaul of the state-owned bank system.
- *High unemployment* In China, there are 6 million young people entering the workforce each year, when 12 million will be laid off after restructuring the state enterprises. In 1998, China National Petroleum Corporation also announced plans to dismiss 250,000 workers in the next few years. China faces a challenge to create jobs for those laid-off workers and the new workforce, when economists estimate that one in three state industrial workers is redundant.
- *Soaring inflation or deflation* Inflation has emerged as the scourge of economic life and may threaten steady economic growth in the

---

[13] Also see *World Development Report, 1998,* the World Bank.

future. The annual rate of inflation had been held between 4 and 6 percent annually until 1991, but then rose to 16.6 percent on average from 1993 to 1995. In large cities, the rate reached 25 percent in 1994. The central government plans to hold the inflation rate between 1 and 15 percent per year. Monetary constraints are the most promising policy instruments available to central authorities, and in 1993, the PBC, the central bank in China, increased interest rates significantly and tightened state credit. However, local governments and enterprises always resist anti-inflationary measures, since they nearly always depend heavily on central government credits and investment to finance industrial and infrastructure expansion. Nevertheless, if the inflation rate in major cities jumps or threatens to stay high, the central government will have no choice but to tighten monetary policy, slowing economic growth.

China successfully slowed down its overheated economy and brought down the inflation in 1996 and 1997. Soon after that, however, China found itself with another round of economic problems: deflation and possible economic sluggishness. In the first half of 1998, retail prices fell at an annual rate of 3 percent, due to fear of lay off which forced people to save rather than to spend. China started to struggle to boost demand, and halt deflation.

- *Excessive investment in construction* Economic expansion in China in the last decade has been powered by a sustained increase in investment in construction. Growth led by fixed assets investment instead of productivity improvement has contributed greatly to rising inflation. Over-investment in the industrial sector at the expense of energy and infrastructure may be popular in many provinces, but it threatens to imperil overall economic productivity, and therefore growth, in the future.
- *Crumbling "bottleneck" infrastructure* China's economic boom is exerting huge pressures on dilapidated or deteriorating "bottleneck" industries such as energy and transportation, which are increasingly failing to meet the needs of China's growing economy. In the oil sector, for example, refinery throughput increased by only 4 percent in 1996, far below the 12 percent growth rate of industrial production. In 1993, the door has opened to direct foreign investment and involvement in China's infrastructure,[14] but it is not clear at present to what extent the central government is willing to liberalize and decentralize infrastructure maintenance and development.

[14] See *Xinhua News Agency*, "China to Boost Infrastructure", January 25, 1993.

- *Monetary chaos* Central government revenues are declining and central government spending is increasing. During the past three years (from 1994 to 1996), according to Chinese officials, China's budget deficit was maintained above 55 billion yuan (about US$ 7 billion), almost doubling the annual deficit level in the early 1990s. Increased government deficits have put a large strain on the financial sector, which is unable to absorb this debt without putting inflationary pressure on currency and drawing down foreign reserves.

To summarize, even with all these problems, past successes are fueling dynamic forces that will continue to move China into high levels of economic growth. The overall economic outlook is bright because the growth in certain key sectors, especially nonstate enterprises, will continue. Some obstacles remain, however, such as inflation, high debt, and poor infrastructure. The implications of Asian crisis for China is clear: the easy days of export-led economic growth may end. China needs to establish a healthy economy. The consequences of this economic growth outlook for oil production and consumption are explored in subsequent chapters. Before considering them, however, China's success at coping with its economic problems also depends on the scope and pace of political reform.

1.2 Political Outlook

China is entering an extremely sensitive period, as its leadership faces a highly uncertain political transition into the post-Deng era. Maintaining political and social stability must still be the number one priority for the leadership. China's economy, and especially its oil market, is inextricably linked to government policies and particular regimes. Much of the economic growth scenario sketched out previously hinges on how political developments unfold. Two issues require close examination. First, what form will domestic political change take over the next 10 years? We offer three plausible scenarios: moderate change, turbulent change, and chaos. Second, how will China's foreign relations evolve, especially with key players in the oil business?

1.2.1 Petroleum politics

What does the departure of paramount Chinese leader Deng Xiaoping mean to the oil sector of the world's fourth largest oil consumer? To answer this question, we should first understand the relationship between Chinese politics and oil policy. Three layers of decision-making affect the oil sector in China: the political system, macroeconomic policy, and oil policy. In China, oil policy is a function of macroeconomic policy, which is related to larger political issues. In other words, the oil industry and market are primarily sensitive to government *oil* policy, which emanates largely from a greater macroeconomic policy, which, in turn, depends on the character and intentions of those holding power. As a result, there is little *direct* linkage between the oil business and strictly political events, at least in the short run. However, Deng's death certainly has a direct and a long-term effect on China's economic policy. After Deng, macroeconomic choices will directly determine the fate of the oil industry and market. We have observed that oil policy has been subject to frequent changes as the central government has tried to determine the role oil should play in the economic transition.

Decisions regarding China's economic policy are often determined by the political needs of those in power. There are three things worth noting when considering the development of Chinese macroeconomic policies. First, before Deng's departure, the current leaders kept a low profile. There was no open policy shift until Deng died. Deng's eventual departure strengthened the current leaders' position in keeping a tight rein on the economic and political situation. By their nature, the three "giants": President Jiang Zemin, former Premier Li Peng, and Premier Zhu Rongji, are advocates of a centrally planned economy. The "interference" in economic affairs by Deng during the early 1990s, particularly his policy of "growth-at-all-costs", risked Jiang, Li and Zhu's political lives because they became the targets of criticism due to the economic chaos which some reforms have created. Criticism has come from both sides — reformers complaining that reforms have been "too slow" and conservatives arguing that the reforms were "too liberal". During the Deng "deathwatch", weaker directives from Deng resulted in an increased freedom of decision-making for the current leaders. Because they consider political and economic stability to be their number one concern, they would vastly prefer to see more central government measures to control the market and the economy, and to drag their feet on any further economic reform.

Second, Deng's departure may result in a power struggle. Unlike most countries, political groups fighting for leadership in China are divided by personality rather than ideology. Conservatives and reformers may exist in the same group, if they align themselves with the same potential leader. Therefore, the people who gain power will adopt whatever policy fits their political ambitions. China's economic future then depends on who grasps power and what kind of political propaganda they employ to weaken their rivals. Consequently, there will be great uncertainty regarding the direction of long-term macroeconomic policy and economic reform after Deng.

Third, the next leader's ability to enact major changes in economic openness will be limited because the Chinese people generally favor the economic openness of the Deng era. Attempts to reverse economic openness will meet strong resistance from the regions and the people that have benefited from the movement towards a free market economy. If the political system survives for a few years after Deng, it is likely to be stable enough to prevent any dramatic changes in China's macroeconomic policy.

1.2.2 Domestic political change: moderation or turbulence?

Deng Xiaoping's departure left an enormous gap at the top of the leadership structure and may usher in a period of political uncertainty. The disappearance of a strong ruler such as Deng — such "strongman" rule having deep roots in Chinese history — could substantially weaken central power, thereby creating opportunities for restless provinces to become more independent. It is certainly premature, however, to predict a breakup of the country or even de facto independence of the provinces.

After Deng departed, China remained and will continue on its path during Deng's era: rapid introduction of market reforms with little political change. There is little chance that China would step back to a Maoist state of isolation from the world economy and politics, because the few hardliners who would still support such a move are passing away, both ideologically and physically. There are three possible directions domestic politics could take after Deng: (1) peaceful leadership transition and gradual changes in political power along the lines of South Korea or Taiwan; (2) violent, tumultuous change involving coups and swings between authoritarianism and reform as we are seeing in Russia; and (3) societal disintegration, factionalism on regional lines leading to chaos and a high degree of violence (as seen in several republics of the former Soviet Union (FSU) and especially Yugoslavia).

Some political observers predicted that Deng's departure might lead a certain degree of turbulence and perhaps even political violence during the succession process. Because no strong, obvious successor was in view; it was possible that a clique of leaders would attempt to run the country, allowing for the gradual emergence of a head figure. The long time Deng held power, however, reduced the possibility of severe political violence after he departed. As economic growth takes hold and education and economic development progress, it becomes more likely that Deng's succession can unfold without political violence.

Descent into Yugoslavia-type chaos is unlikely for several reasons. First, after witnessing the collapse of the FSU and Yugoslavia, the people in China do not want to repeat such economic disaster and social unrest.

Most Chinese believe that only a stable political situation can keep the economy on a fast track. Second, economic growth has facilitated greater interdependence between regions. Third, the cohesion and structure of the People's Liberation Army (PLA) have improved. Without a split in the PLA, there is little possibility that China will break up.

Could China break up into different republics, as did the FSU, even with a minimum of violence? While both China and the Soviet Union encompass large land areas with independent-minded borderlands, we do not think that geographic breakup is likely, even though secession activities will continue in Tibet and Xinjiang. First of all, there are some important demographic differences between the two cases. Unlike the FSU, an ethnic minority dominates no region in China. Even in minority regions (with the exception of Tibet), the minority nationalities still account for a small percentage of the population, while the majority is composed of Han nationals in virtually all areas. Secondly, the FSU was around twice as long as Communist China. The FSU has more than 70 years' history, and Gorbachev was the fourth generation in its communist leadership. But socialist China is less than 50 years old, and its paramount leader still belongs to the first generation of communist revolutionaries. Therefore, China could theoretically rebuild its political structure with less resistance.

There is a consensus among China's leaders on the importance of "sustained, rapid, and healthy" economic growth. Political instability and social unrest are obviously undesirable because they breed unstable economic conditions. Political survival requires both political and economic stability. The struggle for power between the central government on the one hand and local governments and SOEs on the other is likely to intensify in the coming years. But in the past the increase of autonomy in enterprises and provincial regions has depended more on the central government's willingness to decentralize than on the will of the enterprises and provinces to fight for independence. In the post-Deng era, the periphery will have become accustomed to increased autonomy, and the central government will increasingly rely on political maneuvering, persuasion, deal making, and compromise. Future relations could evolve into a kind of "centrally managed pluralism" in which the central government will rely on political skills, rather than commands, to balance the interests of different enterprises and provinces.

The PLA represents another force for stability, as well as economic reform. The army is the ultimate guarantor of the state and has played an especially important role, as shown in the two-digit increase in the military budget during the past few years. The military has also started to cash in on economic reform and has an interest in seeing it continue. In keeping with the business fever running through China, the military actively involved itself in commercial operations, and built its own business connections with SOEs, local governments, and foreign partners. If the military can be even more independent, such a system of cooping the military's interest in economic growth could prevent China from a chaotic breakup or military conflict. However, on July 22, 1998, President Jiang Zemin, also as a commander-in-chief ordered the PLA, as well as police and judicial organs, to withdraw from all commercial activities, where they had enjoyed special privileges such as tax breaks for a long time. The undercutting of PLA-run business aimed at reducing the PLA's economic and political autonomy.

To summarize, China will continue on its present course of economic reform and slow political change after Deng. China has entered a phase characterized by political maneuvering and uncertainty, but there is little chance for violence, a total disintegration of the society or widespread and prolonged internal war. China's culture and history point to the reemergence of a strong figure as head of government. With the army standing for stability and supporting economic growth, political change will resume on a moderate course toward the end of the decade.

1.2.3 China's foreign relations: impact on the oil business

China's foreign relations with certain countries in the region and around the world are crucial determinants of how political and economic change develops. In some cases, foreign relations will have a direct bearing on oil market developments. First, they will help to shape the region's geopolitical environment and hence limit or enlarge the range of options open to the Chinese government to achieve growth and stability. Second, they will determine investment levels, which will be a key element in achieving a smooth growth path. Finally, they will influence oil import and export policies.

Below, we evaluate China's relations with these countries, which will have a determinate influence on one or more of these factors. China's relations with these key countries remain uncertain. In our view, however, they will probably evolve in a manner that is broadly conducive to regional stability, economic growth, and oil market intervention. The chapters that follow assume that no major or cataclysmic events disrupt regional stability, investment, or trade for any extended period of time.

1.2.3.1 Japan

China has had a long relationship with Japan where oil trade is concerned. Crude exports to Japan peaked at about 350,000 b/d in 1985 and averaged 230,000 b/d during the years from 1995 to 1997. In addition to crude oil, there are also some petroleum product exports to Japan, but the quantities are small. To secure and diversify its oil supply, Japan's Import-Export Bank has provided China three times with jumbo loans at low interest rates to develop energy projects in China, primarily in oilfield exploration. These loans have helped China buy technology and equipment to upgrade the Daqing and Shengli oilfields.

Japan has a greater economic interest in China than does either the US or Europe. China's huge market is extremely important for Japan's export-oriented economy and, while Japan sided with Western countries on the embargo on China due to the Tiananmen incident, Japan became the first country to break its commitment. When many foreign companies closed or at least delayed new business investments in China after the Tiananmen incident, most Japanese companies stayed and took advantage of decreased competition to expand their business. Thus, with their largely symbiotic relationship, both countries will continue to cooperate closely in overall economic development, including the oil sector.

1.2.3.2 North Korea

Like the client states of the FSU, North Korea has become dependent on China for virtually all its energy and food supplies, even though China established formal diplomatic relations with South Korea in 1989. China is North Korea's largest trading partner, with commerce between them estimated at $850 million in 1996. Three-quarters of North Korea's total energy and food imports come from China.[15] According to the Chinese Customs Authority, China's crude oil exports to North Korea ranged from 10,000 b/d to 22,000 b/d during the period of 1990 to 1997 (Table 1.4) (China also exported crude to South Korea, 63,000 b/d in 1996 and 91,000 b/d in 1997). In November of 1993, oil exports to North Korea rose to 26,000 b/d, and in December 1993 they jumped again to 30,000 b/d, in part to replace declining deliveries from Russia, which before 1992 used to supply as much as 10,000 b/d. North Korea consumes close to 50,000 b/d of oil, almost half of which comes from China. The volume of crude exported to North Korea is unlikely to rise further, because the tight market in China will leave little for export. In addition, severe cash shortages have made it more difficult for North Korea to buy Middle Eastern crude; China has also urged North Korea to pay cash instead of barter.

The issue of international inspection of North Korean nuclear facilities increased tensions in the Korean Peninsula, and China is the key to

[15] See John Burthon, "Self-Reliance May Soon Be Put to the Test", *Financial Times*, March 22, 1994, p. 6.

Table 1.4 Oil exports from China to North Korea

	1995				1996			
	000 b/d	As a % of total exports	Million US$	US$/bbl	000 b/d	As a % of total exports	Million US$	US$/bbl
Crude oil	20.43	5.4	130.9	17.55	18.72	4.6	118.7	17.37
Gasoline	0.25	0.6	2.6	28.11	0.06	0.2	0.8	35.30
Kerosene	0.76	9.4	6.5	23.48	0.84	5.9	8.4	25.82
Light diesel	0.24	0.9	2.4	27.60	0.27	0.8	3.2	32.10
Lubricant	0.18	3.0	5.7	88.15	0.08	3.1	2.5	81.42
Fuel oil	0.07	2.1	0.6	23.60	0.09	1.7	0.6	20.10
Others	0.05	0.8	0.3	18.48	0.01	0.2	0.1	50.39

Source: Chinese Customs.

any international approach. China wants stability on the Korean peninsula so that its fast-growing economy can benefit from the increasing prosperity of the region. China also fears turmoil on its borders if North Korean refugees should flood into its territory. China has thus made efforts to persuade the North to accept nuclear inspections. In the final analysis, China will reluctantly support or at least not actively oppose international efforts to halt North Korea's nuclear program. China would try to prevent North Korea from provoking war, since this would in no way benefit China's economic progress.

1.2.3.3 United States

The annual renewal of China's most-favored-nation status (MFN) is Washington's weapon to pressure Beijing on human rights.[16] However, China's duality of economic openness and political tightening complicates US policy making toward China. An overemphasis on extending American values to China, including human rights, military sales, family planning, Tibet, etc., may conflict with US interests in maintaining and expanding its economic and trade relations with China. Washington must balance these two potentially conflicting objectives and search for a more coherent China policy.

Chinese leaders have insisted that China will not be persuaded by US attempts to use trade as a weapon, stating emphatically that "China can live without it (MFN)". While Washington insists that economic reforms must be accompanied by political openness, Beijing believes

[16] In May 1993, President Clinton decided to attach conditions to renewal of MFN in 1994, but in May 1994 he withdraw human rights condition from linkage with MFN.

that political tightening ensures stability, which in turn guarantees economic prosperity. It is not in the Beijing regime's interest to loosen political controls, with or without MFN.

If the US restrains trade, how will it affect the oil market? It was reported that China relies on the US market for about 12 percent of its total exports and about 17 percent of its total imports (about 15 percent of its total value of imports/exports) in both 1995 and 1996.[17] In contrast, less than 5 percent of total US exports and imports are related to China. The World Bank estimates that preferential low tariffs affect 96 percent of China's exports to the US. If China's preferential trade status were eliminated, its trade surplus with the US, which was about $25 billion in 1996 including those Chinese goods shipped through Hong Kong, would dramatically shrink, jeopardizing a cash flow that is crucial for China's burgeoning economy.

During the years from 1993 to 1997, China exported only 45,000–60,000 b/d of crude to the US, almost nothing when looking at the magnitude of total US crude imports.[18] As China's market booms, its surplus of crude has shrunk, leaving ever fewer barrels to export to the US and other markets. More importantly, troubled relations between the US and China would restrain the involvement of American oil companies in China's crude oil exploration, refining, and marketing activities. A pragmatic administration in Washington could extract concessions from Beijing. Washington holds the cards regarding MFN. Rather than making an "all or nothing" decision, the Clinton administration might impose trade sanctions only on certain goods or industries controlled by the Chinese central government as a compromise solution that would satisfy both business groups and human rights groups. Selective sanctions might backfire, however, because China could easily convert its state produced products into private sector products, giving US. Customs no way to make distinctions.

Selective sanctions applied to the oil sector would hurt the Chinese state oil business worse than any other state-controlled industry, if such a targeted approach were to channel foreign investment away from the state sector and into the private sector. We believe that targeted sanctions aiming to encourage state enterprises in the oil sector to become private, as laid out in the

[17] There are statistical discrepancies in the calculation of US–China and China–US trade. US calculations of its imports from China include those Chinese goods shipped through Hong Kong, but its accounting of its exports to China excludes those American goods trans-shipped through Hong Kong. While including most imports from the US, Chinese calculations do not count those Chinese goods trans-shipped through Hong Kong as exports to the US.

[18] According to Energy Information, US Department of Energy, the US imported 8.9 million b/d of oil (on a net basis) in 1997. See Petroleum Section, EIA web site, April 1998.

congressional bill, will not work. The oil industry is vital for the government, and it will retain a firm grip on activities in this sector.

US involvement in Asia is important for that region's stability and continued economic growth. Trade in oil is a more marginal concern, since the current and projected oil trade is quite small. As noted above, one area that does impact directly, however, is US investment in China's oil industry. If the US imposes selective sanctions on state enterprises as a consequence of its MFN compromise, it could hurt the Chinese oil industry because the state has an almost exclusive stake in it. But even in this case, other countries can provide the necessary capital and technology China needs to modernize its oil sector. The most pragmatic and beneficial approach is for Washington to keep investment flowing between the two countries.

1.2.4 Perplexing dispute over oil in the Spratly Islands

Unresolved skirmishes between China and the Philippines over the Spratly Islands have stoked the fires of controversy.[19] These islands in the South China Sea have increasingly been the focus of international attention after emerging as huge potential reservoirs of as-yet untapped oil and natural gas. The regional balance was significantly altered after the last US military base in the Philippines was closed in 1992, leaving a vacuum in the region. A military confrontation in the future could escalate into a regional and international crisis.

In the South China Sea there are two archipelagos in dispute: the Parcel Islands and the Spratly Islands. Only China and Vietnam have claimed sovereignty over the former. The latter have been wholly claimed by China, Vietnam, and Taiwan, and claimed in part by Malaysia, the Philippines, and Brunei. It was reported that the Spratly Islands are divided as follows: Vietnam holds more than 20; the Philippines, 14; China, seven; Malaysia, three; and Taiwan, one. Only the island held by Taiwan has fresh water, an indication that the occupancy of these islands is more important as an issue of geopolitical strategy rather than for the value of the territory itself. All of the claimants except Brunei maintain a military airstrip or guard post on at least part of their occupied area. The dispute became even more complicated after most of these countries accepted a 200-mile exclusive economic zone for their sea territory.[20] The

[19] See Reuters North American Wire, May 10, 1997, "China Protests Philippine Flag in Disputed Shoal". Also see Inter Press Service, May 6, 1997, "Southeast Asia: New Chinese foray into Spratlys Jars Nerves".

[20] The UN reached an agreement in November 1994 to give each ratifying state a 200-mile Exclusive Economic Zone around its land border.

claims of the various countries are based on either of the following two reasons or both: (1) an ancestral discovery, and (2) the conventional law of the sea under the UN charter. In 1991, foreign investment in oil exploration and development began to pour into the South China Sea area, which further intensified competition among the islands' claimants.[21] All the countries are attempting to use the presence of foreign oil companies under contract to them to support their claims to their respective waters.

1.2.4.1 Disputes over oil

What levels of natural resources are present in the Spratly area? Geologists have provided varying estimates of the amounts that could be extracted. So far there is no clear picture of the oil and natural gas reserves in the region, and all the estimates are speculative. According to the Chinese press, the South China Sea as a whole has 10 times more oil and seven times more natural gas then China's onshore reserves. However, these estimates are unreliable because there was no identification of the types of reserves compared. Another estimate from a Chinese newspaper, *Zhongguo Qingnian Bo* (Chinese Youth's Newspaper), reveals that in the Spratly Islands (called Nansha Islands by the Chinese), there is estimated 1 billion tons (73 billion barrels) of oil reserves and 1 trillion cubic meters of natural gas. It means that the discovered oil reserves in the Spratly Islands are equal to as much as 40 percent of China's onshore reserves.

According to Chinese sources, E&D activities in the Spratly Islands are conducted in eight sedimentary basins: Zengmuansha Basin, Brunei Shaba Basin, Wananxi Basin, Xibalawang Basin, Andutan Basin, Zhenghe Basin and Sino–Vietnamese Basin. About 200 oil-gas structures and 61 oilfields and 34 gasfields have been found. The proven reserves have increased to 8.6 billion barrels of oil and 3296 billion cubic meters of natural gas. So far, 34 oilfields and 4 gasfields have begun operation and are producing about 520,000 b/d of oil and 17.74 billion cubic meters of natural gas annually (Table 1.5). Although the crude output in the Spratlys accounts for only 10 percent of total production in the Asia-Pacific region, continuing increases in oil production from the Spratly Islands will accelerate the growing economic development in Southeast Asia (Table 1.6).

In 1993, China changed from being a long-time net oil exporter to being a net oil importer (a combination of crude oil and petroleum products). In 1996, for the first time in the past quarter century, China

[21] For instance, Mobil of the US is working in waters claimed by China under a contract with Vietnam. China has hired Crestone Energy of the US to explore for oil in the area disputed with Vietnam.

Table 1.5 Crude oil production of the Spratly claimants

	1985	1986	1987	1988	1989	1990	1991	1992	1993	1994	1995	1996	1997
Production (000 b/d)													
Brunei	170	165	145	139	149	150	162	169	158	162	175	170	167
China	2498	2614	2683	2733	2754	2766	2820	2828	2898	2928	2967	3145	3244
Malaysia	442	489	461	515	589	580	623	634	647	644	683	706	691
Philippines	10	7	5	5	6	5	3	18	9	7	15	11	11
Taiwan	2	2	3	2	2	3	2	1	1	1	1	1	1
Vietnam	0	0	0	0	0	55	80	111	128	142	166	165	174
Total Asia-Pacific	5659	5803	5794	5882	6072	6274	6443	6474	6550	6730	6864	7079	7201
Share in Regional Total (%)													
Brunei	3.0	2.8	2.5	2.4	2.5	2.4	2.5	2.6	2.4	2.4	2.5	2.4	2.3
China	44.1	45.0	46.3	46.5	45.3	44.1	43.8	43.7	44.3	43.5	43.2	44.5	45.1
Malaysia	7.8	8.4	8.0	8.8	9.7	9.2	9.7	9.8	9.9	9.6	10.0	10.0	9.6
Philippines	0.2	0.1	0.1	0.1	0.1	0.1	0.0	0.3	0.1	0.1	0.2	0.2	0.2
Taiwan	0.0	0.0	0.1	0.0	0.0	0.0	0.0	0.0	0.0	0.0	0.0	0.0	0.0
Vietnam	0.0	0.0	0.0	0.0	0.0	0.9	1.2	1.7	2.0	2.1	2.4	2.3	2.4
Sub-total	55.2	56.5	56.9	57.7	57.6	56.7	57.3	58.1	58.6	57.7	58.4	59.3	59.5

Note: Crude oil production in Asia-Pacific excludes US PADD V.
Sources: International Energy Agency, OECD, State Statistical Bureau, Energy Commission in Taiwan, and Energy Security Analysis, Inc.

almost all of its economic sectors, including aviation and telecommunication, while China still prohibited foreign investment in these sectors. In some surveys conducted by foreign investors, the liberal aspects of the Vietnamese market are greater than in the Chinese market. The economic liberalization is expected to rescue the Vietnamese economy from its unsuccessful attempts to be self-sufficient. Since 1987, when Vietnam opened its door to the world, its market has become a competitive field to those from the Americas, western Europe, and the Asia-Pacific region.

The oil reserves under Vietnam's continental shelf are estimated at 3–5 billion barrels. In 1997, Vietnam pumped 174,000 b/d of crude oil from its three producing offshore fields in the South China Sea — gaining an annual growth of 20 percent since 1990, the fastest growing crude supply of any country in the region. Vietnam received more than a quarter of total export revenues from its crude oil exports, generating more than $1 billion in 1996. Vietnam expects to produce 400,000 b/d of crude oil and to build two refineries by 2000.

In 1986, VietSovPetro, a 50–50 joint venture between Vietnam and Russia, produced its first barrel of oil. Beginning in 1988, oil companies from the US, France, Japan, Australia, South Korea, and Malaysia have entered into contracts with Vietnam for oil exploration in the waters of the South China Sea. Direct foreign investment in the oil sector amounted to $1.2 billion, accounting for 12 percent of total foreign investment in the country. As of early 1995, about 160 wells have been drilled. Most of Vietnam's crude is shipped to refineries in Japan and Singapore.

Thus far, Vietnam is the only country which has faced a military confrontation with China in the South China Sea. Vietnam became a full member of ASEAN in July 1995, which has strengthened its position in the dispute with China. Therefore, it is unlikely that Vietnam will soften its position with China regarding the Spratly Islands, even though these two countries have normalized relations after a border war in the late 1970s.[22] More importantly, Vietnam's economy has become reliant on the oil it has developed in the South China Sea.

The Philippines produces a very small volume of crude oil product compared to other claimants of the Spratlys. In recent years, crude oil production in the Philippines has ranged from only 10,000 b/d to 20,000 b/d. The Philippines' main concern in the Spratly Islands is one of economic advancement. Manila hopes its economy will take off with the discovery of oil under the sea territory it claims. Oil income is viewed as a means of eradicating poverty and strengthening the country's economic base. The 1951 Philippine–US Mutual Defense Treaty, in theory,

[22] See Reuters North American Wire, April 11, 1997, "Vietnam Says Deadlocked with China in Oil Talks", and April 10, 1997, "China Sees no Quick Fix for Vietnam Oil Dispute".

applied to the country's land territory. The Philippines are making efforts to extend this Treaty to cover the new territory in the 200-mile exclusive economic zone in order to use a US military umbrella to defend its ownership of potential oil reserves.

Like all major oil producing countries, oil income has pushed Malaysia up into the ranking of upper-middle-income countries around the world.[23] Malaysia's crude production has steadily increased during the last decade, from 440,000 b/d in 1985 to approximately 700,000 b/d in 1997, averaging a growth rate of 4 percent per annum. This growth was primarily due to an increase in the pace of oil exploration and development in the Spratly archipelago. These wells were drilled by and are operated by a consortium including Malaysia and several foreign countries. Consequently, Malaysia is not interested in the proposal of shelving the dispute and developing oil jointly. Its policy is to lay aside bilateral disputes but to develop oil individually.

The Chinese Petroleum Corporation(CPC), a state-owned company in Taiwan, has reportedly explored oil in the area under its control in the Spratlys, in order to alter the situation in which there is virtually no domestic crude output (there was only 1000 b/d crude production over the past six years in Taiwan). However, the Spratly Islands are not the only territorial flash point for Taiwan. The debate on Taiwan's independence and the unification with mainland China has made the Taiwanese government very cautious regarding any action they may take regarding the South China Sea quarrel. Involvement in this dispute will present Taiwan with a dilemma: the "one China policy", on which both sides of the Taiwan Strait continue to focus, will not lead Taiwan to side with China in any territorial dispute. Moreover, protecting territory or developing oil in the disputed area may worsen Taiwan's relations with other South Asian countries upon which Taiwan depends for help in breaking China's diplomatic restraints. Furthermore, Taiwan's focus is on admission to the World Trade Organization, the International Monetary Fund, and ultimately the UN. Therefore, we anticipate that Taiwan will be surprisingly quiet over the Spratly dispute.

The geological location of Brunei gives it a great advantage despite its small land area. The country's sizable oil supply reminds the world of its importance as an oil supplier in the region. In Brunei, oil production peaked at 250,000 b/d in 1979, but slid in the 1980s to bottom out at about 140,000 b/d in 1987. Since then, its crude production has grown slowly but steadily to reach 170,000 b/d in 1997. So far, Brunei has not quarreled with others regarding territory in the Spratly Islands, and will probably continue to maintain its crude operation. Despite its territorial claims to

[23] See *World Development Report, 1996,* the World Bank.

part of the Spratly Islands, Brunei is unlikely to become involved in any military conflict due to its small size and other claimants unwillingness to take aggressive action against it.

1.2.4.2 Who tips the power balance?

None of the countries which claim sovereignty over the Spratly Islands have sufficient surveillance capabilities and military vessels to patrol all of the waters they are claiming. Over time, diplomatic efforts will help each claimant in the dispute, but are not likely to settle the issue decisively. The final determination will depend heavily upon the military and economic power backing individual claims. Any military action (such as new occupancy) must be backed by the armed forces (Table 1.7). There is a great possibility that the dispute over territory may well lead to an armed conflict.

Table 1.7 Defense expenditures and armed forces of the Spratly claimants in 1995

	Brunei	China	Malaysia	Philippines	Taiwan	Vietnam
Military expenditures (billion US$)	0.27	63.51	2.44	1.15	13.14	0.54
Armed forces (000s)	5	2930	122	110	425	550
Military expenses as % of GNP	6.0	2.3	3.0	1.5	5.0	2.6
Military expenses as % of central government expenditures	n.a.	18.5	12.4	8.5	34.9	10.9
Military expenses per capita, US$	919	53	125	16	618	7
Armed forces per 1000 population	17.1	2.4	6.2	1.5	20.0	7.6

Source: *World Military Expenditures and Arms Transfers, 1996,* US Arms Control and Disarmament Agency, July 1997.

Of the various claimants, China is the only country that has the military power to alter the existing occupancy balance. China has a modern range of destroyers, frigates, and submarines, as well as thousands of jet fighters. Furthermore, China has increased its military orders for aircraft and warships supplied by Russia, which further exacerbates the vast disparity in comparative military strengths. Undoubtedly, there will be an increased Chinese vessel presence as well as more construction activity in these disputed islands. Continuing high economic growth will strengthen China and will increase its capability for further military operations.

The Philippines have limited military capacity and economic power. Philippine military forces are ill equipped, with only nine elderly F-5 jet fighters (seven F-5A and two F-5B). Its navy includes only one small patrol craft and eight small gunboats.[24] Therefore, the Philippines will rely primarily on diplomatic operations in conjunction with the US and ASEAN. First, the Philippines have lobbied the US Congress to revise the 1951 Philippines–US Mutual Defense Treaty. If the US feels that the freedom of the South China Seas is being threatened by the presence of Chinese warships, Congress may revise its 1951 Treaty with the Philippines to include territorial waters. If so, the Philippines may be willing to take more aggressive action. A conflict between China and the Philippines over the Spratly Islands will force the US to make some difficult policy choices. In addition, the Philippines have brought the issue of the Spratly Islands to the attention of ASEAN and other international organizations. ASEAN has taken a unified position on the Spratly Islands dispute, indicating its serious concern over recent developments. However, ASEAN's power as a regional organization is still weak. On the one hand, most ASEAN member countries fear China's expansion in the South China Sea. On the other hand, they do not want to see any break in normal relations with China. Finally, without assistance from superior military forces, the Philippines have no card to play in the Spratly Islands dispute. Any outright military confrontation with China will force the Philippines to retreat from the islands they claim.

Vietnam also has a limited and technologically inferior navy and air force. In the confrontation with China, Vietnam is not a serious threat to China in terms of military capability. But it has the capacity to maintain a persistent and effective harassment effort toward China if it were to lose its islands.

Malaysia has a relatively modern navy and air force. In a confrontation of short duration, it has the capacity to engage China militarily. But its military capability will weaken in a longer conflict. More importantly, no superpower supports Malaysia's position.

Japan has significant investments in Vietnam, and has made efforts to settle the South China Sea dispute. Japan needs to maintain stability in the South China Sea to ensure steady oil transportation in this corridor. Increasing oil supplies from the South China Sea will significantly reduce the need to ship large amounts of crude from the Middle East to Japan. However, Japan's political and military position is not strong enough to allow it to play a key role in the resolution of the dispute.

US influence in the region was diminished when it left its military base in the Philippines. The US has indicated that it will not become embroiled in the Spratly wrangle. However, this does not mean the US will take a hands-off stance. In fact, US policy towards the South China Sea issue

---

[24] The discovery of a Chinese presence on its claimed island has accelerated a $2 billion armed forces modernization bill passed by the Philippine Congress in early 1995.

may eventually shift. In any case, US involvement in the Spratly dispute will likely result in a direct political and diplomatic rift with China. Although economic ties between the US and China have grown rapidly over the past few years, the relations between these two countries are still delicate. For the US, a new rift over the Spratlys may simply add another problem onto the list of its policy conflicts with China. However, the US could also possibly concede on this issue in order to make progress on other more critical issues such as nuclear sales, arms proliferation, and cooperation with China in the UN Security Council.

1.2.4.3 A solution in sight?

The evolution of the political and territorial balance will be determined primarily by the policies of the US and China. Essentially, the dispute over the Spratly Islands is one between the US and China, because no other countries, including Vietnam, Malaysia and the Philippines, can compete with China for military influence in the region. Both the US and China consider the South China Sea as a strategic area. Any policy towards the Spratly Islands dispute will be determined in view of strategic considerations. US concerns are not only the safety of navigation through the Strait of Malacca, but also the strategic balance in the western Pacific Ocean, and US interests in this region into the next century. China's concerns focus on two issues: (1) energy supplies and security in its economic development; and (2) safeguard of its access to the Indian Ocean in the future. Nevertheless, China is a more critical player in this game than the US.

Any possible solution of the Spratly Islands dispute will depend decisively upon China's next action. After its offer to shelve the dispute and develop resources jointly was rejected by other claimants, China faces three options. The first is to maintain a low profile in diplomatic efforts but to expedite the expansion of occupancy of the Spratly Islands, including the building of so-called "fisherman's shelters" to mark China's ownership. However, this strategy will be transparent to other countries, and will likely be resisted by ASEAN member nations. In this case, the US and Japan will stand behind the ASEAN countries, and the organization will push for a strong stand against China. In this scenario, some military conflict is unavoidable. China has indicated that if its offer of discussion on joint development continues to be rebuffed, it will take control of the islands forcibly.

The second option is that China makes a great concession to other claimants in order to ensure a peaceful and stable environment to develop its economy and increase its national strength. China's decision to accept this particular solution would depend on pressure not only from the external situation but also from internal conditions: political and economic turbulence and military tensions in the Taiwan strait could push Chinese policy in this direction. In this scenario, the parties would retain control of their current possessions. A stalemate over the disputed Spratly archipelago is expected to remain of long duration. Even with a peaceful solution, boundary disputes over oil exploration are likely to occur and will affect drilling zones. Joint development of the Spratly Islands' resources remains questionable due to the difficulties of determining exactly how each country would cooperate in oil exploration in this area. Ironically, China has maintained its levels of crude imports from Malaysia and has increased the imported volume of crude from Vietnam during the past years, some of which has originated from the South China Sea region. Does this imply that China acquiesces in the ownership of these oil producing area or does it reflect an inconsistency in China's territorial claims and its oil import needs? Its softer line on crude imports, particularly from Vietnam, may well explain its difficulties in settling offshore boundary disputes.

The final option faced by China is to use military means to capture part or most of the islands, which are now occupied by others. Past international practices indicate that as time goes by, the possession of any territories will legalize ownership and establish individual countries' control over the relevant land and water. China is concerned that the current occupation of its claimed islands will translate into permanent sovereignty by others. However, any possible military conflict will have a heavy cost for all of the parties involved. China may pursue a strategy of steadily overcoming each claim individually in order to minimize domestic costs and international criticism. Of all other claimants, China may first challenge Vietnam, because so far Vietnam has no major power backing it. The Western countries would probably stay out of the dispute and leave the issue of conflict resolution to ASEAN member nations.

In any event, China faces high expenses for maintaining occupation of the islands. Meanwhile, China also runs a risk of increasing diplomatic confrontation with ASEAN, as well as with nations outside the region. The Philippines, Malaysia and Vietnam will take a more aggressive diplomatic stance against China, and are likely to ask for assistance from ASEAN and even the UN in order to receive the backing of the international community in their efforts to maintain their territorial claims. China may in fact choose some or all of these options at different points in time, depending on the circumstances of the moment and the motivations of its leaders.

In conclusion, an easing of political tensions could lower potential risks and benefit all the parties, including the commercial interests of foreign oil companies. However, an escalation of conflict over the oil-laden waters could jeopardize normal diplomatic relations between China and the other nations involved, as well as foreign oil companies' activities in what appears to be a promising field. The ultimate resolution of the dispute over the Spratly Islands could set an example for the future handling of rival claims, but whether the approach will be one of cooperation or conflict is unclear. In this case, it is hoped that China and the other claimant countries will find a fairly cooperative approach to South China Sea oil exploration.

1.2.5 Political deadlock in the Taiwan Strait: economic relations help bridge the gap

With Taiwan's first democratic presidential election, tensions in the Taiwan Strait were exacerbated by increased military pressure from mainland China. After Taiwan's presidential election, tensions in the Taiwan Strait will not easily subside because the basis of the conflict remains unresolved. Thus, the threat of a clash remains. At Taiwan's presidential inauguration, newly elected President Lee offered to fly to Beijing on a "journey of peace". While some viewed this as a hopeful sign, it is more realistic to view it as primarily a diplomatic gesture rather than as a concrete step that can be taken to solve the deadlock between China and Taiwan.

1.2.5.1 "One China" policy

Although the "one China" policy (which holds that both mainland China and Taiwan are part of China) is agreed upon by both governments in theory, in practice it reflects two very different views. For mainland China, "one China" indicates itself, the People's Republic of China. Its "one state, two system" policy can be interpreted to refer to the coexistence of capitalism and socialism under the supervision of a communist central government. (Hong Kong integrated into this system on July 1, 1997.)

However, for Taiwan, "one China" represents a historical and geographic reality. Within its acceptance of the "one China" policy, Taiwan emphasizes that both current *political* entities in the Taiwan Strait must be recognized. The interpretation of the "one China" policy is that it does not conflict with the existence of two separate governments or political systems. Taiwan seeks to break its diplomatic isolation and desires international recognition at a level corresponding to its economic power. (It is the third largest holder of hard currency reserves and the 14th largest country in international trade.) Interestingly enough, as the mainland asserts that "one China" is the People's Republic of China, Taiwan has increasingly shifted its view about the "one China" policy from an abstract term of one Chinese nation having two political systems toward a definition of it as the "Republic of China" (i.e. Taiwan).

Despite some conciliatory speeches on both sides, there have not been any signs of a dramatic change in relations between China and Taiwan. Taiwan continues to insist that the mainland make two commitments: (1) not to exercise its military threat toward Taiwan, and (2) to regard Taiwan as a separate political entity in international diplomacy. There is virtually no chance that the mainland will concede to either of these two demands, because doing so would risk giving the green light to Taiwan's independence movement.

In the years when Chiang and his son were in power, Taiwan safely stood back as an observer rather than a player in international affairs. But President Lee has been moving Taiwan toward active participation in world affairs, and in doing so may be leading both sides of the Taiwan Strait from a systematic competition toward a national confrontation. The election of Taiwan's independence-minded President Lee to another five-year term is likely to continue this trend toward greater international visibility, and toward increased resistance to the mainland Chinese government's efforts to impede progress in this area. A "peace agreement" with China promoted by President Lee during and after his presidential campaign was more a campaign slogan than a sincere gesture of conciliation.

Political relations between China and Taiwan are unlikely to return to the momentarily-friendly status existing before Mr. Lee made a private trip to the US in June 1995. Since China has targeted President Lee as an advocate of secretly promoting independence, his reelection is bound to further antagonize the mainland's political leaders, and may push China further in the direction of a military solution to the stalemate. In addition, bilateral talks (suspended after Mr. Lee's US visit) will not take place unless this atmosphere of mutual distrust is remedied.

Given that the political balance in the Taiwan Strait has been disrupted, China and Taiwan are unable to maintain the relatively cordial relationship they enjoyed prior to President Lee's visit to the US in 1995. With unification in the foreseeable future unlikely, there remain four possible scenarios for further development: (1) a fragile balance and continued peace with some compromises on both sides; (2) substantial improvement in relations if the mainland changes its interpretation of the "one China" policy; (3) further tension if Taiwan increases its efforts toward international recognition or if the mainland promotes a hard-line policy; and (4) escalation of tension leading to military action. We regard the first scenario as the most likely to take place over the next few years.

In order to ease the tensions in the Taiwan Strait, substantial concessions need to be made by both sides: (1) both must fundamentally reassess their respective views of the "one China" policy; (2) Taiwan must limit its efforts to seek international recognition and discourage the movement toward independence; and (3) China must promise not to use force against Taiwan.

1.2.5.2 Likelihood of military conflict

While it is impossible to guess the precise circumstances that may cause a conflict to begin, there is an increasing likelihood that military action will viewed as a viable policy alternative by the mainland. During the coming years, military action could be triggered by: (1) official recognition of Taiwan by the international community, or (2) China's domestic political pressures and its desire to visibly assert dominance in the region. Any Taiwanese activities perceived to be additional steps along the road toward independence are likely to bring about swift action on the part of China.

If China takes any military action against Taiwan, China will face three significant consequences: (1) international condemnation of the action and pressure to cease aggression; (2) an increased chance of direct conflict with the US; and (3) a halt of investment flows from Taiwan. Contrary to the opinion of many political commentators, we believe that if China senses that Taiwan is on the verge of declaring independence, military action against Taiwan will be unavoidable.

The Taiwan issue has been a factor for instability in East Asia. Tensions between China and Taiwan could spread throughout the entire region, threatening political stability as well as fragile economic relationships. Given its potential for disruption, the Taiwan question could surpass its scope of definition by China as an internal issue, and thus become a concern for neighboring countries. However, the possibility of internationalizing the Taiwan issue (which has long been promoted by the advocates of Taiwanese independence) is an outcome that Beijing would not like to see. Yet, any military action will undoubtedly result in greater international attention and possible censure, both political and economic. Beijing has to carefully consider the negative consequences that would result from military action, and to determine whether its policy goals warrant such severe measures.

The US role is critical in this game. The Taiwan Relations Act of 1979 provides for US military support of Taiwan, but this creates a dilemma for the US. On the one hand, if the US accepts the responsibility of defending Taiwan, it will encourage Taiwanese advocates of independence to irresponsibly declare their separation from China, which would certainly result in military action by China. On the other hand, if the US indicates the intention to remain uninvolved in any potential conflict,

Chinese hard-liners will see few factors constraining their increasing propensity to use strong-arm tactics or to actually launch a military attack.

The tensions between China and Taiwan are caused by many factors, including historical resentment and ideological differences, and could be further complicated by the US presence in the area, thereby introducing a third party into the conflict, one whose active participation could lead other nations possessing pretensions to superpower status to also become involved, further disturbing the delicate strategic balance in Asia. Furthermore, as Taiwan's first democratically elected president, Lee is likely to continue to pursue greater international visibility, and may accept an invitation to visit the US again. This could develop into another round of political and diplomatic confrontations between Taiwan and China, as well as between the US and China. Before this occurs, each side must cautiously consider both the political, economic, and military policy implications.

1.2.5.3 Growing economic linkage

Economic relations between China and Taiwan have been strengthened substantially in the past years. The fundamental economic equation of this relationship is as follows: the mainland needs capital, technology, and managerial skills from Taiwan, while Taiwan needs the mainland's resources, cheap labor, and huge market. Total trade between China and Taiwan reached about $19 billion in 1996, up more than four times from 1991 levels. In both 1995 and 1996, Taiwan was the third largest source of direct investment in mainland China following Hong Kong and Japan (see Table 1.8). Capital flows from Taiwan into China have the potential for even greater acceleration if the Taiwanese government further loosens its restriction on investment in China.

Table 1.8 Foreign direct investment in China

	1991	1992	1993	1994	1995	1996
Direct investment ($US billion)						
Hong Kong	2.58	7.71	17.44	19.82	20.19	20.85
Japan	0.81	0.75	1.86	2.09	3.21	3.69
Taiwan	0.47	1.05	3.14	3.39	3.17	3.48
US	n.a.	n.a.	n.a.	n.a.	3.08	3.44
Total	4.37	11.01	27.52	33.77	37.81	42.14
Share in total direct investment (%)						
Hong Kong	59.0	70.0	63.4	58.7	53.4	49.5
Japan	18.5	6.8	6.8	6.2	8.5	8.8
Taiwan	10.8	9.5	11.4	10.0	8.4	8.3
US	n.a.	n.a.	n.a.	n.a.	8.2	8.2

Sources: State Statistical Bureau and *China Statistical Yearbook, 1997*, China Statistical Publishing House.

Deteriorating conditions for domestic investment, combined with international competition, have increasingly forced Taiwanese businesses to transfer their labor- and resource-intensive industries to the mainland. During the past few years, approximately 25,000 Taiwanese businessmen have invested an estimated US$ 20 billion in China, both with and without government approval, in order to cash in on cheaper labor and raw materials there. This trend gives every indication of continuing, assuming that the political situation does not dramatically disintegrate.

Economic linkages have been assisted by a loosening of previous regulations. The Taiwanese government began liberalizing China-bound investment in late 1990, while it has continued to ban official or direct commercial links with China since 1949. The Taiwanese government has liberalized indirect commerce with China by raising its investment ceiling of mainland-bound investment by local manufacturers from approximately US$ 1.5 million (NT$ 40 million) to approximately US$ 2.3 million (NT$ 60 million). The Taiwanese government also limits Taiwanese investment to an amount less than 20 percent of the company's capital. However, to bypass this restriction, some Taiwanese investment in China has been made through specific companies' overseas subsidiaries. The Taiwanese government is also considering altering its policy of indirect investment (which requires Taiwanese companies with a subsidiary in a third country to invest no more than US$ 1 million in China) by raising the maximum investment to US$ 2 million.

Both sides have derived substantial benefits from these growing economic ties. The cut in capital flows from Taiwan to China that would result from a political or military conflict could negatively impact Taiwan's economic growth as much as China's, given the close linkages between the two economies.

1.2.5.4 Consequences for oil

Should tensions escalate, the Chinese Army has four potential military options: (1) invading the smaller islands belonging to Taiwan; (2) a selective missile strike on Taiwan; (3) enforcing a blockade around Taiwan; and (4) a full military invasion and occupation. While it is difficult to assess the actual feasibility of these various options, it is important to examine the possible effect of military action on the oil market.

Due to geography, China may have the capability to completely blockade Taiwan's international shipping lanes. Chinese missile tests near Taiwan's two major ports, Kaohsiung and Keelung, which oil tankers use to approach Taiwan and off-load shipments, were indicative of China's intention to disrupt the sea navigation lanes if necessary.

Taiwan lacks domestic energy resources and imports 96 percent of its energy. Oil accounts for more than a half of total energy consumption. In 1996, Taiwan imported more than 700,000 b/d of crude oil (Table 1.9), of which approximately 75 percent was shipped from the Middle East. This is a potentially precarious situation, especially if China were to begin a blockade of the island.

Table 1.9 Energy imports in Taiwan

	1991	1992	1993	1994	1995	1996
Imports (000 b/d)						
Oil	534	579	610	630	712	730
LNG	39	41	44	55	63	65
Share of imports in total energy consumption						
Energy	95%	95%	96%	95%	96%	96%
Oil	53%	53%	53%	52%	54%	53%
LNG	4%	4%	4%	4%	5%	5%

Source: Energy Statistical Data Book, 1996, Energy Commission, Ministry of Economic Affairs, Taiwan, March 1997.

Taiwan realizes the importance of diversifying its sources of energy supply, and has been taking action in this regard. Taiwan purchased crude from several other sources. For instance, the island received its first shipment of Alaska North Slope (ANS) crude in 1996. North Sea crude was also shipped to Taiwan in 1996 with the first crude term contract of 10,000 b/d. A deal of another 10,000 b/d crude was completed with Argentina in late 1996. In addition, Taiwan will also import about 30,000 b/d of crude from Zaire, making it one of the few Asian term customers for West African crude oil.

Taiwan also took steps to increase its strategic crude stocks as tensions with China grew. According to Taiwan's Energy Management Law, the CPC, a state-owned oil company, must have at least 60-day reserves of crude oil, 45-day reserves of gasoline and diesel fuel, and 25-day reserves of liquefied natural gas at all times. During the particularly tense period immediately prior to the presidential election, CPC increased its strategic crude reserve to more than 60 days of stocks.

If military pressure from China escalates into a blockade of Taiwan's main island, its 60 or 70 days crude inventory could be drained very quickly. Meanwhile, a sudden crude surplus of 700,000 b/d will also cause jitters in the oil market. Prolonged disruption of imported supplies could effectively cripple any Taiwanese resistance, a fact of which both governments are well aware. Any blockade would, however, also increase international pressure on China, and would force its Asian neighbors to take sides. Thus, any news that China is beginning an actual military drive against Taiwan will not only affect prices in the Asian crude oil market, but it will also have significant strategic geopolitical effects whose unpredictable outcome could have repercussions on the world oil market as well.

1.2.5.5 Oil cooperation: an exception?

Ironically, there have been some improvements in the oil business between Taiwan and China. Although Taiwan's government prohibits its state-owned companies from doing business with China, it is reported that the CPC has been granted an exception and will buy mainland crude and may also organize a joint venture for oil exploration in China's offshore regions.[25] Lifting the restriction on mainland oil purchases indicates that Taiwan has given priority to its goal of diversifying its energy supply sources.

China National Offshore Oil Corporation, a state-owned offshore exploration company in the mainland, has invited CPC to join offshore oil and gas exploration in the Pearl River Mouth basin and a subsidiary of CPC in the East China Sea. If CPC accepts, this will be the first joint endeavor between mainland and Taiwanese oil businesses, as well as a sign that Taiwan might be willing to lift its restriction on state companies investing in the mainland. In addition, if the Taiwanese government still enunciates a "one China" policy, they may, in the future, attempt to use this policy to justify efforts to try to exert some authority over natural resource development on the Chinese continent. Oil exploration, by its nature, is thus an element of Taiwan's interest in expanding its sovereignty to the mainland. Increasing involvement by Taiwanese companies in the mainland secures both parties' interests, and should cement evolving economic cooperation.

For the last 40 years, Taiwan and China have had no direct oil trading links. Taiwan bought a limited number of crude oil barrels from China mainly through Hong Kong during the last few years. At the same time, Taiwan occasionally sold gasoil through a third party into mainland China. However, in the past, if CPC knew in advance that a cargo of gasoil, which was sold to a trader, might go to China, CPC would cancel the deal. As a state-owned oil company, CPC has to abide by the rules set by the government, but it cannot always be certain of the final destination of the gasoil it sells. Lifting the ban on purchasing China's oil will clear the way to oil trading between Taiwan and China. Specifically, CPC wants light sweet Daqing crude, and may purchase as much

[25] It was reported that CPC withdrew an oil and gas exploration vessel from the Taiwan Strait before Chinese missile tests.

as 10,000 b/d over the next few years, as long as the economic relationship between China and Taiwan does not completely disintegrate. However, due to the shortage of petroleum products in the South and the East, China may not be able to provide Taiwan with any significant amount of petroleum products in the immediate future.

While economic relations between China and Taiwan will continue to improve as trade ties grow, the political situation will remain unchanged unless one side or another makes a significant concession. However, this is something that at the present neither appears willing to do. With tied economic relations between both sides of the Taiwan Strait, China will certainly gain more than Taiwan. In addition, during the economic race, time is on China's side if there is no more political turmoil domestically. The economic strength of China could overshadow the economic miracle that has happened in Taiwan. As time goes on, Taiwan may lose its bargaining chip in political talk with China.

1.3 Conclusion

In the past decade, the Chinese economy has evolved away from a centrally planned economy toward a more hybrid one. It is hoped that economic reform will improve the central government's ability to manage a multiple-level economy through improvements in areas such as enterprises, taxation, finance, and pricing. While the economic "go–stop" cycle of the past is unavoidable, the prospects of economic growth are bright. Economic growth has averaged over 9 percent for the past decade, and it is expected to continue to average 9 percent in the late 1990s and the 20th century. Growth will remain cyclical, however, rising well over 12 percent in some years and falling below 7 percent in other years.

The reform and openness policies initiated by Deng Xiaoping in the 1980s have advanced to the point of no return. Economic liberalization will continue and may even accelerate during the post-Deng era.

China is also entering an extremely sensitive period of political transition. But there is little chance that China will either retreat into Maoist isolation or fly apart like Yugoslavia. Political change will resume on a moderate course over the years to come.

import quotas restricted the volume of product imports in 1994. Inflexibility in oil allocation blocked product flow between regions, as well as crude flows among the refineries. The refineries in the south could not received adequate crude supplies from the central government to produce products, while surplus products stockpiled in the north and could not be shipped to the south due to a transportation bottleneck. Therefore, total petroleum product demand showed virtually no growth in 1994, compared to about a 300,000 b/d increase in 1993.

We have touched on how macroeconomic and political reform has affected China's economy and its energy sector. In this chapter, we will examine in greater detail the specific developments that are of acute concern to the oil industry. A summarized evolution of China's oil policy is presented in Fig. 2.1.

This chapter focuses on oil industry reforms and oil policy in China. Section 1 concludes the recent reforms in the oil industry. Section 2 details oil pricing, distinguishing two stages: price deregulation before May 1994 and price reregulation after May 1994. Section 3 reviews the oil distribution system. In particular, this section discusses the reorganization of oil distribution in China. Finally, Section 4 reviews and summarizes oil import licenses, quotas, and tariffs in China and addresses the impact on the oil market of changes in these areas.

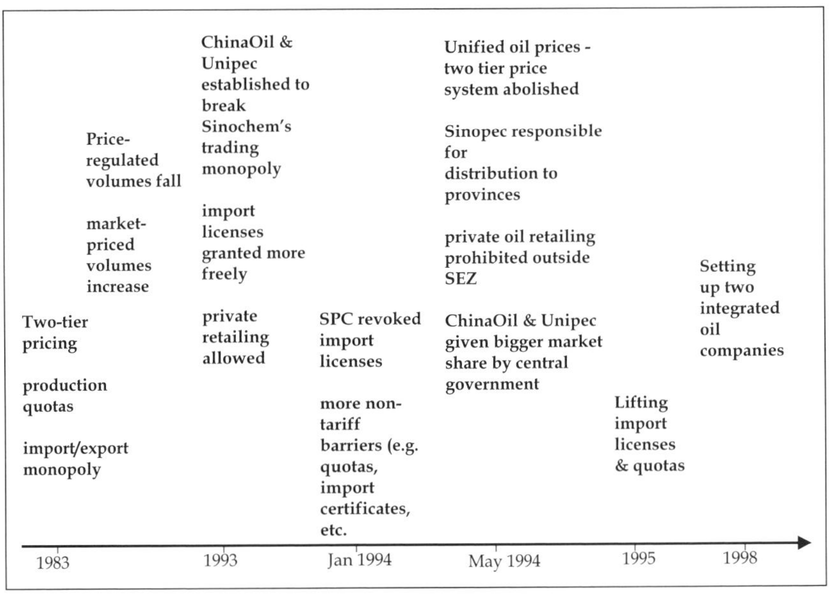

Fig. 2.1 Evolution of China's oil policy

2.1 Oil Industry Reforms

China's oil industry and market has experienced significant change. But the oil industry still has little autonomy over production and pricing decisions. The year of 1994 showed that, fearing inflation and an overheated economy, the central government stepped back in to reregulate oil prices and centralize the distribution system, which has prevented the oil industry from enjoying the boom that market mechanisms created in many other sectors of the economy. The main facets of oil industry reform are: (1) establishing petroleum exchanges; (2) forming joint stock sharing companies; (3) eliminating guaranteed profits for refineries; (4) industrial integration; (5) trade liberalization; and (6) exchange rate deregulation.

2.1.1 Establishing petroleum exchanges

As part of the government's efforts to move to a market-oriented economy, petroleum exchanges were set up in Shanghai, Nanjing, and Beijing, where trade was conducted in local currency. The exchange consisted of domestic oil producers, refiners, traders, financial institutions, and investment houses. Crude, gasoline, gasoil and LPG were traded in the exchanges. Oil companies dominated trade, however, with speculators playing only a marginal role. The petroleum exchanges facilitated oil price reform by providing the central government with important market information, but 90 percent of China's crude distribution was still determined by the state, leaving little to offer at the petroleum exchange houses. Trade in petroleum products, less regulated by the central government than crude, surpassed crude trading.

China's fledgling oil futures exchanges were suffering from a lack of liquidity due partly to government rules against foreign participation. If China lifted the ban on crude import licenses, imported crude was expected to be traded freely at the exchanges, and then foreign traders might be allowed to participate in the exchanges. In addition, the Chinese securities exchanges were planning to let foreign securities firms become members. But there was no sign that the government would allow foreign investors to be members of the petroleum exchanges.

However, the trading in exchanges faded after the central government reregulated the oil prices in May 1994. Moreover, petroleum exchanges have been merged into other commodities futures exchanges, but the activity on oil disappeared. The regulated oil prices are not expected to be abandoned in the near future. As a result, no trading volume in oil will be seen in any commodity exchanges in China.

2.1.2 Emerging stock-sharing companies

While privatization has gathered much momentum in the global oil industry, China is embracing a form of it in line with the transition in its economic system. Despite central government intentions to change many state-owned companies into joint-stock companies during the period from 1998 to 2000, actual progress may come slowly. However, there are some precursors in the oil industry. One example is the Petroleum and Petrochemical Company of Shanghai (PPCS), which is no longer receiving any state subsidies. PPCS, a subsidiary of Sinopec, has been a downstream experimental model in raising its own funds for the expansion of its refining operation. PPCS put its stock on the New York Stock Exchange at the end of July of 1993,[1] and raised $342.5 million by selling 30 percent of its assets. At the beginning of the stock issue, PPCS was selling 18 million shares at $20 per share, reaching a peak value in December 1993 of $48.125 per share. Meanwhile, Zhanghai Petrochemicals in the eastern coastal province also became a publicly listed company. There are similar plans to list more coastal refineries.

The first stock-sharing upstream company is based on a small and new oilfield, Toutai Oil Company in Heilongjiang Province. It is owned by four state-owned enterprises (SOEs) and began operations in 1993. Toutai Oil Company is still in the initial stage of operating differently from the other state-owned crude producers, and its level of success remains to be seen.

Stock issuing in SOEs, which are assured a constant source of raising funds, highlights the softer approach taken by central government toward SOEs. When funds come from most nonstate-controlled sources, SOEs enterprises are likely to provide output without receiving planned indicators from a central plan, neither for their production nor for their purchases and sales. However, loosening the reins of control over the state-owned oil industry seems to be on a trial basis. Whether the move away from state control continues will partly depend on the success of the first experiments mentioned above. It is still too early to be more specific how to conduct a change in the total oil industry. Because reshaping the huge state-owned firms in the oil industrial sector will be among the most difficult of institutional changes in a centrally planned economy, no one knows for sure which approach is the best.

The decision in 1992 to set up an upstream conglomerate, China National Petroleum Corporation (CNPC), replacing the Ministry of Petroleum Industry, may improve planning and coordination under the existing system, but is unlikely to advance the cause of industrial competition. In addition, the losses in the oil industry will continue to be

[1] As of 1993, seven Chinese state-owned companies have stock traded on Wall Street.

subsidized by central or local government budgets. Along with industrial reform, the fundamental trend in the oil industry is likely to accelerate the progress of the market-orientation movement. But it seems unclear whether the central government wants to abandon its large share in and control of the oil industry, and whether stock-sharing firms will become the dominant approach to reform.

2.1.3 Elimination of guaranteed profit for refineries

The reform of the tax system will have the most far-reaching impact on state-owned refineries. Under the previous contract-based system of revenue sharing, refineries were allowed to keep any profits exceeding the amounts they had agreed to remit to the local or central government. Such a system benefited refineries with low crude input costs and high petroleum product prices controlled by the central administration, leading to guaranteed profit for these refiners. The tax-sharing system, which took effect in January 1994, required refineries to pay three major categories of tax: value-added tax (VAT), enterprise income tax, and sales tax. The combined effect of these three taxes has been, in many cases, to eliminate the after-tax profits of the refineries. This change in the tax base will also be a factor in price setting for both crude oil and refined products.

The crude and product pricing reforms have also eliminated the guaranteed profit for refineries. Prior to the price reregulation in 1994, the ratio between crude oil, gasoline, and gasoil was roughly 1:2.9:2.7 in China, compared to the ratio of 1:1.6:1.3 between Dubai, unleaded gasoline, and gasoil in Singapore in 1993. 1994's price reregulation has lowered the ratio down to 1:2.0:1.7,[2] but this is still higher than that in the larger Asian market. Nevertheless, mandated reductions in refinery margins have shifted excessive profits from refineries to previously loss-making crude producers.

2.1.4 Vertical integration

China's oil industry focuses on expanding investment by diversifying decision-making and attracting outside financial sources. Upstream, downstream and trade companies have joined together to set up refinery and petrochemical works, or cooperate with provincial governments and international oil companies to expand downstream activities. For instance, with the expected additional revenue from its crude sales to Sinopec following the deregulation of crude prices, CNPC intends to expand its downstream sector (Table 2.1). CNPC plans to create five

[2] The ratio is taken from an average of nine city prices in May 1994.

refining–petrochemical integrated centers before the end of the century.[3] Daqing, as one of these centers, is promoting a number of downstream projects, which are expected to increase the petrochemical share in its industrial output. Sinopec has started oil trading to import crudes for its coastal refineries. Sinopec is also expected to move further into the retailing business, while Sinochem will play a larger role in the refining business. Early in 1993, Sinochem was also stripped of its monopoly on the import and export of crude and refined products, and its market share was cut, which created two joint-venture trading companies: ChinaOil joined with CNPC, and Unipec joined with Sinopec. Joint-venture agreements, either with domestic oil companies or with foreign companies, have become the preferred route for vertical integration.

When CNPC expanded its participation in foreign upstream projects,[4] Sinopec and Sinochem also began looking for upstream and downstream investment opportunities in South Asia and in other places around the world. China's major oil companies are seeking equity interests in oil projects around the world to broaden their sources of crude and petroleum product supply because the gap between domestic supply and demand is expected to widen after the year 2000. Similar steps are being taken by China National Offshore Oil Corporation, which is planning to set up petroleum-trading subsidiaries overseas.

Table 2.1 Restructuring of China's oil industry in 1998

	Crude oil production (million b/d)	Refining capacity (million b/d)	New assets
CNPC	2.14	1.86	19 companies from Sinopec including Daqing Petrochemical Works 23% stake in joint-venture refinery with Total of France in Dalian 44% stake in acetic acid joint venture with British Petroleum in Sichuan
Sinopec	0.72	2.42	12 companies from CNPC including Shengli oilfield

Source: "Slick Maneuvers: China's Petroleum Shake-Up Dashes Competition Hopes," *Far Eastern Economic Review*, July 2, 1998, p. 61.

[3] They are Karamay and Dushanzi in northwest Xingjiang, Daqing in northeast Heilongjiang province, Zhongyuan in central Henan province, and Bohai Bay in north China.
[4] Sapet Development Corporation, a subsidiary of CNPC, signed an oil and gas production contract with Peru's state oil company Petroleos del Peru along Peru's northern coast. Sapet is committed to invest at least $25 million during the years from 1995 to 2000, signaling the country's first oil investment in Latin America.

All the efforts of integration made progress in 1998. Setting up two integrated oil conglomerates initialized the preliminary industry reorganization. CNPC started to operate a number of refineries which were previously under Sinopec's control. Meanwhile Sinopec also took over several oilfields, which belonged to CNPC before. CNPC's territory is in the north and the west, while Sinopec is in the south and the east. These two conglomerates are given the power to set the retail prices of refined products and issue permits for oil imports.

This vertical integration is expected to limit overlapping investment in the oil industry, protect domestic markets from imported foreign products, and create some efficiency gain. However, this vertical integration is unlikely to induce domestic competition at least in the near future, and will not solve the problems of outdated technology and lack of management skills in the industry. Moreover, this market-oriented industry restructuring will not pave the way for more foreign investment downstream, because China's oil policy that governs the foreign investment in the oil industry shows no sign of change yet.

2.1.5 Trade liberalization

There are two official hurdles to crude and petroleum product imports: (1) international trade licenses, which were previously issued exclusively to Sinochem but in 1993 were extended to two others, ChinaOil and Unipec; and (2) import quotas. Chinese oil buyers have to apply for import quotas and work through licensed trading companies to import barrels (for a more detailed discussion, see Section 2.4.1, Import licenses and quotas).

Despite strict regulation of international trade, trade liberalization in China has been centered on two key aspects. First, China's four special economic zones have been given the right to import crude oil and petroleum products by themselves without strict constraints on import licenses and quotas. Second, Sinopec, provincial refiners and traders were occasionally assigned import licenses, which caused them to act independently of Sinochem, and led to more organizations becoming involved in the oil market.

It was reported that the central government has abolished crude import licenses and quotas (for details, see Section 2.4.3, Lifting import licenses and quotas for crude: what next?) If market competition happens, it may increase the linkages between the Chinese oil market and the world oil market, bringing them ever closer.

2.1.6 Impact of foreign exchange rate on import–export

A new foreign exchange system is in place, with an exchange rate of about 8.5 yuan against 1 dollar. If the rate goes to 10:1, imported oil measured in domestic prices will be significantly higher; if it sinks to below 8:1, imported oil prices will fall accordingly. The unified two-tier foreign exchange rate (i.e. a state-controlled and a swap market exchange rate — for more detailed discussion, see Section 1.1.2) has eroded some of Sinochem's profits, which were made by purchasing domestic crude and refined products at low state-controlled exchange rates, then selling the crude and product on the world oil market. Earnings by crude import-oriented refineries have also been negatively impacted, reflecting an increased cost for imported feedstock. Beneficiaries of the lower yuan have encouraged the oil industry to export, but higher oil demand resulting from the economic boom has diminished the overall amount available for exports.

2.2 Oil Pricing

Because the growth rate of the oil supply is dependent on capital input, the major obstacle in the development of the oil industry in China has been insufficient investment. Before 1980, oilfield development and refinery constructions were primarily financed by central government budget allocations as part of the fixed capital investment program. With state-controlled low prices and soft-budget constraints, there were large deficits between revenues and costs. The funds needed for self-development in the oil industry was therefore insufficient. Even though most offshore oil projects have been permitted to sell their crude at world market prices, onshore upstream development has suffered from a shortage of funds. In addition, pricing systems also partly contribute to the paradox of high-energy intensities per unit value and low energy consumption per capita.

2.2.1 Price deregulation

The major achievement of price deregulation was to phase out subsidized prices for petroleum products, except in a few sectors such as agriculture and the military. The price scheme introduced in 1985 allowed upstream and downstream industries to adjust their output prices in relation to the fluctuation of input costs. Freeing product prices became the key to future financial success. In upstream industries, there were three different prices: (1) "in-quota" low prices (300–500 yuan/ton, depending on the quality of oil, US$ 4.95–8.25/bbl equivalent at the exchange rate to 8.3 yuan per dollar); (2) "in-quota" high price (about 700 yuan/ton in 1993, US$ 16.82/bbl equivalent, and 964 yuan/ton in 1997, US$ 15.91/bbl equivalent); and (3) market prices. As Table 2.2 shows, since 1988, "in-quota" prices have been incrementally raised by three times for low prices and by 90 percent for high prices. Under this system, CNPC sold a declining percent of its production (about 80 percent in 1997) to Sinopec at state-controlled low prices (both low and high in quota prices). In return, the central government provided billions of yuan annually in annual subsidies for profit-losing oil producers. However, free market prices were also permitted, for as much as 20 percent of CNPC's total production.

Table 2.2 Crude oil prices in China

	Low in-quota (regulated low price)		High in-quota (regulated high price)		Exchange Rate (yuan/US$)	
	Yuan/ton	US$/bbl	Yuan/ton	US$/bbl	Official	Market
1988	115	4.23	500	18.41	3.72	n.a.
1989	144	5.25	500	18.22	3.76	n.a.
1990	174	4.99	500	14.33	4.78.	n.a.
1991	210	5.99	530	15.13	4.80	n.a.
1992	210	5.05	530	12.74	5.70	n.a.
1993	300	7.21	700	16.82	5.76	n.a.
1995	700	11.48	1200	19.69	8.35	8.35
1997	880	14.52	1200	19.81	8.31	8.31

Note: Both official and market foreign exchange rates are used before 1993.

Sources: *China Statistical Yearbook, 1997*, China Statistical Publishing House, the World Bank, China's National Petroleum Corporation (CNPC), Shanghai Petroleum Exchange, and *China OGP*, Vol. 5, No. 2, January 15, 1997 p. 7.

Price decontrol in petroleum products is outpacing crude price reform. Since 1982, there have been two different price levels: "in-quota" low prices and market prices for products once the refinery has met the quota assigned by the state. The proportion of market-priced refined products increased rapidly from less than 10 percent in 1983 to about 65 percent in 1993. That year, only one-third of total refined products were being sold at state-controlled low prices. Unlike crude oil prices, two-thirds of petroleum product retail prices in China are at or higher than world market prices. For example, prices for gasoline and diesel fuel are substantially higher than world prices, despite their generally poorer quality. Between 1983 and 1993 the market price was applied to even more product sales, and government subsidies were left on only a few consumer categories, notably diesel fuel for farmers, fuel oil for certain enterprises in the strategic sectors, and refined products for military use.

Since 1986, the central government has allowed regions with surplus hard currency to import petroleum products, which were in severely short supply. This resulted in product imports rising at an annual growth rate of 33 percent since 1989, which also caused a great upward pressure on domestic product prices.

Elimination of price controls is seen as a sign that the central government is anxious to pull the oil industry out of the red. Higher crude prices have allowed onshore oilfields to invest the additional earnings into increased exploration and development, greatly increasing the activity of this sector. Increased crude prices did not cause the government to transfer subsidies from crude producers to refiners, due to the fact that prices for most petroleum products are already high enough to keep refinery margins profitable. However, price decontrol has also forced refineries to pay more for feedstock, while retail product prices are still centrally set for sales to farmers and the military. As long as policy-protected consumers (such as the military and farmers) remain short of cash, centrally set refined product prices will not be abolished in the near future.

2.2.2 Price reregulation: one step back, two steps forward?

The effort to control inflation and cool down China's overheated economy has reversed the course of oil price reform. On May 1, 1994, prices for both crude and refined products were re-controlled by the state along with a broad range of consumer goods and key commodities. It was also up to the state to decide differences between wholesale and retail prices for individual cities and provinces. These government-mandated price guidelines have been combined with new tax regulations and a unified foreign exchange system.

Three different crude price levels have been set up: (1) a regulated low price (RMB 700/ton, or US$ 12.00/bbl equivalent); (2) a regulated high price (RMB 1200/ton, or US$ 20.00/bbl); and a few market-level prices. Crude priced at the regulated low price makes up approximately two-thirds of total market share, while the crude supplied at the regulated high price makes up the remaining third, with little crude supplied at market prices.

Under these rules, the state allocates about 80 percent of total annual crude production to be sold at the price of about 754 yuan/ton ($11.87/bbl) for Daqing crude and 684 yuan/ton ($10.77/bbl) for other crude.[5] Crude oil sold at these prices is supplied to refineries operating under state plans, most of which come under Sinopec's control. The other 20 percent of crude, known as nonstate-allocated crude, has prices ranging from 1160 yuan/ton to 1310 yuan/ton for Daqing, Shengli, and other crudes (see Table 2.3). These crudes are supplied to refineries

[5] Daqing crude always has a price about 70 yuan/ton higher than others.

processing volumes beyond their centrally planned volumes, namely refineries which are controlled respectively by CNPC oilfields, local authorities, and the Ministry of Chemical Industry. The new price system is seen as a major boost for CNPC, as more crude is sold at significantly higher prices than before.

Table 2.3 Crude oil prices after price re-regulation

	Yuan/ton	US$/bbl
Daqing	1310	20.63
Shengli	1220	19.21
Gudao	1160	18.26

Notes:
(1) Crude prices in major oilfields set by the central government.
(2) Foreign exchange rate of 8.7 yuan per US dollar is applied.
Source: State Statistical Bureau.

Different wholesale prices for refined products are set for three different regions: Xinjiang in the west, Qinghai-Gansu-Yumen in the central region, and the rest of China in the east. For petroleum products coming from refineries located in remote areas such as Xinjiang and Qinghai (where transportation is difficult and costly), wholesale prices are relatively low compared with those from refineries in coastal regions. Both wholesale and retail prices for petroleum products will be determined by the central government (the State Planning Commission (SPC)) and local governments (local pricing administrations). Thirty-five major cities had their retail prices decided by the SPC. For other cities, local pricing administrations calculated retail prices for their respective regions according to a formula designed by the state. Nevertheless, price recontrol did not move the price back down to the previous state-controlled price system. As Table 2.4 shows, the reregulated prices remain well above those paid in the international market, although direct comparison of domestic and international prices is difficult due to differences in exchange rates and product quality. As of the completion of this book, however, not all new prices have taken effect.

Another part of these new regulations establishes that when joint-venture refineries sell oil products processed with imported crude on the domestic market, prices will also be set by the state. Foreign-funded enterprises (both singly owned and joint ventures) are not allowed to market their imported petroleum products in China's domestic market.

There are several negative effects resulting from these new price regulations. First, the decision to fix oil prices has plunged the country's fledgling oil futures markets into confusion. Price ceilings have killed futures

Table 2.4 Refined product prices after price re-regulation

	#70 Gasoline		#0 Diesel oil	
	Yuan/ton	$/gallon	Yuan/ton	$/bbl
Wholesale prices set by the central government in three major regions after May 1994				
Xinjiang	2500	0.80	1755	27.04
Qinghai-Gansu-Yumen	2180	0.70	1830	28.20
The rest of China	2290	0.74	1900	29.27
Retail prices in selected cities in June 1994				
Tianjin	n.a.	n.a.	2000	30.82
Shenyang	2000	0.64	1850	28.50
Shanghai	2400	0.77	2349	36.19
Hangzhou	2635	0.85	2212	34.08
Qingdao	2180	0.70	2000	30.82
Wuhan	2400	0.77	1950	30.05
Guangzhou	2700	0.87	2270	34.98
Chongqing	2550	0.82	n.a.	n.a.
Xi'an	2500	0.80	2050	31.59

Note: Foreign exchange rate of 8.7 yuan per US dollar is applied.
Source: State Statistical Bureau.

trading in crude and petroleum product contracts, and the exchange may not open again unless the price ceilings are removed. Second, while the new state prices of some petroleum products range around international prices, the fixed prices of diesel and gasoline are well above the spot market. That means that restricting imports, on the whole, will raise the cost of energy for domestic consumers. Third, the new price regulations will not be able to identify which refineries need to be replaced over time to increase efficiency. Unlike price controls on coal and grain, which were intended to help consumers cope with inflation, the oil price controls appear to have only one obvious beneficiary: the state oil monopoly. Finally, price regulations are prompting a wait-and-see attitude about large-scale foreign investments, particularly those involved in downstream production and distribution.

At this point in time, it is not clear how long the price regulations will last. These regulations are expected to be eliminated after China's economy becomes healthier and inflation is under control. These administrative measures give China only short-term stability, not the structural improvements needed for long-term economic growth. In the foreseeable future, China will return to market-driven pricing while maintaining the option of again imposing short-term price controls if they are deemed necessary.

China's tightly regulated oil pricing system is facing a direct challenge.

First, over the next few years, more major refining and petrochemical plants will be using imported crude oil, since domestic crude supply is suffering an increasingly serious shortfall. It will be difficult for these refiners to financially reconcile their constantly fluctuating import costs with regulated domestic product prices. Second, some onshore oilfields, which have foreign investments, were allowed to sell crude into the domestic market at international prices in 1997. This complicates and possibly destabilizes the regulatory price structure in China. Third, there is an urgent need to obtain additional funds (through higher crude oil pricing) to increase exploration in remote western areas.

Under these circumstances, the central government is being pressured to enact price reform. In a segregated oil industry with separate upstream and downstream companies, upstream companies, and a few refiners who use imported crude, have an incentive to push price deregulation as a means of improving their profitability. However, because of their inefficient operations, the other refiners are not in favor of this change. Deregulated oil prices would subject refiners to margin volatility. Nevertheless, the emergence of a new oil pricing system is expected to further open China's oil market and eventually encourage additional foreign investment.

2.3 Oil Distribution Policy

There is an imbalance in the distribution of petroleum product between northern and southern China. Supply in the north usually exceeds demand, while the south has to depend on imports of crude and products from the north and from the rest of the world. Meanwhile, new product distribution centers have been built or are under construction in the southern and eastern provinces, and service stations are springing up quickly in large cities and along new highways. Product transportation logistics still rely heavily on railroads and waterways, however, so some restrictions and bottlenecks will persist.

2.3.1 Distribution decentralization (1991 to early 1994)

The distribution of oil products in China is done by outlets operating under three major sales organizations: the sales company under Sinopec, the sales company under CNPC, and the sales company supervised by provincial authorities. But emerging agents and organizations, even private, were actively involved in product wholesaling, an activity normally prohibited to private firms. The dominant role is played by Sinopec's sales company, which is supposed to balance distribution among different regions and supply the major consumers:

railways, public transportation, civil aviation, the army, the state oil products reserve, the petroleum industry itself, and export companies.

Along with the expanding price decontrol in 1992 and 1993, refineries and sales companies were allowed to sell products not only to end-users, but also to unlicensed re-sellers, who sold the products with a high price markup. Meanwhile, private petroleum retailers were booming around the country, and within only two years they made up about three-quarters of all gas stations. These private retailers even took advantage of the liberalized economic environment and diversified into the wholesale business, which was prohibited. The decontrol of petroleum prices and distribution increased availability and released a huge pent-up demand for oil products in China. This in turn led to a situation where, even with imports surging, product prices spiraled upwards, because supplies were not imported quickly enough to meet the new, higher level of demand. The inflationary effects of product prices rocketing upwards and the foreign exchange balance effects of the much higher level of imports were too much for China's economic planners to bear. As part of a program to reduce inflation and cool the economy, product distribution, imports, and prices were recontrolled in May of 1994.

2.3.2 Distribution recentralization: who wins and who loses?

In line with the price and tax reforms conducted before 1994, the oil distribution system is in a process of transition. In May 1994, the central government started to reorganize domestic product distribution in an effort to maintain unified prices and increase the consistency of product quality. Under the new distribution system, the central government designated Sinopec as the exclusive agent to run the country's wholesale business. Six regional sales companies under Sinopec (i.e. the regional branches in the north, northeast, east, south-central, southwest, and west) have been given responsibility for supervising and coordinating product sales, but they may not sell products directly to consumers (prior to the distribution re-organization, the six major regional Sinopec subsidiaries could sell directly to consumers). Provincial sales companies receive refined products directly from Sinopec (usually at a refinery gate) at state-controlled prices, and then resell products to local retailers and the major end-users. Benchmark retail prices are set by the central government for 35 major cities, while wholesale prices are not fixed and could be 4–6 percent lower than retail prices, depending on region (for more details, please refer to Section 2.2 in this chapter).

In 1994, the central government also took back crude allocation authority, part of which had been assigned to oil producers in 1993 (i.e. the crude producer could sell crude output above its production quota to refineries at market prices). Under such a recentralized allocation system, Beijing provided crude equally to refineries based on their capacities. When the central planners saw high inventories, they cut the crude supply of every refiner equally, regardless of the demand situation which particular refiners faced. This inflexible crude allocation caused refineries in the east and south to received inadequate crude supplies to meet their burgeoning regional demand, while product inventories in the northeast (where demand growth was relatively slow) remained high. As a result, national refinery runs in 1994 were held in check for the first time in 30 years.

The new distribution system constrains the provincial sales companies in trading with other provincial sales companies by requiring them to take full responsibility for their local product supply. The provincial sales companies have been warned that if they run out of a product and have previously sold some to another province, the penalties will be severe. Adequacy of supply to meet provincial demand will have to be guaranteed before sales to another province will be considered. Provincial sales companies are also forbidden to sell products to any resellers. Also, product imports are allowed by only three state trade companies: Sinochem, ChinaOil, and Unipec. Refineries and provincial governments are no longer allowed to import themselves, despite their import quotas. (For details of import license and quota, see Section 2.4). The net effect of these changes is to dramatically reduce the flexibility of the distribution system and to firmly consolidate control over prices and volumes of petroleum products in the hands of the central government.

After this recentalization, profit margins for gasoline station operators throughout the country shrank dramatically from 20 percent to about 5–7 percent. It was reported that central authorities would close or take over many small-scale retailers and distributors who had thrived on selling low-cost imports. State distributors will guarantee that CNPC and Sinopec sell their stocks of crude and refined products before imports are allowed, relegating imports to a role of supplementing domestic supplies, rather than competing with them.

This distribution policy will have several negative effects. First, decreased flexibility in the distribution system could force provincial sales companies to hoard supplies. Second, the new distribution policy will reduce interprovincial trade and undermine economic growth in the region where there is great demand growth. For instance, reassertion of control over the distribution network creates potential supply problems for consumers in southern China. China produces and refines most of its oil in the north, but China's greatest economic growth, as well as oil demand growth, is in the south. Distribution regulations will limit local government and companies' short-term flexibility in meeting demand increases.

This distribution monopoly will force foreign oil companies to work through Sinopec to sell oil products in China, because China does not want retail networks set up by foreign firms. There is still a ban on foreign-owned gas stations in mainland China. But there is an exception. Up to 1997, in Shenzhen Special Economic Zone (SEZ), China allowed 28 foreign-owned gas stations selling 150,000 tons of gasoline a year, most of it imported. Such foreign-owned gas stations are allowed to pay half the standard import tariff (discussed in greater detail in Section 2.4.4).

In September 1996, a provisional regulation was approved by the State Council that allows foreign companies to set up joint-trade ventures for the first time with Chinese partners in Shanghai's Pudong area and in Shenzhen, both of which are SEZs. This could be regarded as a preliminary step in breaking the monopoly in China's international trade long held by state-run companies. With the increasing flow of oil into the domestic market, this policy change is expected to lead to more competition among trading companies and lower prices for Chinese end-users.

It was reported that Japan's Idemitsu opened a service station in northern China's Dalian Economic and Technology Development Zone. The station is operated through a joint venture between Idemitsu (70 percent) and ChinaOil, a Chinese petroleum product trading company (30 percent). The latter has agreed to supply gasoline to the service station from Daqing, Dalian, and other domestic refineries. The retail business partially owned by foreign companies in SEZs will serve as a test case for Chinese policy-makers in determining if foreign investors will be permitted to establish a retail network beyond SEZs. To gain entry into this potential market, foreign investors should seek business opportunities that are in the form of joint ventures or are located in an SEZ, or both.

Joint-venture refineries are designated to sell their products in the world market. However, the policy stipulates that they can market a certain percentage of their products in the domestic market. How large is the percentage depends much on the needs of the Chinese market and the negotiations with the related government authority.

2.4 Oil Import Policy

The Chinese government has established three different categories of commodities, each with a different level of regulation. For the first type of commodity, import/export privileges are given exclusively to designated state agencies, and the volume of trade is based on quotas assigned by the central or provincial government. For the second type of commodity, imports and exports are not chartered, but the volume is still based on quotas assigned by the central or provincial government. The third type of commodity consists of those not considered very important, and their import or export requires neither chartered agencies nor an import/export quota.

2.4.1 Import license and quotas

Crude oil is included in the first type of commodity.[6] Gasoline, diesel oil, kerosene, and fuel oil have been in the second category since 1983, but were moved into the first category on May 1, 1994, as part of a program to control inflation and to guarantee receipt of the taxes levied by the central government (for a more detailed discussion about tax and price reform, see Sections 1.1.2 and 2.2.2). Other petroleum products, such as lubricants, belong to the third category.

Inclusion in the first category of commodities has important ramifications for trade in crude oil and the principal petroleum products. China's SPC[7] has limited the number of import agencies, and it strictly oversees the volume of oil imports by issuing a restricted number of import licenses on a quarterly basis. Until the early 1990s, Sinochem held a monopoly in the import/export of crude oil and all petroleum products. Attracted by the profits Sinochem was earning, China's upstream company, CNPC, entered into a joint venture with Sinochem to form Unipec. Similarly, China's national refining company, Sinopec, formed a joint venture with Sinochem to create ChinaOil. Both of these offspring companies have been licensed to import and export products and crude. In addition, some other state-owned companies and local governments have been designated to trade either on a case-by-case basis or for a specified time period.

In 1994, the central government issued new regulations governing export of crude and products to clarify responsibilities among importing and exporting companies. While crude designated for export by the SPC continues to be handled by Sinochem, about 35 percent of that total export volume is to be exported by ChinaOil, with the proceeds going for repayment of foreign loans. Refined products allocated to Sinopec and its affiliated refineries for export are to be handled jointly by Unipec and Sinopec. Meanwhile, four refineries under Sinopec — Guangzhou, Maoming, Zhenghai, and Gaoqiao — were given the right to import crude oil and export refined products for their own purposes. In general, few licenses have issued directly to end-users, such as power stations.

[6] Identified by the SPC of the Foreign Economic and Trade Ministry (FETM), formerly named the Ministry of Foreign Trade and Economic Cooperation, decree No. 1 of December 29, 1993 (published in FETM "Gazette" No. 24 of January 18, 1994).

[7] The name of State Planning Commission (SPC) was changed to State Planning and Development Commission (SPDC) in 1998.

Under the new regulation, oil import and export licenses for refiners, some state-owed companies, and some end-users have been withdrawn by the central government for an unknown period.

In addition to the SPC, two government bodies play a role in administering China's import quotas: the State Economic and Trade Commission (ETC) and the Foreign Economic and Trade Ministry (FETM). The former manages import quotas, and the latter oversees quotas for a variety of general commodities.

All import channels, including general trade, trade by foreign funded enterprises and border trade, are subject to import quotas set by the state. The total import of petroleum products is primarily planned and managed by the SPC, according to the import applications presented by the provincial governments. The SPC assigns the next year's quota to each province at the end of the current year. The SPC sets import quotas based on the sometimes conflicting principles of achieving a balance between imports and exports and remaining in line with projected demand and historical provincial quota volumes and usage, as determined by the ETC. This system rewards oil gluttony; the greater the consumption in the previous year, the greater the chance of obtaining a high quota this year. The quotas are given mainly in terms of monetary value and in a few cases by quantity.[8] It is essential to remember that requesting and receiving a quota is only the first hurdle that would-be consumers must clear. Another, perhaps more important task is finding an agency with the license and ability to actually import the product.

After getting their total quota from the central government, provincial governments allocate shares to the various companies and local governments in the region, based on the same principles applied by the SPC. The exception to this subsequent allocation at the regional level is for the import of crude oil. Crude imports are tightly controlled by the SPC, which assigns quotas to provincial governments and Sinopec. There is no further quota assignment for crude import to local governments and other companies.

After quota allocation, import certificates are also required. The import certificates can be granted by three government agencies: the central bureau of FETM, one of the regional offices set up by FETM, and the provincial Bureau of Foreign Economic and Trade (BFET). The FETM provides certificates for crude oil and petroleum products, while the FETM's regional offices and BFET grant certificates for petroleum products only.

---

[8] The bill setting the import quotas must be approved by both the ETC and the State Council before submission to the National People's Congress.

The government makes two lists. The first consists of the various import quotas of crude oil and product imports for every region. The second list details the imports of products for which no quota is necessary (e.g. lubricants). The government keeps an eye on these imports since they consume foreign exchange. The makeup of the two lists is determined jointly by the ETC and FETM. Tendering is encouraged but not required for all imports, and products not on either list can also be opened for tender at the request of a client. Contracts below US$ 100,000 are not usually tendered. Tenders are usually open to either internal or international bidding and are usually carried out by the local branches of the relevant importing agent.

Third party processing arrangements can gain tariff and quota exemptions, but they are subject to import/export licensing regulations. In general, refineries that are engaged in third party processing (at the time of publication of this book, there were four such refiners in China) usually have some spare capacity and are located close to port facilities. They access their crude traders with import licenses. Besides the government bodies mentioned above, the crude imports and corresponding refined product exports resulting from third party processing are supervised by Chinese Customs authorities. Government officials verify third party processing agreements by ensuring that a certain quantity of refined products, often specified as a certain proportion of crude imports, is exported.

There is some pressure in China to allow greater imports of refined products. In line with the centralization of distribution (see Section 2.3.2), large-scale SOEs in coastal and river areas and coastal provincial governments are no longer granted the right to import oil. Only designated state companies (i.e. Sinochem, ChinaOil, and Unipec) are allowed to import crude oil and the four principal products (gasoline, diesel oil, kerosene, and fuel oil). This new regulation has encountered some resistance from the SEZs, where some local distributors and gas stations are financed by foreign investors. In addition, despite the stringent state control on oil imports, import sources other than state-designated trade companies exist, largely because more influential state-owned companies and local governments have found ways — e.g. border and barter trades — to include crude oil and refined products in their overall import lists.

The import quota system for crude and refined products is not expected to weaken in the near future. However, it is important to note that the Chinese government's ban on oil imports was most effective in its initial phase, and that its effectiveness faded with the passage of time. The rule seems to be: the stricter the policy, the shorter the policy's real life.

At the same time, policies that try to transform the industry often find stiff, entrenched opposition at several levels. For example, the implementation of ambitious reforms in taxation, finance, and SOEs planned for 1994 faces a variety of obstacles and challenges, making achieving of these reforms doubtful. Thus a corollary to the above rule might be: the greater the change, the longer it takes. The upshot of these two phenomena is the gradualism will characterize change in the Chinese oil sector over the next decade.

The far-reaching new rules issued in 1994 raise some trade issues. China has pledged under its Market Access Memorandum of Understanding with the US that it will eliminate all quantitative restrictions on refined product imports. It means that oil imports cannot be restricted by license or quota. Controlling imports must therefore be a short-term expedient if China is to continue to move toward harmonization with international trading practices. The full set of new rules affecting trade in the oil business will inevitably conflict with China's desire to pursue admission to the World Trade Organization (WTO).

2.4.2 Periodic import ban

During the years from 1994 to 1995, the central government banned petroleum product imports twice: the first four months of 1994 and the second quarter of 1995. This act reflected a still weak domestic oil market amidst the continuing macroeconomic fight to control inflation, as well as a concern for domestic refineries' survival in the face of pressure from the cheap imported products.

Import prospects depend on three variables: domestic demand, price differential to other Asian markets, and oil policy. Chinese oil policy seems volatile and disorderly, but it is the most important factor influencing the oil trade in China. The trade press often reports on Chinese "import quotas". While the quota announcements deserve attention, foreign companies and analysts should be cautioned not to equate quotas with future imports. The quotas tend to work in only one direction: that is, import volumes have trouble exceeding quotas, but high quotas are not completely met.

1993 was the year when the central government loosened its control to promote a new round of economic reforms. Consequently, the market boomed and the regulations on import quotas and licenses were relaxed, resulting in soaring product imports. The quickly expanding imports of the first half of 1994 were a consequence of 1993's liberal policy. But when the government steeped back from this decentralization policy in mid-1994, it dramatically curtailed it import quotas and actual Chinese imports plummeted. Product imports have remained low since then.

In 1994, the ban on crude and petroleum product imports was only partially effective. The first four months' figures revealed that there was enough leeway in the restrictions for product importers to move material into the country. Product imports increased in February, March and April, after declining in January 1994. The increase in product imports was primarily prolonging a continuous demand increase in China since 1993. The Chinese government's ban on oil imports in 1994 was effective only in its early phase. Its effectiveness faded with the passage of time.

In 1995, quotas were scheduled to be assigned semi-annually; and the quotas for January to June 1995 were higher than the same period of 1994. However, the government suspension of product imports in the second quarter meant that many importers who held quotas could not use them. 1995's "suspension" of imports was different from 1994's "ban", but its effect on Singapore margins was similar.

Three additional points merit mention. First, an import ban or suspension does not mean that China's imported barrels will drop to zero. Certain provinces and companies can always obtain special treatment, and military imports are never restricted by any regulations. Second, due to the higher value-added tax (VAT) after the tax reforms of 1994, oil smuggling has flourished. The volume of smuggled refined products is increasing in terms of absolute barrels, but still accounts for a small percentage of total product imports. The smuggling is concentrated in Guangdong and Fujian (where demand is rising), and diesel and fuel oil are the main products favored by smugglers. It was reported that the government is cracking down on smuggling, but oil smuggling will continue to provide a relatively minor share of total oil product imports. Third, Chinese policy allows local governments and refineries to import crude and petroleum products using their own hard currency. With that any central government ban on oil imports cannot be very effective for long. As long as regional economic growth places more hard currency in local government coffers, regional markets will require and find more refined product imports. The ability of the central government to limit imports is, therefore, subject to some constraints.

However, the central government will continue to use bans or suspensions for product imports when necessary. Furthermore, the import quotas of products for the next few years, assigned quarterly or semi-annually, are unlikely to be high.

2.4.3 Lifting import licenses and quotas for crude: what next?

Under pressure to open its internal market as a precondition for entering the WTO, China scraped import licenses and quotas for 367 commodities in 1995. Crude oil was on the list. In the negotiations with WTO, however, China claimed that if any obstacles continue to prevent China from becoming a member of WTO, China would retreat from its trade policy shift. In short, China has made this conditional trade policy change reluctantly, and it presents a political and economic tradeoff for admission into the WTO, rather than any domestic desire or demand for such an action.

Crude oil was selected because it is not consumable by end-users. Refineries' crude intake and production are completely administrated by the central government, and the change will have virtually no impact on the domestic market. To promote exports and to attract foreign funds, Beijing has made some sacrifices. In 1982, China first started its policy of trading its petroleum resources for foreign capital in offshore oil exploration. This policy has further expanded to inland Tarim oil exploration. Importantly, however, a reliable policy allowing foreign capital into the domestic oil market has not yet emerged.

After liberalizing crude trade, two developments are possible. First, China has indicated that a new "crude registration system" will be created. In its nature, such a "registration system" is de facto regulation replacing the licenses and quotas. Although we do not know the details of the new restrictions, they would function in the same way as the old crude regulation. The "registration system" may simplify the old regulations (which posed three hurdles in their procedure — requiring licenses, quotas, and import certificates), but it will probably serve as a more ambiguous and flexible tool to control the market. The central government can use such a system with fewer complaints about trade barriers, and can change the requirement and content of the new system at will.

Dropping the regulations on crude import licenses and quotas does not mean that foreign crude will be more accessible to domestic traders and refiners. There is no refinery capacity outside of Sinopec and CNPC to run additional imported crude; therefore lifting the requirements to have licenses does not present any new opportunities. Sinochem (crude importer and exporter), Unipec (crude importer) and ChinaOil (crude exporter) can continue to monopolize transactions in the international crude oil trade. Other importers will have difficulties competing with them, not only because these three have trading experiences and credit, but also because without the government's permission refiners will not refine crude from other sources.

The second possible development would be for the central administrative control over crude imports and exports to become too weak to constrain the trade transaction (if crude trade can be made profitable). This is more likely, if the political situation evolves in such a way that the central government has less political and economic power to keep local governments and enterprises within bounds. In the worst case, the sources of crude supply for refineries could be disrupted, and refinery output fall due to disarray of imports and exports. If such disorder threatens, the central government will tighten its restriction to take a tougher stand than before.

Will China abolishing licenses and quotas for petroleum products as it did for crude oil? The answer is no. The lifting of license and quotas for crude oil seems to be a gesture made for political and economic reasons. Scrapping import licenses and quotas for petroleum products would indicate that China would like to deregulate its oil market. That will happen only if and when the macroeconomic environment demands a more free market. Sharing its oil market with foreign investors is likely to be the *last* step in China's agenda. The import quota system for refined products is not expected to weaken in the near future. In fact, recent events suggest that the regulations could tighten further if the government felt it necessary.

2.4.4 Tariffs

While China's oil policy remained essentially stable, a number of new measures were proposed and adopted. Their implementation is likely to have a far-reaching impact on the Chinese oil market in the future.

In an effort to gain admission to WTO and as a policy change to reduce import costs in line with high oil prices in the world oil market, China has made a commitment to reduce its overall level of tariffs from the current 27.5 percent to 15 percent, in order to approach the general standard of WTO's member countries. From 1992 to 1997, China reduced tariffs several times on imported commodities, including crude and petroleum products (Table 2.5). In 1997, crude import tariffs fell from 3 percent to about 1.5 percent, and the gasoline import tariff from 15 percent to 9 percent. In 1997, crude import tariffs continued to fall from 2 percent to 16 yuan per ton. Similar cuts in import tariffs on naphtha and gasoline may take effect in the near future. Tariffs on other refined products were pared from 15 percent to 6 percent.

In line with these tariff reductions, China applied its first ever universal VAT while simultaneously tightening product import licenses. The tax changes are widely seen as a move by the Chinese central government to recover the revenue it forfeited through import tariff cuts. Withdrawing import licenses from some companies and traders, who held product import licenses before, would restrict import activities — especially those carried out by local authorities and end-users. The anticipation of the relaxation of tariff restrictions and the tax, both enforced for January 1, 1994, helped to cause the import boom at the end of 1993.

This is a further step toward the integration of China's oil market into

Table 2.5 Import tariff (%) of refined products

	1992	1993	1994	1997
Crude oil	3.0	2.0	1.5	16 yuan/tone[a]
Motor gasoline	15.0	n.a.	9.0	9.0
Naphtha	15.0	10.0	6.0	6.0
Kerosene	n.a.	n.a.	9.0	9.0
Light diesel	n.a.	n.a.	6.0	6.0
Heavy diesel	n.a.	n.a.	6.0	6.0
Fuel oil	15.0	15.0	12.0	12.0
Propane	15.0	12.0	9.0	6.0
Butane	15.0	12.0	9.0	6.0
Liquid paraffin wax	n.a.	15.0	12.0	9.0
Other heavy oils	n.a.	15.0	12.0	12.0
Liquefied natural gas	n.a.	15.0	12.0	12.0
Paraffin wax	35.0	33.0	30.0	9.0
Petroleum asphalt	25.0	23.0	20.0	8.0
Liquefied propane	n.a.	n.a.	n.a.	6.0
Liquefied butane	n.a.	n.a.	n.a.	6.0
Other LPG	n.a.	n.a.	n.a.	6.0

[a]Crude import is subject to this specific tariff starting from July 1, 1997.

Sources: *China's OGP*, October 15, 1997, p. 28, December 15, 1993, p. 2, and July 15, 1993, p. 5; and other sources.

the world market at a time when China has become a net importer of both crude oil and refined products. By reducing their costs, Chinese importers will have the additional flexibility of purchasing in the spot and/or futures market, or otherwise hedging their positions — choices, which are increasingly important, given high world oil market prices. With lower tariffs, China will be a more aggressive buyer in the international oil market. In addition, the import tariff cuts on petroleum products will squeeze the profits of oil smugglers, thereby reducing the (as yet undetermined) amount of oil smuggled into the country.

Despite the tariff cut, limiting the issuance of import licenses has been the dominant influence on the market and has effectively curtailed imports in 1994. Even for entities with officially sanctioned quotas, the import ban has crimped the business of importing fuel into China through legal channels. Smuggled fuel continues to seep into China (particularly from Hong Kong), but the quantities are rather small in the context of import volumes.

In June 1994, the State Council decided that foreign-funded enterprises in the SEZs) can import tax-free oil for their own production purposes and that foreign-funded gas stations serving those regions need only pay half the standard tariff. But there are already indications that some of the refined products imported by foreign-funded enterprises are flowing to inland markets beyond the SEZs. Due to this tariff immunity for foreign companies, the Chinese oil industry has raised concerns regarding product dumping, which slashes the revenues of state-owned oil sales companies. This has resulted in an effort, unresolved at this point, to impose restrictions on the importation of oil by foreign-funded enterprises in SEZs, to force them to conform to national policies.

2.5 Conclusion

Reforms in the oil industry are far from complete, and the legacy of central control remains. Investment decisions in oil enterprises remain strongly influenced by government intervention, competition remains largely absent, and the oil trade is still administered by government-controlled agencies internally and externally.

The cyclic growth of "stop–go" in the oil industry emanates from a fundamental tension in the oil sector. On one side lie the forces for liberalization, deregulation, and rapid petroleum product consumption growth, the booming economy, cheap imports, decentralization, and the growing political and economic clout of provincial authorities and businesses. On the other side are forces that promote more affordable growth: the central government, inflation watchers, and those charged with managing foreign exchange concerns. Neither of these two forces is completely in control. Each will try to find a way to counteract or sidestep the activities of other, and this dance is likely to characterize the oil sector over the next decade.

In 1994, the government re-imposed central control over its oil industry by fixing the price of crude and petroleum products, and channeling virtually all sales through state agencies. This move marks a temporary retreat from the ongoing experiment with oil deregulation, which has picked up speed since 1992. For as long as it lasts, this renewed central recontrol over oil prices and distribution will create increased uncertainty for foreign participation.

Chapter 3
Crude Oil Production

China's largest developed oilfields have already passed their point of peak productivity.[1] China's success in meeting its goals for production during the next 10 years will be largely determined by the extent of new discoveries and the speed with which they can be developed. Crude oil exploration on such a scale is inherently uncertain and future strategies for crude production and petroleum product consumption will have to take this uncertainty into account.

China's crude oil production grew at double-digit rates between 1960 and 1980, from 100,000 b/d to 2.1 million b/d. But since then, crude production has dramatically slowed down, growing by 2.5 percent per year during the 1980s and 2 percent per year from 1990 to 1997, reaching 3.2 million b/d in 1997. In 1979, based on erroneous estimates of potential reserves, political leaders issued forecasts that production would reach 10 million b/d by the year 2000. In early the 1990s, official forecasts called for 4 million b/d by 2000, but continuing correction was soon made, reflecting that this figure may be still overly optimistic. Official forecasts of crude production ranged from 3.2 million to 3.4 million b/d by 2000. Nevertheless, China's reserves and expanded production capacities, particularly in the western provinces, give it enormous potential after 2005.

In the short term, new discoveries in the northeast corridor will barely offset the depletion of older fields. Most of the increased production will come from remote and relatively unexplored fields in Xinjiang and Qinghai, with lesser additions from offshore oilfields. Offshore development has been disappointing thus far, and it will take more than 10 years (and significant investment) before any significant volumes are produced by the promising Tarim Basin. Additional volumes could be recovered from existing oilfields with infill drilling and other expensive improvements in equipment, technology, and practices. The crude slate

[1] In China, most oilfields are subsidiaries of CNPC, to some extent operating independently.

will remain predominately light and sweet, with some increase in light medium-sulfur crudes.

This chapter is divided into three sections. Section 1 considers issues regarding crude oil production, including an overview of oil reserves and production by oilfields. Section 2 provides a brief overview of the prospects for the exploration and development (E&D) of the Tarim Basin. Section 3 presents a methodology for forecasting crude production, as well as our projection of Chinese crude output over the next decade, including production constraints, the extent of exploration activities, and the adequacy of exploration and production techniques.

3.1 Crude Production Levels, Onshore and Offshore

Before 1998, five organizations were involved in crude production in China: China National Petroleum Corporation (CNPC), China National Offshore Oil Corporation (CNOOC), the Ministry of Geology and Mineral Resources (MGMR), local governments, and joint ventures between CNOOC and foreign oil companies. However, the MGMR and local governments only take very small share in total crude output (see Fig. 3.1). After 1998, China Petrochemical Corporation (Sinopec) jointed in the crude oil operation after taking over several oilfields from CNPC.

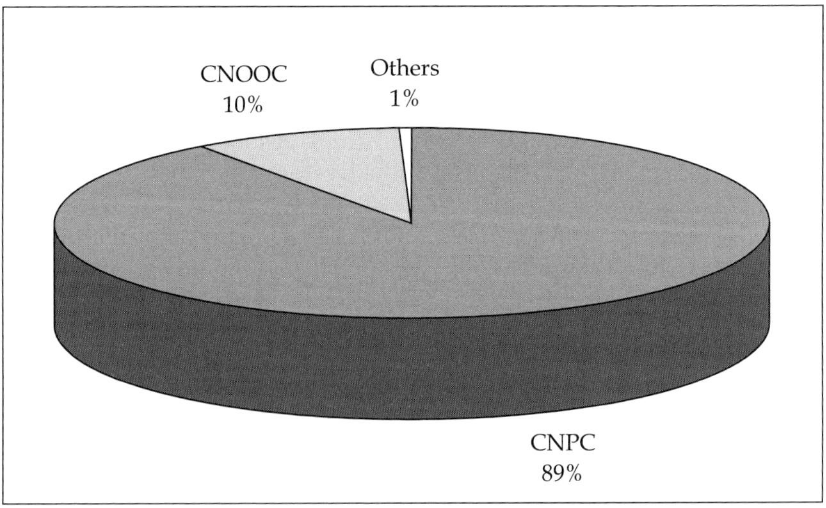

Fig. 3.1 Crude oil production by organization in 1996
Source: *China Petroleum Industry, 1996,* The China Petroleum Information Institute, CNPC, 1997.

The Constitution of the People's Republic of China occupies the highest position in the legal framework of natural resource exploration and exploitation. Article 9 of the Constitution provides that "[M]ineral resources, waters, forests, mountains, grassland, unreclaimed land, beaches and other national resources are owned by the State, that is by the whole people, with the exception of forests, mountains, grassland, unclaimed land and beaches that are owned by collectives in accordance with the law". In Article 9 of the Constitution, it also provides that "the State ensures the rational use of national resources and protects rare animals and plants. Appropriation or damage of natural resources by any organization or individual by whatever means is prohibited".[2] According to these provisions, the Chinese government has the right and power to regulate the exploitation and development of petroleum resources. The law sets forth the guiding principles and basic regulatory structure for activities of CNPC and CNOOC. The exploiters of petroleum resources must pay resource taxes and fees for resources extraction.[3] Table 3.1 lists the resources taxes for major oilfields.

Table 3.1 Resource taxes for major oilfields since 1986

Oilfield	Resource tax rates (yuan/ton)
Daqing	24
Shengli	8
Dagang	8
Henan	6
Huabei	3
Zhongyuan	3
Jilin	3
Liahe	1
Jianghan	0
Jiangsu	0
Yuman	0

Source: State Taxation Bureau.

In August 1995, to attract more foreign investment in China's oil E&D, the Ministry of Finance and the State Taxation Bureau announced a new cut in royalties which were paid by joint ventures of oil and gas exploration. The royalty amendments come as China starts its third round of

[2] PRC Constitution Article 9 (adopted December 4, 1982), published in Beijing Review, December 27, 1982, p. 13.

[3] Article 5 of the Law of PRC on Mineral Resources provides that "... the entity which extracts mineral resource must pay a resources tax and a fee for resources compensation". See Chen Mingyi, et al., *Selections of Environmental Laws and These* (Chinese), Wuhan University Press, 1989, pp. 94–101.

Table 3.2 Royalty rates for foreign companies operating (as percentage of production)

Annual production per field	New royalty in regions		Previous royalty
	Xinjiang, Qinghai, Tibet or shallow waters	Other areas	All regions
Crude Oil (10,000 tons)			
100	0.000	1.000	6.40
200	2.500	3.000	9.45
300	4.330	4.670	10.47
400	5.750	6.000	10.98
500	7.100	7.300	11.28
Natural Gas (100 million cubic meters)			
20	0.000	0.500	6.40
30	0.333	0.833	8.43
40	0.625	1.125	9.45
50	0.900	1.300	10.06

Source: Knight Ridder Money Center, p. 5306, August 23, 1995.

international bidding for exploration of 12 blocks in the Tarim basin and Junggar basin in Xinjiang. These new royalties divide the regions into those which have different E&D costs. In contrast to the previous policy, it requires less royalties in the remote western part of China (Xinjiang, Qinghai, and Tibet), which have high investment risk (see Table 3.2).

Table 3.3 shows how China's total crude production has climbed during the past 48 years. China produced barely 100,000 b/d of crude oil in 1960, but its 3.2 million b/d in 1997 placed China fifth among the world's oil producers. The average growth rate was 19.6 percent from 1960 to 1970, 13.2 percent from 1970 to 1980, 2.8 percent from 1980 to 1990, and 2.1 percent from 1990 to 1997. These growth rates exhibit the extent of the problems in the Chinese upstream industry. The rate of growth has dropped from double digits to 2 percent or less, and the addition of new oil production capacity has not been enough to meet expanding crude oil demand. The volume of discovered oil[4] accounts for less than a third of the oil-in-place actually found by extensive drilling. The remaining two-thirds or more of the oil cannot be extracted with ordinary technology.

The Chinese oil industry made great strides in the early 1960s, and China became a major world producer by the mid-1970s (Fig. 3.2), when oil began to play a major role in the development of the energy sector. During this period, the Daqing oilfield was developed, but there was very

[4] For various reserve definitions, see Figs 3.3 and 3.4.

Table 3.3 China's crude oil production history, 1950–1997

	Crude production (000 b/d)			Cumulative crude production (million bbl)		
	Total	Onshore	Offshore	Total	Onshore	Offshore
1950	4	4	0	2	2	0
1955	19	19	0	24	24	0
1960	104	104	0	125	125	0
1965	226	226	0	398	398	0
1970	614	614	0	1106	1106	0
1975	1541	1541	0	3154	3154	0
1976	1743	1743	0	3790	3790	0
1977	1873	1873	0	4474	4474	0
1978	2081	2081	0	5233	5233	0
1979	2123	2123	0	6008	6008	0
1980	2119	2119	0	6782	6782	0
1981	2024	2024	0	7521	7521	0
1982	2042	2042	0	8266	8266	0
1983	2121	2121	0	9040	9040	0
1984	2292	2292	1	9877	9877	0
1985	2498	2496	2	10,789	10,788	1
1986	2614	2606	8	11,743	11,739	4
1987	2683	2688	15	12,722	12,713	10
1988	2733	2718	15	13,720	13,705	15
1989	2754	2730	24	14,725	14,701	24
1990	2766	2740	26	15,735	15,701	33
1991	2820	2748	72	16,764	16,704	60
1992	2828	2756	72	17,796	17,710	86
1993	2872	2780	92	18,844	18,725	120
1994	2913	2784	129	19,908	19,741	167
1995	2968	2800	168	20,991	20,763	228
1996	3137	2838	299	22,136	21,799	337
1997	3198	2875	323	23,303	22,848	455

Sources: CNPC, State Statistical Bureau, and *The Petroleum Resources of China*, Energy Information Administration, US Department of Energy, 1987.

little subsequent investment in exploration or development of other known fields. In the late 1970s, when Daqing's output declined, so did China's. In 1979, based on false information about China's crude oil reserves, political leaders announced that China would open 10 more giant oilfields the size of Daqing by 2000, and the annual oil output would then reach 500 million tons per year (10 million b/d). Expectations that China would become a "second Middle East" made China's critical energy shortage during the early 1980s even more discouraging.

Chinese oil production started to stagnate in 1979. Output dropped by approximately 4.5 percent in 1981 and remained at the same level in 1982 (Fig. 3.2). During the three-year period between 1979 and 1982,

Table 3.4 Main contracts and agreements signed by CNPC in 1994 and 1995

Company	Signing date	Acreage (km^2)	Location	Classification
Phillips Petroleum Int. Corp. Asia & Arco China Inc.	Feb. 4, 1994	3700		Geophysical survey agreement
Phillips Petroleum Int. Corp. Asia & Arco China Inc.	Feb. 4, 1994	5400		Geophysical survey agreement
Kerr McGee China Petroleum Ltd. & Murphy Pacific Rim, Ltd.	Aug. 17, 1994	2280		PSC
BHP Petroleum (China) Pty. Ltd.	Aug. 19, 1994	3640		PSC
Oriental Petroleum (South Sea) Ltd.	Nov 8, 1994	5200		PSC (Promotion)
Phillips Petroleum Int. Corp. Asia	Dec 7, 1994	9200		PSC
Primeline Petroleum Corp.	Dec 12, 1994	4500		PSC
Arco China Inc.	Dec 19, 1994	510		PSC (Promotion)
LVR	Feb. 27, 1995		Dagang	Petroleum agreement
LVR	Feb. 27, 1995		Liaohe	Petroleum agreement
Esso (China) Ltd.	Apr. 7, 1995	29,900	Daqing	Petroleum contract
Esso (China) Ltd.	Apr. 7, 1995	11,591	Qinghai	Petroleum contract
Maxus	Apr. 1995		Zhoukou	Joint study agreement
Kerr-McGee, EDC, & Estsco	Jul. 21, 1995	308	Bohai Gulf	Petroleum contract
Shell	Jul. 1995		Ordos	Joint study agreement
Amoco	Nov. 1995		Sebei	Joint study agreement
Texaco	Dec. 21, 1995	14,059	Sichuan	Petroleum contract
Texaco	Dec. 21, 1995	10,105	Sichuan	Petroleum contract

Sources: *China Petroleum Industry, 1994*, The China Petroleum Information Institute, CNPC, 1995, p. 25, and *China Petroleum Industry, 1995*, The China Petroleum Information Institute, CNPC, 1996, p. 62.

Table 3.5 Main contracts and agreements signed by CNOOC in 1994 and 1995

Company	Signing date	Acreage (km^2)	Share	Classification
Agip (Overseas) Limited	Jan. 8, 1994			Petroleum contract
BP Exploration Operation Co. Ltd.	Mar. 22, 1994			Geophysical
ARCO China Inc.	Nov. 9, 1994			Joint study
Oriental Petroleum (Bohai Sea) Ltd.	Nov. 18, 1994			Exploration
Anadarko China Co.	Oct. 1994			Joint study
Energy Development Corp.	Sep. 22, 1994			Joint study
Pennzoil Co.	Jun. 1994			Joint study
Marathon Petroleum (China) Ltd.	Jan. 18, 1995	4800	100	PSC
Triton (China) Inc. LLC	Feb. 8, 1995	3200	100	PSC
Chevron Overseas Petroleum Ltd.	Feb. 24, 1995	25,200	100	PSC
Cluff Oil (China) Ltd.	Mar. 10, 1995	3430	100	GSA
Triton (China) Inc. LLC	Apr. 6, 1995	7900	100	Joint study
Chevron Overseas Petroleum Ltd.	Jul. 26, 1995	3300	100	Geophysical
Occidental Eastern, Inc.	Nov. 8, 1995	2340	100	Geophysical
Shell Exploration (China) Ltd.	Nov. 9, 1995	10,000	100	Geophysical
Caim Energy PLC	Nov. 28, 1995	2945	100	PSC

Sources: *China Petroleum Industry, 1994*, The China Petroleum Information Institute, CNPC, 1995, p. 27, and *China Petroleum Industry, 1995*, The China Petroleum Information Institute, CNPC, 1996, p. 67.

Table 3.6 Crude production by major oilfield

(000 b/d) region/field	1986	1987	1988	1989	1990	1991	1992	1993	1994	1995	1996	1997
Northeast:	1359	1395	1425	1446	1455	1455	1455	1465	1483	1495	1491	1497
Daqing	1111	1111	1111	1111	1112	1112	1110	1115	1117	1117	1117	1117
Liaohe	200	227	251	267	272	274	276	283	300	310	300	300
Jilin	47	57	63	68	71	68	68	67	66	68	74	80
North:	278	242	206	191	184	176	174	173	177	179	180	180
Huabei	200	159	122	110	107	100	96	92	93	93	93	93
Dagang	78	83	84	82	77	76	78	81	85	86	87	87
East:	727	781	823	823	813	810	800	779	731	700	682	662
Shengli	590	632	664	667	670	671	667	652	616	598	581	559
Zhongyuan	126	136	144	140	126	122	116	110	96	82	80	80
Jiangshu	11	13	14	16	17	17	17	17	18	20	21	23
South-central:	71	71	72	70	67	62	61	59	58	55	55	53
Henan	50	50	51	51	50	48	46	43	41	38	37	37
Jianghan	21	20	20	19	17	15	15	16	17	17	17	16
Northwest:	160	168	177	189	204	226	248	284	308	343	395	444
Xingjiang	110	115	123	128	136	140	146	152	158	158	166	174
Changqing	29	29	28	29	29	31	33	36	39	44	55	66
Qinghai	7	12	13	15	16	20	21	22	23	24	28	32
Yuman	11	11	11	11	10	10	9	9	9	8	9	8
Yanchang	2	2	2	7	8	9	10	11	13	15	18	21
Tarim	0	0	0	1	3	11	18	32	39	50	62	84
Turpan-Hami	0	0	0	0	1	4	12	23	28	44	58	60
Others	12	11	16	10	17	18	18	5	12	11	17	29
Onshore	2606	2668	2718	2730	2740	2748	2756	2780	2784	2800	2838	2875
Offshore	8	15	15	24	26	72	72	92	129	168	299	323
Total	2614	2683	2733	2754	2766	2820	2828	2872	2913	2968	3137	3198

Note: Other oilfields include Jidong, Sichuan, Yunnan, Guizhou/Guangxi and Anhui.
Sources: CNPC, State Statistical Bureau, and *China Energy Statistical Yearbook, 1991*, China Statistical Publishing House.

China's crude supply policy, i.e. "stabilizing the east, developing the west", seems to be working well in the east, but has thus far shown few signs of success in the west. The three largest oilfields in the east had mixed results regarding crude production during the recent years: Daqing maintained output, while Shengli's production fell, and Liaohe's supply rose somewhat.

Regional oilfield output statistics underscore the problems facing the nation's crude production. As Table 3.4 shows, northeastern China produces 50 percent of the country's crude oil, yet production of all the eastern fields, accounting for as much as one-quarter of China's total output, has fallen by approximately 1 percent per year since 1988. China's northwest (with the Tarim Basin), which accounts for less than 15 percent of total oil output, remains virtually unexplored and undeveloped.

The main obstacle preventing Chinese crude production from meeting growing demand is the fact that output from China's older eastern fields is declining, while the vast western and offshore basins have not yet been explored and are years away from coming on-stream. China has high hopes that these eastern fields will experience some growth or, at the very least, maintain their present output level until the vast reserves in the northwest are explored and developed.

In response to this challenge, China has formulated a three-pronged policy for oil development: (1) employ new techniques to maintain production at older fields; (2) step up exploration to find new sources of oil in the south and the northwest, as well as in the peripheral areas of the older oil producing basins; and (3) employ the latest equipment and technology to effectively and efficiently bring the newly confirmed oil reserves into commercial production (Table 3.7).

To ensure enough oil supply, China has moved to find overseas reserves to compensate domestic crude production decrease. CNPC is carrying out oil E&D projects in the Middle East, Canada, Peru and Thailand, and oilfields are being evaluated and considered in Papua New Guinea, the Middle East, Russia and the US. For instance, a 26-year agreement between China and Iraq was signed. The agreement covers the development of the Al-Ahdab field west of Al-Kut, which is around 100 miles south of the capital Baghdad. Plans call for CNPC to secure 200,000 b/d (or 10 million tons) of crude capacity from its overseas ventures by the year 2000.

Table 3.7 Well and footage drilled by CNPC in 1995

	Numbers of wells			Drilling footage (1000 m)		
	Total wells	Exploratory wells	Development wells	Total footage	Exploratory footage	Development footage
Daqing	2851	100	2751	3632	198	3434
Shengli	1672	166	1,506	3375	448	2927
Liaohe	1288	143	1145	2087	285	1802
Huabei	357	80	277	742	154	5388
Xinjiang	348	29	319	520	76	444
Dagang	219	32	187	534	111	423
Zhongyuan	137	26	111	401	48	353
Henan	191	22	169	323	48	275
Jilin	991	51	938	1110	84	1026
Changqing	532	41	491	832	100	732
Sichuan	78	16	62	227	55	172
Others	815	234	583	1547	583	930
Total CNPC	9479	940	8539	15,330	2190	13,106

Source: *China Petroleum Industry, 1995*, The China Petroleum Information Institute, CNPC, 1996, p. 28.

The following review of the particular developments and priorities at major onshore and offshore oilfields will give a clearer picture of how the national policies and plans are being implemented in particular regions. It also outlines some of the unique problems and obstacles faced by China's upstream industry.

China's upstream activities can be grouped into three key geographic regions for discussion purposes: onshore in the eastern regional basins, offshore in the eastern regional basins, and the western regional basins (all onshore), each of which has its own distinct resource potential and development time frame. Within these general regions, the major basins are shown in Table 3.8. Selected fields and basins within these three general regions are discussed below with regard to the production quantities, proven reserves, anticipated ultimate recovery volumes, and prospective levels of production.

Table 3.8 Major basins in three general regions

(1) Eastern Region (Onshore):	(3) Offshore:
Songliao	South Yellow Sea
Huabei	North Yellow Sea
Subei	East China Sea
Sichuan Basin	Pearl River Mouth
Ordos Basin	South China Sea
Bohai Gulf Basin	
(2) Western Region:	
Shan-Gan-Ning	
Junggar	
Zhaoshui	
Turpan-Hami	
Jiuquan	
Tarim	
Qaidam	
Yanqi	
Sichuan Basin	
Ordos Basin	

3.1.1 Onshore exploration in Eastern Regional Basins

The onshore eastern corridor, which extends from the Hainan Island to the northern border of Heilongjiang province, and includes the major oilfield complexes — Daqing, Liaohe, Dagang, Shengli, Huabei, and Zhongyuan, which provided about 90 percent of China's total crude oil output in the early 1990s and contain about 80 percent of the nation's proven reserves. Production has declined in eastern fields, with year-to-year output dropping 12,000 b/d in the early 1990s, compared to annual increases of 50,000 b/d in the 1980s.

China has implemented an extensive program of infill drilling, water injection wells, and well completion onshore in the eastern regional basins (see Table 3.9). These measures have contributed to the production gains of recent years. Without these measures, production from these fields would have declined. However, additional production gains from these efforts cannot last long, despite that Chinese oilmen believe that their technology to improve production is of world standard.

Table 3.9 Wells and production of various onshore fields in 1995

Fields	Oil wells	Water injection wells	Gas wells	Annual oil production (000 b/d)	Annual gas production (million m^3)
Daqing	20,784	9528	75	1120	2290
Shengli	14,367	5518	274	600	1290
Huabei	2886	1001	23	93	310
Liaohe	9808	1295	400	310	1750
Xingjiang	6904	1724	44	158	880
Dagang	2137	794	43	86	380
Henan	1205	410	3	38	40
Zhongyuan	2499	1465	48	82	1100
Jilin	5695	1528	7	68	180
Changqing	2213	632	18	44	100
Jianghan	461	173	10	17	80
Yumen	1139	428	0	8	10
Jiangsu	539	146	6	20	20
Qinghai	351	88		25	60
Sichuan	243	0	954	3	7180
Dian-Qian-Fui	144	45	14	2	20
Jidong	193	38		10	140
Tarim	118	19		50	120
Tuha	496	270		48	
Others	73	10		27	880
Total	72,255	25,212	1919	2809	16,150

Source: *China Petroleum Industry, 1995,* The China Petroleum Information Institute, CNPC, 1996, p. 21.

3.1.1.1 Daqing field — Songliao Basin

When relations between China and the USSR deteriorated in the late 1950s, the Soviets withdrew financial and technical aid to China and cut oil exports as well. China earnestly began to look for oil to replace the Soviet imports. These efforts paid off with the discovery of Daqing, the country's largest field, in 1959. The Daqing field has the additional advantage of being relatively near the population centers of the eastern coast, where most oil is consumed. Daqing oil is sweet low sulfur crude, but has low gasoline yield. It can produce high quality kerosene and diesel oil, particularly light diesel. Its residual fuel oil, with a low content of metal, can directly input into cracking units.

In the late 1970s, most foreign and Chinese experts predicted that Daqing would experience a severe decline in output by the mid-1980s. However, output has remained over 1.1 million b/d for most of the past 10 years. Ultimately discoverable reserves at Daqing may be as much as 10 to 12 billion barrels, of which at least half had been produced by the end of 1997. With diminishing proven reserves, Daqing has experienced almost zero growth and has been aggressively developed just to maintain its output. By 1997, Daqing has maintained its more than 1 million b/d of crude production for 22 years. Even so, Daqing's share of the nation total crude production fell from 44 percent in 1985 to 35 percent in 1997, as other fields came on line.

Daqing has offered a number of blocks in the field's western region to domestic contractors in order to help boost production, while Daqing is making effort to attract foreign investment in the central region of its field. This policy shift in Daqing signals an important step in diversifying its financial sources.

The most prominent feature of the production technique used at Daqing is the almost universal use of water injection on top of water flooding. The average water-cut at Daqing is in the range of 70 percent and increasing by about 3 percent per year. The application of modern extraction technology has slowed the decline in output at the main field, while helping to find the peripheral oilfields.

New technology that uses polymers to extract oil also helped Daqing to pump out an additional 20,000 b/d each year, and stabilized Daqing's annual output. China was scheduled to introduce tertiary recovery in its major eastern oilfields by applying polymer flooding and alkaline–surfactant–polymer flooding methods. By doing so, it is expected to increase crude production by 140,000–200,000 b/d by 2000[6], partially offsetting the production decline in other eastern oilfields.

Daqing has produced approximately 10 billion barrels of crude oil since it began production in 1959. China has officially announced that Daqing will maintain its 1 million b/d of crude production until 2010, which is an additional five years beyond its previous estimate of 2005[7].

---

[6] See Wang, Tao, "Sustaining Oil industry Development in China and Prosperity Worldwide", *International Petroleum Economics* (in Chinese), 1995, Vol. 3, No. 6, p. 2.
[7] The same as previous footnote.

In addition, the policy to open Daqing oilfield to foreign E&D is likely to help Daqing to achieve this goal.

3.1.1.2 Shengli field — Huabei Basin

The Shengli oilfield, China's second largest field complex, was discovered in 1962 and started production in 1966. It is located just inland from the mouth of the Yellow River and on the shores of the Bohai Sea. Shengli is estimated to have reserves almost as great as that of Daqing. However, Shengli's output had virtually no growth between the late 1980s and early 1990s, and declined rapidly after 1991, with annual output about 600,000 b/d in recent years.

Shengli has substantial reserves of heavy (26.8 API gravity) and medium sulfur content ($0.5\% \leq S \leq 1.0\%$) crude oil. But Shengli has a great number of oilfields, which are significantly different from each other in crude quality. Shengli's oil has a low gasoline yield, but high octane. Its residual fuel oil contains a high sulfur content.

Shengli's reservoirs are unconsolidated formations that will require steam injection. Shengli also has serious sand control problems, which are being addressed with the application of modern gravel pack technology. In addition, modern geophysical technology — both seismic and wireline — provided by foreign contractors has been used at Shengli, putting Shengli at the forefront of China's application of these techniques. The result has been an almost 10 percent annual growth rate in crude output during the mid-1980s. Meanwhile an increase in exploration proved very successful in 1987 and 1988, pushing production up 40,000 b/d and 30,000 b/d per year, respectively.

Shengli oilfield has expended its efforts of oil E&D from onshore to the continental shelf in the shallow water of Bohai Sea. It was reported that Shengli has found one of the largest oilfields of this kind, Chendao Oilfield, with near 700 million barrels of geological reserves.[8] Shengli produced about 10,000 b/d in 1993 and about 50,000 b/d in 1996 from its continental shelf development. It targeted more oil in its offshore oilfields.[9] If it is true, a rise in offshore production will help Shengli to slow down its crude supply declining over the coming years.

In the early 1980s, China declared that Shengli would become a "second Daqing" with a steady annual output of 1.0 million b/d. Based on optimistic discoveries in the mid-1980s, Shengli oilfield was projected to reach 800,000 b/d of crude production by 1990.[10] With the benefit of

[8] Due to lack of comparable information, we assume that the Chinese defined geological reserves are equal to recoverable reserves we define later in this chapter.
[9] See *People's Daily* (Overseas Edition), September 14, 1994, p. 2.
[10] In fact, in the early 1980s, Shengli was scheduled to produce 1.0 million b/d by 1990.

hindsight, it is clear that this original plan was premature. Recent sluggishness in crude production suggests that output may continue to slide, a production target of 800,000 b/d will not be fulfilled in the immediate future.

3.1.1.3 Liaohe field — Songliao Basin

The Liaohe oilfield in China's northeastern Liaoning province and adjacent to the Daqing oilfield is the third largest oilfield in the country. Liaohe holds 730 million barrels of proven crude reserve, which accounts for about 4 percent of national total. However, Liaohe produced about 10 percent of crude oil in national total during the recent years. As more than 20,000 square miles in area have not been tapped, there is huge potential for further exploration in Liaohe. According to Chinese sources, Liaohe has added significant verifiable oil reserves during the past years, more than other eastern oilfields in China.

3.1.1.4 Southern China

In 1985, the MOPI invited foreign exploration companies for the first time to participate in onshore E&D in eight southern provinces (later expanded to 11 southern provinces). The production-sharing contracts that had been signed in offshore bidding provided a general framework for joint onshore exploration. However, because of the geological difficulties and uncertain prospects, the joint onshore exploration projects attracted a few small and medium independents rather than the major oil companies that were active offshore. Although some joint E&D projects are underway in southern China, there is no immediate prospect for new exploration in these areas. China also plans to open a third round of onshore bidding for exploration in the 11 southern provinces in the near future.

3.1.2 Offshore exploration in Eastern Regional Basins

CNOOC undertook oil exploration on China's continental shelf in cooperation with foreign oil companies.[11] The favorable results of early offshore exploration programs have impressed upon China's petroleum planners the limitations of their existing domestic offshore E&P technology. This has resulted in a major shift away from the long-standing policy of "self-reliance".

---

[11] As of early 1995, about 100 agreements and contracts on oil exploration and development had been signed with foreign companies over the past 13 years.

Invitations to bid for offshore acreage previously covered by seismic surveys were issued in early 1982. By the end of 1996, agreements for oil E&D were signed, with over 60 corporations from more than 16 countries and covering 90,000 square kilometers in four offshore blocks: the Bohai Sea, South China Sea, and Yellow Sea, opened to foreign oil explorers in the 1980s, and the East China Sea in 1992. Territorial disputes remain in the East China Sea (with Japan and South Korea,) and in the South China Sea (with Vietnam). By the end of 1996, accumulated investment in offshore E&D was more than $7 billion, in which foreign investment accounted for about 65 percent. By April 1995, there are 12 oilfields and gasfields in production, including seven oilfields and gasfields managed by joint ventures between CNOOC and the foreign oil companies.

China's offshore production, dominated by CNOOC, rose from almost zero in early 1990 to 320,000 b/d in 1997, accounting for more than 9 percent of total domestic production per year. Surprisingly, offshore crude production in 1996 jumped by about 75 percent from 1995 levels, making the most notable contribution to the increase in 1996 domestic crude output. (Correspondingly, in 1996 China produced 2.6 billions cubic meters of natural gas offshore, up seven-fold over 1995.) However, this annual production growth is unsustainable over the long run due to limited production capacities, proven reserves and restricted E&D funding.

Most of these increases came from offshore operations in the South China Sea. The eastern area of the South China Sea pumped out 200,000 b/d of crude, making up two-thirds of total offshore production. The South China Sea was the first offshore area to be opened to foreign oil companies for E&D. Compared to China's other offshore regions, the South China Sea covers a huge area with sizable estimated reserves.

Eighty percent of this crude output was channeled into the domestic market and 20 percent exported to the world market. Although the results of offshore explorations, once touted as the next North Sea, have not been as encouraging as previously hoped, there is no doubt that the number of finds has steadily increased and will raise offshore production from almost zero to a considerable volume.

Although over $4.5 billion worth of foreign investment during the past 10 years has yielded less than expected, China's offshore oil E&D is expected to register fast growth. By 1997, six more offshore oil and gasfields were projected to come on stream (see Table 3.10), in addition to the 12 fields in production. CNOOC predicted that offshore crude production will most likely top out by 2000. Then, oil production in the developed oilfields will gradually move to a declining trend. Maintaining the output in offshore will depend heavily on the new E&D. CNOOC's forecast relied on the expectation that more offshore production capacities would be verified in the South China Sea and Bohai Sea in recent years. If this

Table 3.10 Offshore oilfields and gasfields under E&D

Oilfield/gasfield	Completion date	Type of output	Peak production
Huizhou 32-2	June 1995	Crude	13,000 b/d crude
Yancheng 13-1	December 1995	Natural gas	3.6 billion cm
Caofeidian	May 1995	Crude	10,000 b/d
Xijiang 30-2	September 1995	Crude	34,000 b/d
Huizhou 32-3	June 1995	Crude	16,800 b/d
Liuhua 11-1	April 1996	Crude	50,400 b/d

Source: Wu, Menfei, "The Reform and Development in CNOOC", *Guoji Shiyou Jingji (International Petroleum Economics)*, Vol. 3, April 1995, p. 9.

is true, the offshore output will be a delayed but will be a tangible gain. CNOOC plans to produce total 0.7 billion barrels of oil and 84.5 billion cm of natural gas during the year from 1993 to 2010.[12]

3.1.2.1 Bohai Sea

Offshore oil exploration in China was first conducted in the Bohai Basin, where seismic studies were started in 1966 and an active exploration campaign and production capacity buildup are being completed. Foreign oil companies started to operate in the Bohai Sea, and drilled wildcat wells in the latter part of 1994.

Although several small fields have been discovered, they are modest in size. One of them, Suizhong 36-1, is estimated to have confirmed oil reserves of 1.5 billion barrels, and could start producing 20,000 b/d of crude. The Suizhong oilfield includes Zone A and B. Zone A of the Suizhong oilfield has been bought onstream since 1993, while Zone B was completed at the end of 1995. Another field, Zhao Dong block, which was jointly explored and developed between China and two foreign companies, has had some success, despite its small amount of crude output. Caofeidian 1-6 field has been declared commercial in May 1995, with the output amount is projected at about 10,000 b/d. However, the Caofeidian oilfield only has less than 50 million barrels of crude reserves, and further reserves are being sought. Total Bohai Sea crude production is expected to reach 100,000 b/d.

3.1.2.2 South China Sea

The South China Sea was the first offshore area to be opened to foreign oil companies for oil E&D by the CNOOC. The South China Sea covers a huge area, and much of the region may have sovereignty disputes with South Asian countries (for detailed discussion, see Section 1.2.4, Perplexing Dispute Over Oil in the Spratly Islands, in Chapter 1). Crudes

[12] See Wu, Menfei, "The Reform and Development in CNOOC", *GuoJi Shiyou Jingji (International Petroleum Economics)*, Vol. 3, April 1995, p. 9.

from the discovery in the South China Sea are light (40–44 API gravity) and of low sulfur content (less than 0.5 percent). According to Chinese official sources, by the end of 1996, China had contracted out to more than 30 foreign companies from 11 countries, with a total foreign investment of more than $1 billion, the oil E&D in the South China Sea, which provides about 3.3 billion barrels of geological reserves and has nine oilfields with commercial value. In 1996, crude production from the South China Sea yielded nearly 200,000 b/d, which accounts for almost three-quarters of total offshore oil supplies.[13]

The Pearl River Basin in the South China Sea covers a 131,000 square kilometer area, in which more than 60 percent has been contracted out to 33 foreign companies with a total investment of $1.56 billion. The South China Sea has produced about 100,000 b/d in the Pearl River Basin in cooperation with the ACT-operators group, including Agip, Chevron Overseas Petroleum Ltd., Texaco (each with 16.33 percent shares) and the China Offshore Nanhai East Corporation (CONHE), a branch of the CNOOC (holding 51 percent of the ACT consortium). This ACT-operators group, China's largest offshore producer, is developing two additional fields in the consortium's South China Sea operating area. These fields, named Huizhou 32, have discovered reserves in excess of 400 million barrels with crude quality of 30 to 39 API. Initial oil production started in September 1995, with peak production expected in either late 1997 or early 1998 at approximately 60,000 b/d. In the eastern South China Sea, an oilfield named Liuhua 11-1 is under development by Amoco, and with estimated reserves of 1.8 billion barrels is considered one of China's major offshore discoveries. It was set to came on line in the late 1990s.

CONHE, Phillips Petroleum International Corporation Asia, and Pecten Orient Corporation of the US have jointly explored an oilfield, Xijiang 24-3 in the eastern South China Sea, with estimated geological reserves of 190 million barrels. The field was scheduled to peak its crude production at a 26,000 b/d per year, and to produce more than 12 years in order to tap 30 percent of total geological reserves. Another oilfield, Xijiang 30-2, with estimated geological reserves of 290 million barrels, started its oil output in 1996.

In the South China Sea, natural gas discoveries have been sizable, as CNOOC has announced the discovery of a giant gasfield holding estimated reserves of about 60 billion cubic meters. While prospects for natural gas discoveries are more promising, crude production is not what China and its foreign partners had expected. Even with several commercial discoveries in the South China Sea, oil production will reach a peak

---

[13] See *China Daily*, February 29, 1997, p. 1.

of about 100,000 b/d, a respectable output but far short of initial hopes. However, the volume of natural gas may not be enough to cover the cost of laying the pipelines that would be needed to get the gas to consumers. Further discoveries in the South China Sea may fall in disputed areas heightening claims of sovereignty from a number of South Asian countries.

3.1.2.3 Yellow Sea

So far, the drilling efforts in the Yellow Sea have been only marginally successful, resulting in no significant discoveries. Exploration programs were abandoned in the early 1990s, and future E&P prospects are uncertain.

3.1.2.4 East China Sea

The East China Sea was said to be "one of the world's few remaining virgin regions" for oil exploration. Nevertheless, this exaggerated statement indicated that the East China Sea is the last offshore region in China where potentially large discoveries could be made. Some 460,000 square kilometers of continental shelf basin in the East China Sea is believed to contain 36.8 billion barrels of oil reserves and 2000 billion cubic meters of natural gas.

Foreign oil firms have undertaken extensive prospecting work in the East China Sea. In October 1993, a foreign consortium of Texaco, Agip, and Maersk Oil China signed petroleum contracts with the CNOOC to explore three areas in the East China Sea covering some 8100 square kilometers (2 million acres). These three companies are the first foreign contractors to sign an agreement to explore in this major offshore area. Texaco China B.V. is the operator, with 40 percent interest. Agip and Maersk Oil each participate with 30 percent each in the contract areas. Apart from this area, the results of offshore drilling have been less significant, and the oil companies involved are scaling back their offshore operations.

CNOOC plans to build production capacity of 20,000 to 40,000 b/d in the East China Sea by the year 2000, considerably scaling down its previous ambitious goal of 160,000 b/d production. As a result, the East China Sea production will account for only about 8 percent of offshore total output by the end of the decade.

At present, in themselves financed efforts, few offshore development projects have reached the planning or implementation stage. Without higher world crude prices, development of fields that are deemed to have marginal commercial prospects will be very slow.

3.1.3 Western Regional Basins

China's oil industry began in western China, where the Yuman oilfield in the Gansu province, the Karamay oilfield in Xinjiang, and Lenghu oilfield in the Qinghai province were discovered and developed in the early 1950s. From 1958 onward, oil exploration was shifted to the eastern part of the country, where a number of large oilfields were found and developed. With the depletion of crude reserves in eastern China, China has shifted its onshore exploration efforts back to the west of the country, because substantial additions to production capacity can only be achieved through the accelerated development of the Tarim, Turpan-Hami, Qaidam and Junggar Basins (three of them are located in the Xinjiang Uygur Autonomous Region).

China has discovered 40 oil and gasfields in the Tarim, Jungger, and Turpan-Hami basins in Xinjiang with proven reserves of 14.6 billion barrels of oil equivalent.

The western basins produce just 14 percent of national crude output, but have enormous production potential for the future — the Tarim, Junggar, Qaidam, and Turpan Basins, and numerous mid-sized basins. With several promising high-yield structures identified in western China, the west holds the future of China's oil industry. Xinjiang Uygur Autonomous region includes Tarim, Junggar and Turpan-Hami, and is estimated to hold one-quarter of the country's total oil and gas reserves, according to the CNPC projection. Xinjiang produced 140,000 b/d crude in 1990 and about 330,000 b/d crude in 1997, which exhibits about 13 percent growth rate annually.

3.1.3.1 Qaidam Basin

The Qaidam Basin is located adjacent to the Tarim Basin, about 300 miles away. Qaidam's resource potential is pegged at 15 billion barrels of oil and more than 3.5 trillion cubic feet of gas. Exploratory drilling in the Qaidam Basin has been underway since the 1950s and has yielded a number of small and medium oil and gasfields. The basin's productive capacity is about 20,000 b/d of oil and 97 billion cubic feet per day of gas. Qaidam also completed a 436 kilometer oil pipeline connecting oilfields to the 20,000 b/d Golmud refinery, which was completed in 1994.

3.1.3.2 Junggar Basin

The Karamay oilfield in the Junggar Basin, developed over 20 years ago, is undergoing a production development program that produced about 180,000 b/d of crude oil in 1997. According to CNPC, four new oil-bearing zones were discovered in Junggar Basin since the early 1990s, increasing oil reserve estimates there by 1.5 billion barrels. Crude output in Junggar Basin is expected to be 200,000 b/d by 2000, up considerably from early the 1990s level and representing the highest growth rate among the country's oilfields.

3.1.3.3 Turpan-Hami Basin

In the Turpan-Hami Basin, with a size of 48,000 square kilometer, oil reserves are estimated at more than 700 million barrels, and about 20 exploratory wells have been successful.

3.1.3.4 Shan-Gan-Ning Basin, Sichuan Basin and others

In China's Central Region, more than 30 fields or field complexes have been found in six basins. The central regional basins provided about 5 percent of China's total crude oil output in 1997, but they contain about 15 percent of its proven reserves. Compared to eastern and western regional basins, the central regional basins have less E&D activities.

It was reported that a new oilfield, Horqin oilfield, in North China's Inner Mongolia started to pump oil in late 1994. Hydrocarbon resources in Horqin oilfield are estimated at 2.9 billion barrels and have a potential to be expanded. Crude production is targeted at 25,000 b/d in the late 1990s and reach 30,000 b/d latter on. The crude is planned to feed a 20,000 b/d distillation capacity local refinery.

In addition, China is also taking shares in oilfields in Canada, Peru, and Thailand. Part of crude production there is expected to be shipped back to China. Other crude projects in Papua New Guinea, the Middle East, Russia and the US are being evaluated for the possibility of Chinese participation.

China hopes that by the end of the century the west will be producing 800,000 to 1 million b/d of crude and will have proven reserves of 9.5 billion barrels. It is possible that this latter figure could be as high as 14.6 billion or even 21.9 billion barrels if there are significant new discoveries in the Tarim Basin. However, crude output of 800,000 to 1 million b/d in western China by 2000 is, in our opinion, unrealistic. The growth of production in western China will be determined by the output of the Tarim Basin (addressed in the next section), and we do not believe that it will produce significant volumes of oil by 2000. In the absence of promising new discoveries by foreign exploration in southeastern Tarim, the total crude output in western China is projected to reach only about 700,000 b/d in 2000 and around 1 million b/d in 2005. Western China is the only onshore region where we expect growth in crude production after 2000.

3.2 Tarim Basin Exploration and Development

With existing crude oil production capacity considered insufficient for China's long-term development, the Tarim Basin holds the future for China's indigenous oil supply. As early as 1983, and as a result of joint-exploration efforts between China and two desert-oil-prospecting teams from the US, the Tarim Basin has held great hope.

The Tarim Basin is estimated to be 560,000 square kilometers (138 million acres) in size, more than half of which is covered by the Taklimakan Desert, close to the borders with Tajikistan, Kyrgyzstan, Afghanistan, and Pakistan. With harsh operating conditions, and complicated surface and subsurface structures, the Tarim Basin is estimated to be one of the largest oil-bearing onshore basins in the world. Some oil specialists believe that it may hold nearly as much oil as Saudi Arabia, although these claims are wildly exaggerated. Estimates of Tarim's resources vary, with a reasonable working number falling roughly between 10 and 30 billion barrels along with 8.3 trillion cubic meters of natural gas. A preliminary feasibility study from CNPC indicates that, given the proper technology and reservoir conditions, as much as 35 percent of the oil could be recovered. Chinese geologists have estimated that the reserves of the basin could be as much as 74 billion barrels of oil and 283 trillion cubic feet of natural gas. If true, that would be three times the remaining US proven oil reserves and nearly twice the remaining US proven natural gas reserves. It would place China's oil resources just below those of the largest producers in OPEC, and larger than those of Mexico, Venezuela, Nigeria, Libya or Russia. However, so far the proven reserves of crude oil are about 2.2 billion barrels.

This remote region is about 6000 kilometers from the main consuming areas on the eastern seaboard. Despite that, the region has attracted keen interest from foreign companies because its potential for successful petroleum exploration is considered to be higher than other onshore areas open to foreign activity. However, landlocked frontier oil requires major capital investment, not only in exploration and production, but also transport. When exploration and production techniques were less sophisticated, China was able to develop its oil resource with its large, inexpensive labor pool. But the availability of cheap labor will hardly have an impact on the massive, capital-intensive infrastructure needed to exploit the Tarim Basin.

To a certain extent, political considerations have a bearing on development of the Tarim Basin, which is located in the Uygur Autonomous Zone, a region ethnically different from the Han Chinese. Allowing foreigners into the area could encourage a separatist movement similar to that in Tibet. For 45 years, the Tarim Basin was sealed off to foreign investors to avoid political problems and to maintain China's image of self-sufficiency. Beijing realized that China could not profit from Tarim's treasure without western technology and financial support.

3.2.1 Exploration and development in the Tarim Basin

CNPC and China's MGMR have made several potentially significant discoveries in the northern and central parts of the basin since exploration began in the 1970s. As of the end of 1996, more than 30,000 square kilometers in Tarim had been explored. Proven oil reserves are 2.2 billion barrels. In early 1991, the first Tarim Basin oil made its way to refineries in the Gansu province.

To date, five oilfields have been discovered in the Tarim Basin, and 16 oil and natural gas-bearing structures have been identified. Tarim Basin's oil is moving from the exploration phase to the early development phase. Most of Tarim's production comes from the Lunnan oilfield in the north. But the Tazhong-4 oilfield, which lies in the central Tarim, is estimated to account for 80 percent of the Tarim's total crude oil reserves. It has been under CNPC's development since early 1992. CNPC followed up the Tazhong-4 oil strike by announcing plans to step up exploration work in an area covering 1100 square kilometers in the central part of the Tarim Basin. Tazhong-4 oilfield targets a crude output capacity of 40,000 b/d. Commercial production was bought on stream at the end of 1996.

The Tarim Basin's production averaged 84,000 b/d in 1997. Although recent finds are encouraging, CNPC has revised its forecast of Tarim Basin production. The 100,000 b/d target set for the end of 1994 is expected to be achieved by 1998 or later, due to less investment by the central government.[14] In the late 1990s, exploration in Tarim will accelerate, pushing proven oil reserves to about 6.0 billion barrels. When it is fully up and running, Tazhong oilfield is expected to rank sixth among China's 11 major oilfields in terms of output. In terms of transportation, the next major step will be the expansion of a 195 kilometer pipeline from Tarim to a rail line in Korla to the north. Its capacity will be increased from 20,000 b/d to 100,000 b/d (we address transportation in Section 3.2.2).

The opening up of the attractive Tarim Basin acreage coincides with a slowdown of investor interest in China's offshore acreage and growing concern that China might be changing from net oil exporter to net oil importer. For international oil companies, the arguments for encouraging vigorous exploration activity in the Tarim Basin remain as strong as ever.

---

[14] It was reported that China slashed the exploration budget for Tarim to $300 million from $600 million spent in 1989–1990. See *People's Daily*, "Prospective of the Exploration in Tarim", December, 16, 1996, p. 1.

A total of five blocks in the southeastern sector of Tarim, totaling 72,730 square kilometers, have been offered for international bids organized by the China National Oil and Gas Development Co., a subsidiary of CNPC. Three of the five blocks were taken during the first two rounds of bidding. At least one of the five tracts failed to draw any bids. A short description of the three blocks that were bid upon follows:

Block 1: An international consortium announced the signing of a production sharing contract with CNPC for 9814 square kilometers (2425 million acres) in the Tarim Basin in February 1994. The group is made up of Texaco China B.V., an indirect subsidiary of Texaco Inc., Agip (overseas) Ltd., ELF Hydrocarbures China, Japan Petroleum Exploration Company Ltd. (JAPEX)[15], and Japan Energy Corporation. Each partner in the consortium holds a 20 percent share, with Agip serving as operator. CNPC promised to buy, with hard currency, all petroleum produced by Texaco and its partners, and has an option to participate with up to a 51 percent share of the production.

Block 3: The first Tarim exploration agreement was signed for this block, comprising 15,000 square kilometers, at the end of 1993 by a consortium led by Exxon. The group is made up of Exxon (70 percent), Sumitomo Corp. (15 percent), and Indonesia Petroleum Ltd. (15 percent). The group began seismic activity in 1994, and later conducted exploratory drilling. Exxon drilled a wildcat to a depth of 10,000 feet in Tarim Basin in 1996, and became the first company to do so.

Block 4: This block also covers 15,000 square kilometers. British Petroleum (BP) was set to sign a third seismic contract for exploration in this block. BP bid for exploration rights in a consortium with four Japanese oil and general trading companies: Nippon Oil, Itochu, Mitsubishi, and Mitsui. The BP-led consortium would have the option to exclusively develop the block, depending on the results of the surveys.

Foreign investors, as always, have to make economic choices about where they place their risk capital, even though they may be willing to probe the more costly or technically complex reserves in known regions. The opportunity to access prospective acreage remains as critical as ever. On the other hand, the quantity of oil recovered from Tarim is a function of crude prices in the world oil market. If OPEC capacity utilization reaches 90 percent or more at the beginning of the 21st century, oil prices may move into a high level. This price increase will help to promote investment in Tarim's production capacity. In addition, exploration drilling activity levels are also directly affected by fiscal incentives that government gives.

---

[15] It is reported that Nissho Iwai has acquired about 1 percent of the total shares in Block-1 from JAPEX.

It is still too early to evaluate the Tarim Basin's prospects or how long it will take for significant discoveries to be made and developed. In view of the uncertainties associated with crude exploration in Tarim, until more drilling is done, there is no way to know exactly how much oil and gas there is and how much capital will be required. However, given the remoteness of the basin, it is believed that a 10-year exploration period is needed before there will be a substantial increase in Tarim's output. Consequently, we will not try to predict when and how much of the Tarim oil (foreign E&D only) will come to the market, not only because it is too speculative, but also due to the fact that a span of 10 years, which is the focus on this book, is too short a time frame for substantial production development in the Tarim.

3.2.2 Transporting Tarim's oil

Finding significant oil and gas reserves is necessary but not sufficient for commercialization to succeed. The oil resource of the Tarim Basin are landlocked, so the cost of transporting oil from this region will be much higher than from China's major producing areas, all of which are situated on or close to the sea. The freight components of Middle East Gulf crude account for around 10 percent of the delivered price to Rotterdam or New York (i.e. a couple of dollars per barrel). By contrast, the cost of moving Tarim crude through a pipeline to the Chinese coast 2500 miles away is estimated at $6 per barrel.

Transportation costs are an important element in deciding whether to develop hydrocarbon reserves in the Tarim Basin. The capacity of the railway line (single-track) is limited by the difficult geography, which restricts train weight. Shipping distances are great and development costs of either the rail or pipeline alternative will be large. The first step is likely to be the increase of rail capacity in critical bottleneck sections. The cost of a pipeline would only be justified by the discovery of large oilfields.

There have been two proposals to ship Tarim crude to consumer areas, if there are significant commercial discoveries. One is a $3 billion project that involves building a 3300 kilometer (2050 mile) pipeline from Korla in the north of Tarim to a distribution hub in Luoyang in Henan province of central China. The plan calls for piping Tarim crude through Luoyang to the eastern port city of Lianyungang for export. CNPC has supposedly abandoned this proposal because China is importing more and more products to support its booming economy, and the revenue from crude exports would not offset the money spent on product imports.

Another proposal is to run a 2000 kilometer (about 1250 mile) pipeline from Korla to the highly populated Sichuan province to feed the 200,000 b/d Mianyang refinery in southwest China. According to the most conservative estimates, this plan is expected to cost $1.4 billion (other estimates range from $2 billion to $3.5 billion). This pipeline is designed to move 200,000 to 400,000 b/d of crude oil, combining output from western China's Tarim, Turpan-Hami, and Junggar basins. The Korla–Mianyang line is projected to be completed by the time output from the Tarim Basin reaches 600,000 b/d. The diameter of the pipeline will be determined by the volumes of oil indicated by exploration efforts under the first round contracts.

The Tarim Basin may have the best chance of providing the world with significant additions to reserves. However, Tarim E&D is expensive, given elevated costs, high reservoir pressures and temperatures, and lack of infrastructure. The question is not whether the oil reserves are in Tarim, but whether the oil there is commercially viable. If solutions to the infrastructure problems, capital deficiency, and political issues are found, development of the Tarim Basin could proceed quickly. The Tarim Basin could become Asia's primary source of crude oil in the 21st century. However, the Tarim Basin is probably in too remote an area of China to be developed within the 10-year period that is the focus of this book. We believe that there will be continuing output growth in Chinese-controlled Tarim, but no significant commercial supplies in the foreign-controlled Tarim region until the next century. As a result, crude production in Tarim is expected to increase from 84,000 b/d in 1997 to 140,000 b/d by 2000, to 340,000 b/d by 2005, and then to 760,000 b/d by 2010 (see Table 3.11). If it is true, crude production in Tarim Basin could account for near one fifth of national total at that time.

Table 3.11 Actual and scheduled crude production and reserves in Tarim

	Output (000 b/d)	Proven reserves (billion bbl)
History (Lunnan oilfield discovered):		
1990	10	n.a.
1991	11	n.a.
1992	18	n.a.
1993	32	2.2
1994	39	4.0
1995	55	6.0
Tazhong oilfield started output:		
1996	62	
1997	84	
Projection:		
2000	144	n.a.
2005	340	n.a.
2010	760	n.a.

Sources: CNPC and author's projection.

3.3 Projected Crude Production

Any projection of future oil production is fraught with a high degree of uncertainty. Future production will be determined by a host of factors, including the amount of investment in the oil industry, the speed at which new oil production capacities are opened up, the degree of success in arresting the natural rate of decline in production among the mature oilfields, and the luck in discovering new large oil reserves — either onshore or offshore — in the future.

After a period of stagnation in crude production beginning in 1979, some analysts completely reversed their previous optimistic estimates, predicting that Chinese oil output was likely to remain stagnant until the mid-1980s. The World Bank projected that China might have to start importing oil some time in the 1980s. However, oil production started to grow again in 1984, and crude exports have stabilized since then. Even so, growth in crude production in the last three years has slackened drastically. Is this trend irreversible?

We use several different approaches in answering this question. First, we review the related variables in output projection. Second, we compare reserve estimates obtained from various sources. Third, we use the "top down" and "bottom up" approaches to project China's oil production. The "top down" approach refers to three variables: national total identified reserves, reserve/production ratio, and additions to reserves. The "bottom up" approach is based on projections for each individual oilfield and company with the production estimates obtained from Chinese planners.

3.3.1 Methodology of crude production projection

Because long-term supply forecasts are based on essentially the same information about the available resource base, differences in forecasts of future production levels must be based on different views of the other variables — the political and financial issues — that determine production.

Aside from the sheer size of the petroleum resource itself, the four variables most influential in determining the level of future production are technology, capital availability, the political and legal climate for E&D, and the supply of exploration talent. These are obviously interrelated: the political climate for exploration affects capital availability and explorationist interest, to take but one example.

Given technology, capital availability, a reliable political–legal environment, and a supply of exploration talent, any "bottom-up" production outlook has to be based on three variables: (1) the current estimate of confirmed reserves; (2) the rate at which E&D activity expands those reserves; and (3) the rate at which on-going production depletes those reserves (the reserve / production ratio). For an old oilfield, total amounts recovered would depend on how much oil was originally bypassed.

The conventional categories of the world oil resource base are: proved, confirmed, discovered and recoverable reserves. Long-term production forecasts based on proven reserves tend to be too conservative, but forecasts based on bringing recoverable reserves into production within any specified period of time may be too optimistic. In our forecast, we follow the practice of Energy Information Administration of the US Department of Energy and US Department of the Interior / Minerals Management Service. Resources / reserves are expressed in terms of five categories — total resource, recoverable / unrecoverable resource, discovered / undiscovered under recoverable resource, confirmed / unconfirmed under discovered reserves, cumulative production and proven reserves under confirmed reserves (see Figure 3.3).[16] Confirmed reserves increase as a result of reserve additions (field extensions and new reservoirs) and from infill drilling, improved technology and enhanced recovery techniques, well workovers, recompletions, and longer productive well life due to higher prices. These resource categories — proven, confirmed and discovered reserves — are an expression of the quantity of geologic and engineering data upon which the estimates are based, but are directed to a short-term (10 years) production prediction. In addition, there are another three resource categories — probable, possible and speculative — under unconfirmed and undiscovered reserves. Probable resources are those associated with the further development of fields that have already been discovered. Possible resources are those postulated to exist in new field discoveries. Speculative resources are those believed to exist in new field discoveries not previously proven to be productive. These three categories are compiled with a long-term (30–50 years) projection.

The second variable involved in making a crude production forecast is the rate of addition of reserves. It is difficult to generalize here: prospects range from very large capacity additions in Daqing to small ones in Jiangsu province. In large, mature basins such as Songliao, the historic rate of additions can be a fairly good indicator of future performance, whereas in new regions, such as offshore oil, the past rate of additions may be a very poor indication of things to come.

---

[16] See National Energy Strategy, 1990, *The Domestic Oil and Gas Recoverable Resource Base: Supporting Analysis for the National Energy Strategy*, Energy Information Administration, Office of Oil and Gas, US Department of Energy, SR / NES / 90-05. Also see Cooke, L., 1991, *Estimates of Undiscovered, Recoverable Oil and Gas Resources for the Outer Continental Shelf*, Revised as of January 1990: US Department of the Interior / Minerals Management Service, OCS Report MMS 91-0051.

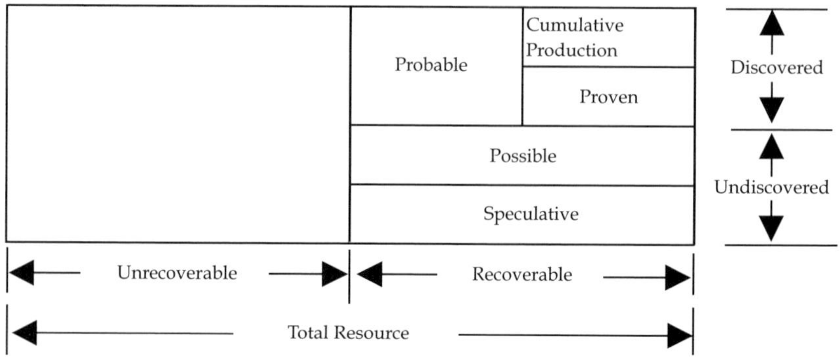

Fig. 3.3 Classification of various resources/reserves

The best prospects for new reserve additions are in western China, including the Tarim Basin. The forecast assumes moderate reserve additions that will enable China to expand its production capacity and prevent it from becoming a big crude-importing country within the decade. This observation, seemingly straightforward, requires many extraordinary things to happen. For example, it requires that oil prices do not collapse (i.e. below $10 per barrel for Dubai, because production in many of China's oilfields is relatively high-cost). It also requires that prices do not explode (i.e. exceed $30 per barrel in 1996 dollars), because this would cause production in the mature oilfields to accelerate. The forecast also depends on China continuing to open its doors to the E&D efforts of foreign companies and investors, and keep them open by offering concession terms at least as good as those in other frontier sectors.

3.3.2 Oil reserves: potential for the future

Crude oil production has been sharply constrained by the level of indigenous hydrocarbon production capacity and the rate of discovery of proven reserve. The first question that needs to be addressed is whether proven and discovered reserves of crude oil are adequate in China. The crude reserve estimates from different sources, domestic and foreign, vary within a large range. Table 3.12 presents estimates from the World Bank, the US Geological Survey, numerous private experts and Chinese official. As always, Chinese specialists usually issue fairly optimistic estimates, while foreign experts are much more conservative. Taking information from both sources may serve to balance the biases out. However, in this book we have chosen to rely primarily on data from the Chinese sources, for two reasons. First, we believe that Chinese published data on proven reserves would be more reliable, because these reserves are realistically targeted for E&D using available technology and at current market prices. Second, Chinese reserve estimates are more recent than foreign estimates we collected, reflecting the introduction of modern geophysical technology (advanced seismic, wireline, and data processing) that has had a dramatic effect on the discovery rate since 1985.

Table 3.12 Estimations of China's crude oil reserves

Name of organization	Estimated reserves (billion barrels)	Estimated year
Discovered Reserves:[1]		
Official Chinese estimates	128	1996
US Energy Information Administration	88	1987
World Bank	40	1985
US Central Intelligence Agency	80–88	1983
Of which onshore:		
Official Chinese estimates[2]	120	1994
Of which offshore:		
Official Chinese estimates[3]	8	1994
Broken Hill Proprietary[4]	13–25	1984
David S. Holland[5]	20–50	1980
A.A. Meyerhoff[6]	30	Mid-1970s
Of which Taim Basin		
Offical Chinese estimate	73	Early 1990s
Foreign specialists	10–30	Early 1990s
Proven Reserves:		
Author's estimates	21	1998
Oil & Gas Journal	24	1994
World Oil	30	1993

Notes:
[1]The estimates from different sources may not comply with our reserve definition.
[2]Data are taken from CNPC.
[3]Data are taken from the *People's Daily* (Overseas Edition), August 15, 1994, p. 1.
[4]Part of the BP operation group.
[5]President, Pennzoil Exploration and Production.
[6]A petroleum geologist.

Although no firm and consistent data on reserves and discoveries have been publicly released by China, which to some extent regards such data as a state secret, fragmentary data have been published on discoveries by CNPC, CNOOC, and other Chinese official sources that enable us to construct a general picture of the reserve situation.

According to former CNPC president Wang Tao, China has crude oil resources of 680 billion barrels, including 460 billion barrels onshore and 220 billion barrels offshore. But discovered oil reserves are less than 20 percent of total crude oil resources, and consist of about 130 billion barrels, of which 92 percent is found onshore and 8 percent offshore.[17] However, two-thirds of discovered oil reserves cannot be extracted, leaving China 21 billion barrels of proven reserves and reserve/production ratio at 18 years.

Another source, Xinhua News Agency, indicated that the country's oil resources (recoverable resource) top about 500 billion barrels (or about 70 billion tons). Only 20 percent of this has been verified (discovered reserves), leaving a huge potential to be tapped through oil exploration.[18]

CNPC data, as Table 3.13 shows, present significant growth in estimates of discovered reserves, up by 42 percent over the past eight years for known fields. In 1996 China's proven oil reserves in onshore areas were estimated at about 17 billion barrels, suggesting about 17 years of production at current onshore crude production levels. The table shows that reserve/production ratios in China's onshore oilfields slid from 1985 to 1991. However, additions to reserves recorded in recent years have been substantial. They offset the production decline in old fields and contributed to the net production increases over the past years. In 1996, the volume of new reserves added to China's oil inventory was reported to be more than 4 billion barrels of oil in place. Much of this increase has come from discovered reserves of already known rather than from new fields. However, recent official reports of new developments in the Songliao and North China basins indicate considerable additional production capacity to be created or tapped by enhanced recovery of existing reserves.[19]

China's continental shelf covers 1,300,000 square kilometers offshore, of which 650,000 square kilometers are thought to be prospective hydrocarbon areas. The estimates for offshore reserves have varied widely and have declined significantly since the early 1980s. China's offshore recoverable oil resources have been estimated in the range of 13 to 50 billion barrels by international experts and rather higher, at 20 to 73 billion barrels, by Chinese experts. China reported that its offshore-discovered reserves are 8.3 billion barrels (or 1.2 billion tons).[20] If we assume that one-third of these reserves can be proved, there are about 3 billion barrels of offshore proven crude reserves.

[17]See Wang, Tao, "Sustaining Oil industry Development in China and Prosperity Worldwide", *International Petroleum Economics*, (in Chinese), 1995, Vol. 3, No. 6, p. 1–2.
[18]See Wen Hui Daily, March 21, 1996.
[19]CNPC has 270 seismic teams and the Ministry of Geology has 60, while most of the oilfields have their own data processing centers. There are also some foreign geophysical teams at work in China.
[20]Data are taken from *People's Daily* (Overseas Edition), "Achievement in Offshore Oil Industry: A Highest Exploration Rate among Oil Producing Countries", August 15, 1994, p. 1.

Table 3.13 Onshore crude reserves (billion barrels)

	1985	1986	1987	1988	1989	1990	1991	1992	1993	1994	1995	1996
Discovered reserves	84.8	91.4	96.5	100.3	103.9	108.6	112.5	116.1	120.1	124.2	128.4	132.7
Cumulative confirmed	27.7	28.7	30.0	30.8	30.6	31.6	32.4	33.7	34.8	35.9	37.2	38.4
Cumulative production	10.8	11.7	12.7	13.7	14.7	15.7	16.7	17.7	18.7	19.7	20.8	21.8
Proven reserves	17.0	17.3	17.3	17.2	16.0	15.9	15.9	16.2	16.4	16.6	16.8	17.0
Reserve/production ratio	18.7	18.2	17.8	17.3	16.1	15.9	15.9	16.1	16.2	16.3	16.5	16.4

Sources: CNPC (before 1993) and author's estimate (after 1994).

Actual reserve potential is almost certainly far higher than these figures indicate, even excluding the western basins or the continental shelf, because: (1) by its own account, China has conducted exploratory drilling on only 20 percent of the area covering its land sedimentary basins; and (2) over 60 percent of the reserves discovered is in peripheral fields outlying existing ones in the northeast corridor. Therefore, estimates of possible oil reserves are highly speculative, particularly for lightly explored areas (such as the Tarim Basin) and for the almost virgin areas offshore.[21] In addition, investment costs and production costs for various sources of crude are a key factor in their E&D.

The growth of China's oil reserves has plateaued, reflecting the fact that new big oilfield discoveries are unlikely, and that there is little possibility of sustaining great growth in proven oil reserves in the next 10 years. But the time for significant oil resource depletion and a drop in annual proven reserves has not come yet. Therefore, during the next 10 years, the increases in annual proven reserves are expected to average 100–200 million barrels per year. If major discoveries are made in the Tarim Basin, it will be difficult to supply a large volume to consuming areas in the near future. In addition, we expect that no significant offshore find will add to oil reserves over the next decade.

In conclusion, as Figure 3.4 shows, China's total discovered reserves are about 150 billion barrels of crude oil, of which 140 billion are located onshore and 10 billion offshore, and this contains an allowance (albeit perhaps too small) for the remote and inaccessible Tarim Basin.[22] China's confirmed reserves were estimated to exceed 44 billion barrels at the end of 1997, and cumulative production was near 23 billion barrels. Total

[21]In early 1980s, the World Bank and the CIA indicated that China's crude production would decline by the mid-1980s.

[22]Chinese geologists have been quoted as estimating that reserves could be 130 billion barrels in Tarim. See "China Stressing Onshore E&D to Spur Crude Output", *Oil & Gas Journal*, July 29, 1991, p. 26.

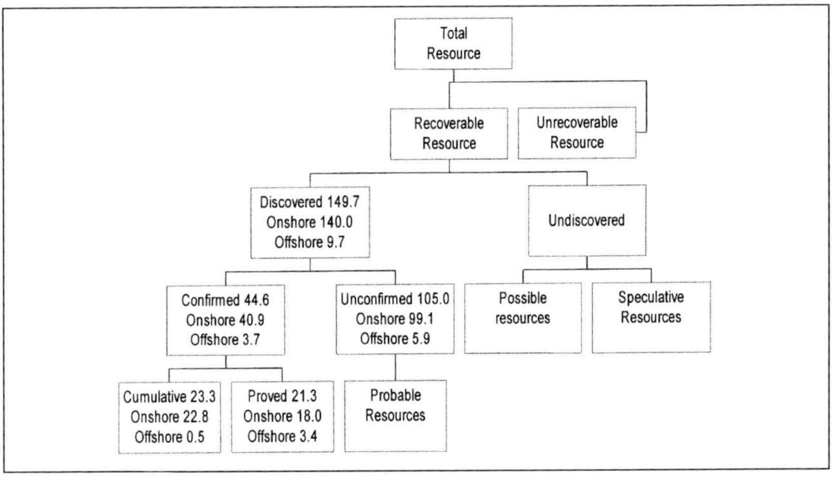

Fig. 3.4 Crude oil reserves at the end of 1996 (billion barrels)
Note: Date is estimated from the partial information taken from Chinese sources in 1994.

proven reserves of crude in 1997 were about 21 billion barrels, representing 18 years of production at the current crude production level (onshore plus offshore).[23] We expect China's crude production capacity to increase by about 1.1–1.5 million barrels per year between 1997 and 2005, which could keep reserve/production ratios between 18 and 20 years.[24]

The estimated undiscovered oil reserves are a matter of complete uncertainty. The mean probable value of the undiscovered recoverable crude oil in China is about twice the present discoverable oil reserves, which we estimate at 132 billion barrels. On the other hand, if half of the discovered reserves are confirmed, Chinese oil resources will be able to sustain the present production level for more than 40 years. In this case, production will probably jump after 2000 with the current growth rate of reserve discovery.

[23]As of 21 billion barrels of proven reserves, China ranks at the 11th largest oil reserve countries around the world, following Saudi Arabia (261 billion barrels), Iraq (100 billion barrels), UAE (98 billion barrels), Kuwait (97 billion barrels), Iran (93 billion barrels), Venezuela (63 million barrels), former Soviet Union (57 billion barrels), Mexico (51 billion barrels), US (24 billion barrels), and Libya (23 billion barrels), according to *Oil & Gas Journal*. *Oil & Gas Journal* estimated China's proven reserve at 24 billion b/d, with a resulting reserve/production ratio of 23 years. See *Oil & Gas Journal*, February 1998.

[24]The government's Ninth Five-year Plan (1996–2000) estimates that oil reserves will increase by 28 billion barrels (or 3.8 billion tons) during the 5-year plan, and that natural gas reserves will rise by 800 billion cubic meters. Xin hua News Agency reported that China will identify proven or probable geological oil and gas reserves of 29.2 billion barrels and 1 trillion cubic meters before 2000 (Xin hua News Agency, April 23, 1996).

Table 3.14 Crude reserves, crude production and their projection ("top down" approach)

	Historical						Forecasts							
	1993	1994	1995	1996	1997	1998	1999	2000	2001	2002	2003	2004	2005	2010
Discovered reserves (billion bbl)	128.4	133.1	137.5	142.4	149.7	156.5	163.6	170.1	176.6	183.3	189.2	195.2	201.4	234.2
Confirmed reserves (billion bbl)	38.3	39.7	41.0	42.5	44.6	46.7	48.8	50.7	52.7	54.7	56.4	58.2	60.1	69.9
Proven reserves (billion bbl)	19.5	19.8	20.0	20.3	21.3	22.2	23.1	23.8	24.5	25.2	25.7	26.2	26.8	29.6
Growth in proven reserves (%)	3.2	1.5	1.2	5.0	5.0	4.0	4.0	3.0	3.0	3.0	2.0	2.0	2.0	2.0
Additions to proven reserves (billion bbl)	0.6	0.3	0.2	0.3	1.0	0.9	0.9	0.7	0.7	0.7	0.5	0.5	0.5	0.6
Cumulative production (billion bbl)	18.8	19.9	21.0	22.1	23.3	24.5	25.7	26.9	28.2	29.4	30.7	32.0	33.3	40.3
Onshore production	18.7	19.7	20.8	21.8	22.8	23.9	25.0	26.1	27.2	28.3	29.4	30.6	31.8	38.3
Offshore production	0.1	0.2	0.2	0.3	0.5	0.6	0.7	0.9	1.0	1.1	1.3	1.4	1.5	2.0
Annual crude production (million b/d)	2.9	2.9	3.0	3.1	3.2	3.3	3.3	3.4	3.4	3.4	3.5	3.5	3.6	4.1
Reserves/production ratio	18.6	18.6	18.5	17.8	18.3	18.6	19.0	19.2	19.8	20.1	20.3	20.5	20.6	19.7

Assumes that (1) total proven reserves in China were 19.5 billion barrels in 1996; (2) annual increase in proven reserves is 7.0 percent before 2001, and 5.0 percent after 2001; and (3) reserve/production ratio will decline slightly.

3.3.3 Crude output projection

Crude production forecasts usually consider 10 years as a short-term projection, 10–30 years as medium term, and 30–50 years as long term. The longer the forecast period, the greater the uncertainty of the projection. For the purpose of this book, we are projecting Chinese crude supply over the medium term or 15 years.

The assumptions we make are: (1) slightly falling reserve/production ratios from the 1996 level; (2) moderate additions to proven reserves each year; (3) a decline in eastern China's output after 2000; (4) a moderate increase in offshore exploration; and (5) no significant commercial barrels of Tarim Basin oil coming on stream over the next 10 years. Based on a wide array of specific assumptions regarding the effectiveness of enhanced oil recovery programs in the mature producing fields of the northeast corridor, and variable discovery and development rates in the western basins and offshore, it is possible to model and project crude oil production curves into the future.

The "top down" crude production projection is presented in Table 3.14. With this approach, we expect China's crude production to rise to 3.4 million b/d in 2000. Over the early 2000s, the growth of crude production is expected to slow down to reach 3.6 million b/d in 2005, without significant production additions in the Tarim Basin. However, crude production could pick up again after more and more Tarim crude is brought on stream. This "top down" projection can be compared to the outlook from a "bottom up" approach, which yields a similar picture of crude output.

As the "bottom up" approach shows (see Table 3.15), incremental output will come mostly from the western part of the country, especially from the four big sedimentary basins: Tarim, Qaidam, Junggar, and Turpan. Production has been increasing and will continue to increase without any significant development in the Tarim Basin. At present, China's three largest producing fields (i.e. Daqing, Shengli and Liaohe) have maintained their present production levels for the past decade but doubts remain about a sustained field flow after 2000. Government policy for crude production is to stabilize output in eastern China, and develop western China, which includes the Tarim Basin. That requires diligent extensions of known reserves and bringing down the reserve/production ratios.

The natural decline of the older fields could be halted through technical improvements. For example, further application of polymer flooding can increase Daqing's light sweet production during the next 10 years by as much as 400,000 b/d over its natural decline. Application of polymer technology, however, was delayed until 1995 due to lack of polymer supplies. As a result, it is unlikely that Daqing will do more

Table 3.15 Crude production projection by oilfield ("bottom-up" approach)

	Crude production (000 b/d)				Annual growth rate (%)				Share in national total (%)		
	1997	2000	2005	2010	1990–97	1997–00	1900–05	1905–10	1990	1997	2010
Northeast:	1497	1491	1423	1389	0.4	–0.1	–0.9	–0.5	52.6	46.8	33.8
Daqing*	1117	1100	1040	1030	0.1	–0.5	–1.1	–0.2	40.2	34.9	25.0
Liaohe	300	296	285	271	1.4	–0.4	–0.8	–1.0	9.8	9.4	6.6
Jilin	80	95	98	89	1.7	5.9	0.6	–2.0	2.6	2.5	2.2
North:	180	175	140	119	–0.3	–1.0	–4.3	–3.2	6.6	5.6	2.9
Huabei	93	93	80	62	–1.9	0.0	–3.0	–5.0	3.9	2.9	1.5
Dagang	87	81	60	57	1.8	–2.1	–5.9	–1.0	2.8	2.7	1.4
East:	662	642	635	618	–2.9	–1.1	–0.2	–0.5	29.4	20.7	15.0
Shengli*	559	542	548	551	–2.6	–1.0	0.2	0.1	24.2	17.5	13.4
Zhongyuan	80	73	62	48	–6.2	–3.0	–3.4	–5.0	4.6	2.5	1.2
Jiangsu	23	26	25	19	4.6	3.9	–0.8	–5.0	0.6	0.7	0.5
South-central:	53	49	41	33	–3.2	–2.7	–3.4	–4.6	2.4	1.7	0.8
Henan	37	34	28	22	–4.3	–2.9	–3.5	–4.5	1.8	1.2	0.5
Jianghan	16	15	13	10	–0.2	–2.2	–3.1	–5.0	0.6	0.5	0.2
Northwest:	444	589	952	1658	11.8	9.8	10.1	11.7	7.4	13.9	40.3
Xinjiang*	174	204	290	467	3.5	5.5	7.3	10.0	4.9	5.4	11.4
Tarim*	84	144	335	766	60.6	19.7	18.5	18.0	0.1	2.6	18.6
Turpan*	60	83	142	242	93.9	11.4	11.3	11.3	0.0	1.9	5.9
Changqing	66	83	94	90	12.2	8.2	2.5	–1.0	1.1	2.1	2.2
Qinghai	32	38	48	60	10.2	5.6	4.8	4.8	0.6	1.0	1.5
Yanchang	21	30	34	26	14.8	12.0	2.3	–5.0	0.3	0.7	0.6
Yumen	8	7	9	7	–3.6	–2.3	3.1	–5.0	0.4	0.2	0.2
Others	38	54	40	29	11.8	12.2	–5.8	–6.0	0.6	1.2	0.7
Offshore*	323	396	329	269	43.3	7.0	–3.6	–4.0	0.9	10.1	6.5
National total	3198	3395	3560	4115	2.1	2.0	1.0	2.9	100.0	100.0	100.0

Note: *means that the projections comply partly or fully with Chinese predictions.

Table 3.16 Comparison of crude production forecasts

(000 b/d)	2000			2005			2010		
	Low	Mid -point	High	Low	Mid -point	High	Low	Mid -point	High
Onshore:									
CNPC[1]	2900	2950	3000	n.a.	n.a.	n.a.	3600	3800	4000
Author		2999			3230			3846	
Offshore:									
CNOOC[2]	340	360	380	n.a.	n.a.	n.a.	200	220	240
Author		396			329			269	
National Total									
CNPC and CNOOC	3240	3310	3380	n.a.	n.a.	n.a.	3800	4020	4240
Author		3395			3560			4115	

Sources: Zhou Yongkang, President of CNPC, *China Daily Business Weekly*, September 23, 1996; *China Daily Business Weekly*, September 23, 1996; and author's forecasts.

is very difficult to assess, and data from recent experience provide an incomplete basis for analysis. Foreign experts' calculations of production costs vary considerably as time goes by and inflation pushes up prices. Despite geographic differences, crude oil costs in China's onshore production are probably around $5.00 per barrel, while costs from offshore production could be up to $10.00 per barrel. As more complicated technology is used and existing oilfields are depleted, production costs could increase significantly. According to Chinese reports, the cost per barrel in the new exploratory oilfields is approaching world oil prices. With this increased rate of cost, China's average cost per barrel could reach or surpass world oil prices by the end of this century.[25] This is particularly true when costs from the Tarim E&D are added to the national average.

In the early 1980s, China hoped to attain 4.0 million b/d of total crude production by 2000, significant volumes coming from offshore oil. It is conceivable that the 4.0 million b/d target will be reached, but only if the pace of investment in EOR is accelerated, and the discovery rate increases through the use of modern geophysical data collection and analysis. In the near term, the key problem facing the Chinese oil industry is that new discoveries in the northeast corridor will be nearly offset by the depletion of older fields. We do not believe that the 4.0 million b/d of crude production can be reached until a significant volume of crude from Tarim can be delivered to consuming areas. However, we do believe that

[25]See Zhang Xisen, "Management System Reforms within China's Oil Industry", Proceedings in Oil and Gas Asia–Pacific Conference, Singapore, September, 1994, p. 7.

China is one of a few countries whose oil reserves and production capacity will give it enormous production potential after 2000.

Finally, Table 3.16 demonstrates several crude production forecasts including the author's projections in order to help reader to make a comparison.

3.4 Conclusion

China's upstream oil industry faces the challenge of a slow-down in the rate of growth of its output. The older existing fields in eastern China, most significantly Daqing, are reaching their production peaks. The vast reserves of the northwest have great potential, but few barrels will come on stream in the foreseeable future.

The exploration effort in China has been relatively limited. China's oil-bearing areas are extensive, but exploration is being undertaken in only 10 of the largest basins, mostly where oil has already been discovered. Even within these basins, exploration activities tend to be concentrated close to existing fields. Secondary recovery operations (i.e. water flooding, polymer injection, polymer flood, huff and puff steam injection, etc.) are in use in China's major fields or field complexes.

Nevertheless, China is Asia's crude expansion giant in the coming decade. We expect a 300,000 to 400,000 b/d increase in crude supplies by 2005. The annual growth rate of crude production should slow down after 1997, reflecting a mature phase of E&D in offshore areas and the pre-mature condition in the west (i.e. Tarim and Xinjiang basins). At present, China's three largest producing fields (Daqing, Shengli, and Liaohe) have maintained their present production levels for the past decade but doubts remain about a sustained field flow after 2000. Although production in mature eastern oilfields is expected to fall, increases in western and offshore fields should more than offset the eastern fields' decline. Western China, and especially Tarim, may have plenty of capacity to meet China's demand for the next 20 years.

Government policy requires both CNPC and CNOOC to be responsible for their own profits and losses. Crude prices remain regulated; however, CNOOC, as an offshore company, has full flexibility to sell crude in the world oil market. In contrast, CNPC, as an onshore firm, has experienced diminished profits resulting from regulated oil prices. Following the successful experiments in Tarim and Turpan-Hami oilfields in northwestern Xinjiang, CNPC will restructure itself so as to improve efficiency, separating the management and budget functions of its oil exploration and production activities from those of technical services, logistics, and other non-oil operations. This is one step closer to western corporate strategies and further from Soviet-style management techniques.

Chapter 4
Refining Industry: Policy and Industry Prospects

With the rapid pace of economic growth and industrialization, China's refining industry faces a constantly increasing demand for petroleum products and petrochemical feedstocks. China has been adding 180,000 b/d of crude processing capacity per year since 1988 and, at 4.2 million b/d, ranks fourth in the world. Demand growth is concentrated in light products, particularly middle distillates, and this has required expensive investments in catalytic and hydrocracking capacity. The country's crude-distillation capacity will be sufficient, while capacity of upgrading units remains insufficient. Secondary units will continue to fall short of meeting increasing demand for middle distillates. However, China faces a regional imbalance between refining capacity (in the northeast) and emerging demand (in the southeast), and problems in the transportation and distribution system are unlikely to be resolved in the near future. In addition, few of China's existing refineries are designed to handle the lighter, higher-sulfur crudes that are being imported in growing volumes.

China's refining system is undergoing a major overhaul. New refineries are being built as existing ones are upgraded and expanded. Most of the new construction is in China's booming southern and eastern coastal regions, where there is easy access to domestic and export demand. Announced refinery construction projects should be sufficient to prevent a shortage of refining capacity over the coming decade. However, it is unlikely that all of these projects will actually be built, and uncertain what their final configuration will be. Because of the need for foreign investment, many of these new refineries will be joint ventures.

There have been four organizations involved in the refining industry in China. They are China Petrochemical Corporation (Sinopec), China National Petroleum Corporation (CNPC), the Ministry of Geology and Mineral Resources, and the provincial governments. Before 1998, however, Sinopec is the major player and the latter two only run a few small refineries. When Dalian refinery started its processing in 1996 a fifth player in the Chinese refining industry was added: the joint-venture refinery. Up to early 1998, of all crude refined in China, about 85 percent of crude was processed by the 34 refineries operated by Sinopec. The rest was processed by other relatively small refineries operated by CNPC and local governments (Tables 4.1 and 4.2).

The first section in this chapter reviews refinery processing capacity and its current problems. Section 2 lays out the procedure for starting a new refinery or expanding an existing refinery, and reviews planned refineries. This section also predicts the refinery processing capacity by 2000, based on the information we have. Section 3 analyzes refinery yields and output, and projects principal product output over the next 10 years. Section 4 reviews the oil transportation system. Finally, Section 5 focuses on refining policy and its recent changes.

4.1 Crude Distillation and Conversion Capacity

China is one of the fastest growing countries in new refinery capacity, with more than 180,000 b/d added each year over the last five years, but capacity utilization has fallen. By the beginning of 1997, Chinese refining capacity was 4.2 million b/d[1], giving China the world's fourth largest crude processing capacity after the US, Russia, and Japan.

4.1.1 Processing capacity

In the 1980s, China's refinery capacity grew slightly more than half as fast as the economy and just over a quarter as fast as industrial output. Atmospheric distillation capacity increased from 3.3 million barrels per day in 1990 to 4.2 million barrels per day at the end of 1996. As Table 4.3 shows, Chinese fluid catalytic cracking (FCC) capacity is relatively high at about 1 million b/d, which is higher than Japan's 700,000 b/d of FCC capacity. Cracking capacity is fairly high relative to crude distillation capacity (Fig. 4.1 and Table 4.3) — 38 percent in China, compared with 30 percent in Asia on the average. China also operates more coking capacity than any other country in Asia.

All existing refineries have been designed and constructed by Chinese firms, though a few processing units have been licensed from international licensors. Most refineries were built between the late 1960s and early 1980s. Overall technical conditions are relatively inferior by international standards, even though technical refitting and modernization has continued. Chinese researchers have recently developed new refining

[1] Of the 4.2 million b/d total, Sinopec, the world's fourth largest refiner, has the share of 3.3 million b/d (or 165 million tons), accounting for 80 percent of total capacity.

Table 4.1 Refinery distillation capacity and throughput in three refining majors (as of 1996)

	Distillation capacity		Refinery throughput		
	000 b/d	Million tons	000 b/d[a]	Million tons per year[a]	Utilization rate (%)[a]
National total	4165	208.3	2791	139.6	67.0
Sinopec	3312	165.6	2,264	113.2	68.4
CNPC	497	24.9	327	16.4	65.8
Local	356	17.8	200	10.0	56.2

[a]Numbers on refinery runs published by Sinopec are different from the data given by State Statistical Bureau, which are frequently used in the book.
Sources: Sinopec and other various sources.

Table 4.2 Distillation capacity changes in Sinopec and non-Sinopec refineries

				Sinopec		Non-Sinopec	
	000 b/d	Million Tons	Annual Additions (000 b/d)	000 b/d	Million Tons	000 b/d	Million Tons
1985	2525	126	–	2020	101	505	–
1986	2525	126	–	2020	101	505	–
1987	2625	131	100	2100	105	525	20
1988	2750	138	125	2200	110	550	25
1989	3000	150	50	2400	120	600	30
1990	3300	165	50	2600	130	700	35
1991	3520	176	50	2800	140	720	36
1992	3815	191	50	3055	153	760	38
1993	3905	195	50	3105	155	800	40
1994	3935	197	30	3135	157	800	40
1995	4015	201	20	3155	158	860	43
1996	4170	209	175	3310	166	860	43

processes, such as an FCC unit with light olefin yields over 50 percent and a combined catalytic cracking–aromatics extraction technology in one unit (the latter can raise the yield of gaseous and liquid products by 1.2 percent with improved quality).[2]

The concentration of demand growth in light products, and middle distillates in particular, has forced heavy investments in cracking capacity to simultaneously meet domestic demand, reduce light diesel imports, and reduce fuel oil yields. Most cat crackers are operated for high distillate yields. Within Sinopec's refineries, there are about 40 FCC units in operation. According to Sinopec, China's FCC units are approaching world technical standards of the late 1980s.

[2] See *China Petrochemical Corporation Annual Report, 1993*, p. 30.

Table 4.3 Refinery distillation capacity comparison

	China		Asia-Pacific (incl US PADD V)	
	Refining processing capacity (000 b/d)	Share in total distillation capacity (%)	Refining processing capacity (000 b/d)	Share in total distillation capacity (%)
Distillation capacity	4165	100	20,450	100
Catalytic cracking	1018	24	3594	18
Thermal cracking	264	6	1428	7
Catalytic reforming	207	5	2584	13
Hydrocracking	201	5	2584	13
Hydro treating	294	7	6530	32
Alkylation	44	1	312	2
Coking	263	6	1428	7
Lube oil hydrotreating	49	1	n.a.	n.a.

Sources: Sinopec and Energy Security Analysis, Inc.

Catalytic reforming is another refining process in which China has made great progress in recent years, after importing modern technology from abroad. If the country moves toward eliminating lead in gasoline, catalytic reforming capacity additions may be expected to continue. Hydrocracking has a relatively small role in the refining system, although China has increased efforts to develop hydroprocessing technology domestically, due to the high cost of importing such technology.

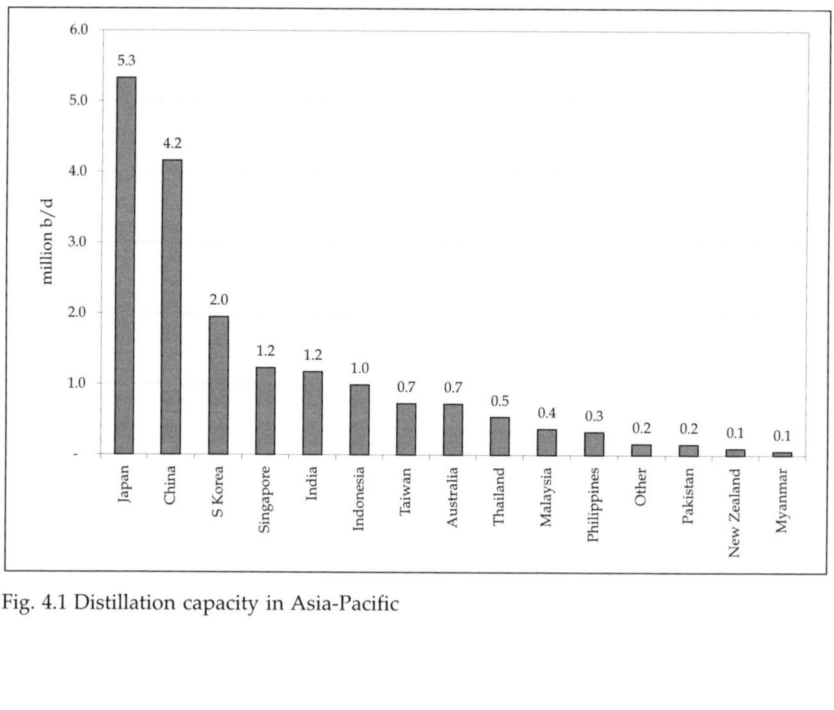

Fig. 4.1 Distillation capacity in Asia-Pacific

Chinese refineries have also added secondary processing capacity at existing refineries, resulting in great yield changes for light products. New construction has focused on hydrocrackers (Daqing, Fushun, Maoming, Nanjing, and Shanghai) for the production of diesel and kerosene; on alkylation units (Shanghai, Tianjin, Zhejiang, Qilu, Fushun, and other smaller refineries) for the octane enhancement of gasoline; and on catalytic crackers for processing the heavy fuel oil fractions.

4.1.2 Refining

The volume of crude refined in China has grown roughly 5 percent per year since 1985. In 1997, it was the first time that China's refinery runs exceeded its crude production level. Crude runs averaged 3.3 million b/d, slightly higher than the crude production of 3.2 million b/d. Of the 3 million b/d of refinery throughput in the calendar year 1996, Sinopec processed about 2.36 million b/d (118 million tons) of crude, accounting for near 80 percent of the total.

The efficiency of Chinese refineries has improved significantly during the past several years. Overall refining utilization fell from 66 percent in 1990 to 76 percent in 1997. China's refining industry is also focusing on the development and efficient operation of technologies for secondary processing, separation and utilization of refinery gases, and formulation and production of advanced refining catalysts and additives, among other areas.

Increasing operational efficiency at existing refineries is a critical issue. Almost all Chinese refineries were built to process domestic crude, which is typically heavy, waxy, and sweet. The supply of domestic crude is increasingly insufficient to meet domestic demand, however, so more crude will have to be imported.[3] However, most refineries cannot handle sour crude, like most Middle Eastern grades, due to metallurgical limitations of the process units. Most of China's existing distillation towers were built with untempered steel, which limits the sulfur content of the crude they can refine without corrosion problems. As a stopgap, refineries may install front-end crude desulfurization units, but that is very expensive, so many complexes will eventually have to be rebuilt.

Furthermore, the topping units usually are combined directly with the vacuum unit. Increasing the gravity of the crude being processed will result in less material being sent to the vacuum distillation unit, leaving

[3] Sinopec concluded at least 80,000 b/d of crude deals with Saudi Arabia, Kuwait and the UAE, with deliveries started in 1995.

vacuum capacity underutilized. Because the heat exchanger systems of the atmospheric and vacuum units are linked, this may cause operational problems. All told, the cost of these modifications may make replacement a more attractive alternative.

There has been little improvement in Chinese product quality in the last 20 years, especially in motor gasoline and diesel oil. Processes such as isomerization and etherfication have not been introduced on a large scale, but reforming capacity has increased. China's refining industry has an intelligent and qualified team in the field of R&D. They have focused their attention on domestic crude, and notable results have been achieved in the past. They are starting to work on processing more foreign crude. Meanwhile, China intends to negotiate with international licensors for advanced technologies. It is believed that technical improvement will be an important aspect in the development of Sinopec's existing refineries in the next decade.

There are 57 refineries in China at this time: Sinopec owns 32, and CNPC owns 25. None of Sinopec's refineries have capacity greater than 200,000 b/d, and only four refineries — Fushun, Qilu, Gaoqiao and Maoming — have capacity above 150,000 b/d (see Table 4.4). In particular, as Table 4.5 shows, CNPC-owned refineries have much smaller capacities than Sinopect's refineries. With its 25 small refineries, CNPC owns a total of 420,000 b/d of distillation capacity near the oilfields (about 12 percent of the national total). These refineries each average less than 17,000 b/d of capacity. This is one explanation for the low energy efficiency and productivity common in Chinese refineries.

As Table 4.4 illustrates, existing refineries were located primarily near oilfields in northern and northeastern China, as well as Shandong province, with a few scattered in various provinces according to local consumption. This resulted in overcapacity in northeastern China and a lack of large refineries in the vast southwestern areas. Refineries in inland provinces usually have relatively small capacities and operate at lower utilization rates. This distribution of refineries has left the southern coastal provinces, which are seeing the most growth in both economic power and oil demand, with few large-scale refineries, while the northern and northeastern regions have more capacity than they need.

1994 was the first year since the early 1960s that China's refinery throughput showed no growth from the previous year, reflecting persistently high inventories and weak demand, which have been hallmarks of the 1994 Chinese oil market. In 1994, China processed 2.64 million b/d of crude, registering a 0.4 percent decrease over 1993. In addition, 1994 third-party crude processing averaged about 24,000 b/d, falling from a peak level of 60,000 b/d in 1993. 1994 refinery operations provide a good example of how the downstream industry works under a centrally

Table 4.4 Refinery processing capacity and throughput by refinery

Name of refinery	Affiliation	Location	Major sources of crude supply	Year of operation start	Distillation capacity (000 b/d)	FCC (000 b/d)
North						
Yanshan Petrochemical	Sinopec	Beijing	Daqing, Huabei	1967	140	49
Tianjin Petrochemical	Sinopec	Tianjin	Dagang	1970	68	31
Shijiazhuang Refinery	Sinopec	Hebei	Huabei	1983	50	18
Cangzhou Refinery	Sinopec	Hebei	n.a.	1975	30	9
Others	Sinopec/non-Sinopec				50	n.a.
Northeast						
Fushun Petrochemical	Sinopec	Liaoning	Daqing	1928	174	49
Dalian Petrochemical	Sinopec	Liaoning	Imported		142	n.a.
Daqing General Petrochemical Works	Sinopec	Heilongjiang	Daqing	1963	120	24
Jinzhou Petrochemical	Sinopec	Liaoning	Liaohe	1937	110	16
Harbin Refinery	Sinopec	Heilongjiang	Daqing	1976	100	6
Jinxi Petro. Proc. & Chem. Complex	Sinopec	Liaoning	Daqing, Liaohe	1941	100	n.a.
Anshan Refinery	Sinopec	Liaoning	Liaohe		50	n.a.
Qianguo Refinery	Sinopec	Heilongjiang	Jilin	1974	50	24
Lingyuan Refinery	Sinopec	Heilongjiang	Daqing	1973	30	6
Liaoyang Petrochemical Fibre	Sinopec	Liaoning	Liaohe	1984	30	2
Others	Sinopec/non-Sinopec				185	n.a.

Continued overleaf

Table 4.4 *Continued*

Name of refinery	Affiliation	Location	Major sources of crude supply	Year of operation start	Distil-lation capacity (000 b/d)	FCC (000 b/d)
East						
Qilu Petrochemical	Sinopec	Shandong	Shengli	1966	160	46
Gaoqiao Petrochemical Co.	Sinopec	Shanghai	Varing	1952	150	20
Jinling Petrochemical	Sinopec	Jiangsu	Jiangshu		140	n.a.
Zhenhai General Petrochemical Works	Sinopec	Zhejiang	Imported	1978	110	28
Shanghai General Petrochemical Works	Sinopec	Shanghai	Imported	1976	106	n.a.
Yangzi Petrochemical	Sinopec	Jiangsu	Jiangsu	1987	60	n.a.
Jinan Refinery	Sinopec	Shandong	Shengli, Imported	1971	60	n.a.
Anqing General Petrochemical Works	Sinopec	Anhui	Varing	1976	56	24
Jiujiang General Petrochemical Works	Sinopec	Jiangxi	Middle East	1981	50	24
Fujian	Sinopec	Fujian	Imported	1992	50	n.a.
Others	Sinopec/non-Sinopec				80	n.a.
South-central						
Maoming Petrochemical	Sinopec	Guangdong	Imported	1955	170	32
Guangzhou General Petro. Works	Sinopec	Guangdong	Imported	1978	104	44
Luoyang General Petrochemical Works	Sinopec	Henan	Zhongyuan	1984	100	40
Baling Petrochemical	Sinopec	Hubei	Jianghan		100	20
Jingmen General Petrochemical Works	Sinopec	Hubei	Jianghan	1975	100	20
Wuhan Petrochemical Works	Sinopec	Hubei	Imported	1972	50	20
Others	Sinopec/non-Sinopec	Henan			25	n.a.
Southwest						
Nanchong Refinery	Sinopec	Sichuan	n.a.			
West						
Lanzhou Petro. Proc. & Chem. Complex	Sinopec	Gansu	Changqing, Yumen	1958	110	24
Dushanzi Refinery	Sinopec	Xinjiang	Karamay	1936	50	
Urumqi General Petrochemical Works	Sinopec	Xinjiang	Karamay/Tarim	1975	50	16
Lanzhou Chemical Industry	Sinopec	Gansu	Changqing, Tumen	1965	14	
Other	Sinopect/non-Sinopec				75	n.a.

Sources: *Sinopec Yearbooks, 1993 and 1997*, and Sinopec.

Table 4.5 Refinery processing capacity in selected CNPC's refineries

Refinery	(000 b/d)
Baoding	8.0
Refinery of Jilin	86.0
Jiangnan Refinery	6.0
Mudanjiang Dongfanghong	4.0
Hangzhou Refinery	3.2
Qingdao	13.0
Dagang	20.0
Asphalt Plat in Liaoning	30.0
Nanchong	3.0
Yanchang Oil	14.0
Yumen	30.0
Qinghai	3.0
Karamay	33.0
Zepu	3.0
Shengli Heavy Oil Plant	20.0
Shengli Refinery	3.0
Refinery of Henan	2.4
Jianghan	2.4
Maling Refinery	6.0
Jajiatan Refinery	2.0
Huabei Chemical	4.0
Daqing Chemical	6.0
Jilin Refinery	3.0
Zhongyuan	3.0
Tianyang	1.0

Source: CNPC.

planned economy. "Excessive" product imports at the end of 1993 — caused by a big price spread between Chinese and world market prices and by the 1993 practice of decentralized import licenses — led to ample product inventories in the first half of 1994. The tightening of policy governing oil distribution in May 1994 restricted demand in the fast-growing areas in southern and eastern China.

However, there was a change at the beginning of 1995 because product inventories receded to normal levels at both the refinery and distributor (an oil sales company in the provinces) levels and because demand picked up again. Since recontrolled imports did not permit product to pour into China at the end of 1994, as happened in previous years, stocks remained under control. The central authorities therefore allocated more crude oil for processing. Expect to see China's refinery runs to have picked up in 1995, increasing moderately to reach 2.88 million b/d in 1995, up by 9 percent from the 1994 level.

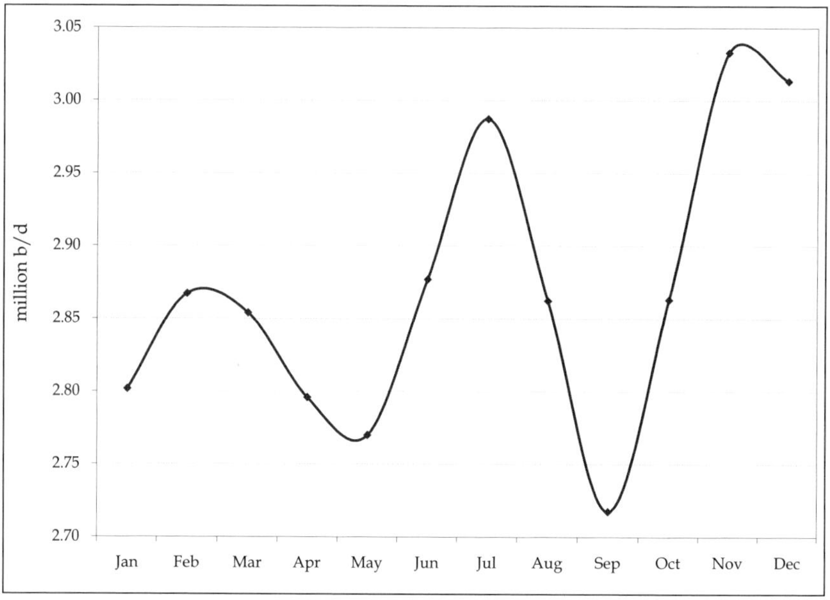

Fig. 4.2 Seasonality of refinery runs (1993–1997)
Source: State Statistical Bureau.

China possesses the second largest refinery capacity in the region, but there is less refinery maintenance information available on an individual refinery basis. However, the seasonality in refinery runs noted during the years from 1993 to 1997 provides some indication that Chinese refineries, on average, undergo turnarounds during May and September (Fig. 4.2).

Chinese refineries have no definite time frame for maintenance. The Chinese oil market is still controlled by the central government's product allocation system, and refineries, except for joint ventures, are constrained by the availability of crude supply. The time of maintenance varies among Chinese refineries. Therefore, refinery turnarounds in China have not historically shown a clear pattern of seasonality. The limited data we received from early 1992 to the first half of 1995 indicated that there was a decline in refinery runs in May and September during the past three years. However, we do not believe that Chinese refinery maintenance will have a significant influence on the Asian market, due to China's regulated oil trade. China affects the Asian market directly through its import/export activities, rather than its refinery operations. Under its centrally planned system, there is no stable, predictable linkage between import/export and refinery operations, and such linkage could be altered at any time.

4.2 Overview of Refining Capacity Plans and Assessment of Actual Changes

The number of proposed new joint-venture refineries in China has grown over time. The highest growth rates are in the south, where smaller refineries are the norm and economic growth is causing oil demand to rise fastest. A number of new refineries and upgrading projects are under construction in coastal areas, and more are being studied. Before we assess the prospects for Chinese refinery construction, two issues should be addressed: changes in the institutional structure in the refining sector and the procedures required for building a new refinery in China.

Before the formation of Sinopec in 1983, the industry was in disarray. Fragmented production facilities were owned and managed by various ministries, national corporations, and local governments. In order to utilize the existing crude supply more efficiently, rationalize production, and coordinate distribution, the government formed Sinopec and gave it responsibility for managing most petroleum and petrochemical processing facilities in the country. At the same time, CNPC was given control over a number of smaller oilfield refineries. In recent years, other state-owned oil-related companies, such as CNPC, China Petrochemical Imports and Exports Corporation (Sinochem), China National Offshore Oil Corporation (CNOOC), and a number of provincial and special economic zone (SEZ) governments began pressing their own refinery expansion programs, with or without Sinopec's cooperation. Many of these efforts are focused on foreign joint ventures, particularly for the construction of new refineries, but also for expansion and upgrading of existing facilities.

Sinopec's share in total refining capacity will decline sharply when others enter the Chinese refining business. Such change does not mean that Sinopec will lose its importance in Chinese refining, but this will break Sinopec's monopoly and introduce competition and a new market environment. Sinopec is involved in most joint-venture refinery projects currently on the drawing board, but the influence they will have over the new enterprises in unclear. The new capacity will, to some extent, compete with Sinopec's existing capacity, but the new capacity will also be more efficient than existing capacity, making its use for domestic consumption more cost-effective than many of Sinopec's other refineries.

The procedure for starting a new refinery or expanding an existing refinery is still strongly influenced by the central planning system in China. The first step is to indicate the intention to build a new refinery and conduct a preliminary feasibility study, with the consent of the provincial government, followed by a more detailed study that can be completed by Chinese initiator(s) or collaboration between Chinese and foreign parties. The normal feasibility study in China requires disclosure

- *Huangdao* (also called Qingdao) refinery in Shandong province — a 200,000 b/d unit — will be the largest refinery in China. It is a joint venture by Sinochem (18 percent), Saudi Aramco (45 percent), South Korea's Ssangyong refining company (15 percent), and Shandong province (22 percent). The project will cost a total of $2 billion.
- *Huizhou* refinery in southern Guangdong Province, with a distillation capacity of 160,000 b/d and 450,000 metric tons per year of ethylene, was expected to built by a group headed by Shell Nanhai (a subsidiary of Royal Dutch Shell). The total construction cost is estimated at $4 to $5 billion (other estimates show $5–6 billion). Shell will take a 50 percent stake in the project. The balance of ownership will be spread among Chinese entities: CNOOC (20 percent), Sinopec (10 percent), the Guangdong government (10 percent), and the Hong Kong Merchants Steam Navigation Co. (10 percent). Huizhou refinery is expected to process heavy crude from the South China Sea. The refinery will also feature a 450,000 ton per year ethylene petrochemical processing plant — the reason for the high cost of the project.
- *Ningbo* refinery (also called Beilum), near Ningbo in Zhejiang province, will have a capacity of 100,000 b/d. It will be owned by Concord Oil, a subsidiary of Hong Kong-based conglomerate Pacific Concord Holdings. It was reported that the Chinese SPC has approved this project.
- *Shanghai* refinery in the Pudong-Shanghai area will be built with a crude distillation capacity of 120,000 b/d, and will be backed by Sinopec, the Shanghai municipal government, and foreign oil companies. The proposed joint venture will be capitalized at $1.4 billion. France Elf withdrew from this project, because Elf objected to its equity stake and the domestic sales arrangement.
- *Shenzhen* refinery in the Shenzhen SEZ, with a 100,000 b/d crude unit, will be built by CNPC, the Shenzhen government, and Yukong with total initial investment of $1.5 billion. Of this, Yukong will contribute 40 percent, and CNPC and the Shenzhen government will kick in the remaining 60 percent. The Shenzhen refinery is scheduled to import Middle Eastern crude and export 60 percent of its products to Hong Kong. It will have eight processing units, including an atmospheric fuel oil desulfurization/hydrogenation complex that is a first for China. It is designed to supply Hong Kong, Shenzhen, and Macao, all of which currently import most of their product from Singapore. The Shenzhen plant could save more than $3.50 a barrel in transportation costs compared with imports from Singapore.

There are other proposed new refineries but less information is available about them, such as Zhanjiang in Guangdong province (70,000 b/d) and

Zhuhai in Guangdong province (50,000 b/d). An outline agreement has been signed with Iran for the study of a refinery project, though no site has so far been selected. In addition, Japanese and Taiwanese companies have shown continuing interest in China's petrochemical sector.

4.2.1.2 Debottlenecking and expansion

In addition to building new refineries, some existing refineries are scheduled to be modernized and revamped. In some instances, it makes more sense to upgrade existing refineries or integrate new development with existing refining infrastructures, because projects for refinery expansion can be achieved relatively easily compared to the plans to construct grassroot refineries, largely due to cost differentials. In addition, revamping the existing plants could also avoid the creation of competition for Sinopec from joint-venture refineries. Many of these expansion plans include units such as reformers, FCCs, hydrocrackers, and alkylation units.

Sinopec has begun a major revamp of its refinery network, debottlenecking capacity and building desulfurization facilities at the major refineries, and plans to revamp and expand seven existing refineries in coastal areas and inland (with access to river transportation) to reach a capacity of 200,000 b/d each by 2000. Sinopec hopes to include foreign companies in these expansion projects, and it is negotiating with foreign oil firms to expand the refining capacity of Sinopec's five major refineries in Maoming, Zhenhai, Dalian, Fujian, and Guangzhou. Expansion for each of them will be from 100,000 b/d to 200,000 b/d.

- *Maoming* refinery in Guangdong province is expected to undertake two projects. The first project, sponsored by Saudi Aramco and Sinopec, is to debottleneck the current 170,000 b/d refinery in order that it may reach a level of 200,000 b/d. The estimated cost is $30 million. This expanded unit will support Vitol–Sinopec's joint-venture 300,000 tons per year ethylene plant. The second project in Maoming is backed by Sinopec and other foreign oil companies (Aramco has indicated interest). It plans to add 100,000 b/d of distillation capacity by 1998 and another 70,000 b/d by the end of the century. This expansion project requires an investment of $1 billion. A new refinery complex is planned to process Middle Eastern crude, which will maintain the position of Maoming refinery as the largest processor of imported crude in the country.
- *Korla* refinery in Xingjiang Uygur Autonomous Region, a joint venture between CNPC and the local government (with a distillation capacity of 20,000 b/d), will be built by CNPC. Infrastructure work has been completed, and the project is waiting for the approval of

the central government. Korla refinery will refine crude from the Tarim basin and supply refined products to southern Xinjiang region.

- *Jiujiang* General Petrochemical in Jiangxi province has a proposed capacity expansion of 30,000 b/d, expanding its crude distillation capacity to 80,000 b/d. It is solely Chinese funded — the consortium members consist of two domestic companies. The project requires an investment of 70 million yuan ($8.2 million). Jiujiang also operates a 24,000 b/d fluid cat-cracker and a 3000 b/d catalytic reformer, and supplies LPG to end-users in the province.
- *Fujian* refinery in Fujian province was built in 1993, with a topping unit of 50,000 b/d. It is planning to expand to 210,000 b/d and develop an adjacent ethylene plant to provide supplies to local mid- and downstream manufacturing industries. The Fujian provincial government expects foreign companies to provide a share of the equity investment in its 18 billion yuan ($2.17 billion) refinery project. The two projects, the refinery and the ethylene plant, are scheduled for completion in the next 7–8 years.
- *Guangzhou* general petrochemical plant in Guangdong province is expected to increase its current topping capacity from 104,000 b/d to 240,000 b/d. Investment in the refinery expansion project is estimated at about $1 billion.
- *Zhenhai* refinery in China's east coast Zhenjiang province has revamped its crude distillation and vacuum distillation units, increasing by 20,000 b/d to 170,000 b/d in 1995. More than half of its capacity can operate with Middle Eastern crude. It also planned to boost the crude processing capacity of its refinery to 300,000 b/d within the next five to 10 years. Zhenhai refinery is one of the few Chinese state-owned companies issuing their stock on the international capital market. By the end of 1994, Zhenhai refinery raised HK$ 1.37 billion ($ 177 million) from the Hong Kong Stock Exchange to fund its revamping program. Its ambitious plans may be met by multiple sources of financing.
- *Cangzhou* refinery in Hebei province scheduled to raise its distillation capacity from the current 30,000 b/d to 80,000 b/d. The refinery expansion will support production of caustic soda and toluene diisocyanate (TDI), and possibly a 300,000–450,000 ton per year ethylene plant. This project will turn Cangzhou city into one of three major chemical center in China. Refinery expansion and TDI will require an investment of approximately 3.6 billion yuan (about US$ 440 million). Refinery expansion project was scheduled to have been completed by the end of 1997.
- CNPC is upgrading three refineries in northern Xinjiang to meet rising production from the Tarim Basin. Debottlenecking and

upgrading will increase refinery capacities at Karamay (40,500 b/d), Dushanzi (71,000 b/d), and Zepu (10,000 b/d) by 20,000 b/d. However, no further information is released for these activities conducted by CNPC.

Aside from boosting topping units, China also aims at upgrading its refining capacities in several refineries. According to various sources, Gaoqiao refinery/petrochemical complex in Shanghai has been proposed for the construction of an integrated normal-paraffin refinery of 1600 b/d and linear alkyl benzene of 1700 b/d. Zhenhai refinery in Zhejiang province was scheduled to set up an additional 16,000 b/d of catalytic reformer capacity to produce more unleaded 90, 95 and 97 RON gasoline in the next few years. It currently operates a 16,000 b/d fluid catalytic cracker. Zhenhai refinery was scheduled to process 40,000 b/d of 1.8 percent sulfur content crude from the Middle East in 1998, compared with the 4000 b/d of sour crude they refined in 1994. Jinzhou petrochemical complex in Liaoning province will complete a hydrotreater with 16,000 b/d processing capacity soon. This hydrotreater is expected to process cheaper sour crude and reduce the smoke point of its jet fuel.

The refinery plans listed above are being actively pursued, but it is unlikely that all of them will come to fruition. The projects listed here are, however, representative of the magnitude and timing of the capacity additions we expect in China. Our output projection for petroleum products will integrate these data and other assumptions.

The refinery plans listed above are being pursued actively, but it is unlikely that all of them will come to fruition. The projects listed here are, however, representative of the magnitude and timing of capacity additions we expect in China. Our output projection for petroleum products will integrate these data and other assumptions.

4.2.1.3 Projection of capacity additions

According to the information available at present, Table 4.7 illustrates our best guess regarding refinery expansion in China by the year 2010. These projections revise the overly optimistic view held by many — Chinese oilmen, foreign investors, and analysts of China's oil market — in early 1994. We saw no new joint-venture refineries built during 1996 and 1997, because all negotiations for new refineries slowed down in 1995. Foreign partners are showing more patience than expected. A delay in the agreements on joint-venture refineries reflects foreign investors' concerns regarding China's oil market regulation and its economic future after Deng. Moreover, a surge in prices of labor and land use in the coastal areas will cut the rate of return on foreign investment.

Table 4.7 Proposed joint-venture refinery projects in China

Refinery Name	Type of Cons-truction	Location	Proposed Distillate Capacity	Estimated Investment	Tentative Foreign Investor(s)	Tentative Chinese Investor(s)	Source of Crude Oil	Status, As of early 1997
Beihai	New	Guangxi	120	$2.6 billion	Kuok Kerry Group (Malaysia)	Three provincial governments	South China Sea	Feasibility study is completed; JV negotiation are under way; Waiting for SPC approval.
Huangdao (Qingdao)	New	Shandong	200	$2 billion	Aramco (48%),	Sinochem (18%), Local government (14%)		JV Ssangyong (20%) negotiation are under way; Waiting for SPC approval.
Huizhou		Guangdong	160	$4–5 billion (with petrochemical capacity)	Shell (50%), Hong Kong MSN Co. (10%)	CNOOC (20%), Sinopec (10%), Local government (10%)	South China Sea	JV negotiation is completed; Feasibility study is in progress; Waiting for SPC approval.
Linggao	New	Hainan	120	$1.67 billion	Hainan Hobond International Petrochem. Co.(100%)			Feasibility study is completed; Construction will start.
Ningbo (Beilum)	New	Zhejiang	150		Concor Oil, Hong Kong (100%)			SPC has approved the project; Refinery is under construction.

Continued overleaf

Table 4.7 *Continued*

Refinery Name	Type of Cons-truction	Location	Proposed Distillate Capacity	Estimated Investment	Tentative Foreign Investor(s)	Tentative Chinese Investor(s)	Source of Crude Oil	Status, As of early 1997
Maoming	Debottlenecking	Guangdong	30	$30 million	Vitol	Sinopec		Feasibility study is in - progress; Negotiation is under way.
Maoming	New	Guangdong	150		Thailand PTT (70%)	Sinopec (30%)		JV negotiation is under way.
Shanghai	New	Shanghai	120	$1.4 billion	Question market after ELF quit		Sinopec and local government	JV negotiations are under way; Feasibility study is in progress.
Shenzhen	New	Shenzhen, ESZ	100	$1.5 billion	Yukong (40%)	CNPC and local government (60%)	Middle East	Feasibility study is completed; Waiting for SPC approval.
Zhenghai	New	Zhejiang	120–140		Amoco	Sinopec	Middle East	JV negotiation is under way; Feasibility study is in progress.

Notes:
This table reflects the author's efforts in collecting Chinese refinery construction information. This information should be updated periodically.
There are other proposed refineries but less information is available, such as Zhanjiang in Guangdong province (70,000 b/d) and Zhudai in Guangdong province (50,000 b/d).
Proposed investment shares are in parentheses with tentative foreign and Chinese investors.
JV = joint venture.

Another round of high inflation could increase foreign investment cost and risk. As a result, a constrained market environment leaves less negotiating leverage to those Chinese businessmen hoping to make a deal.

On the other hand, these projections indicate that if the setting-up of joint-venture refineries is delayed, revamping and expansion of existing refineries may prevent the situation from worsening, and will reduce the current upgraded capacity shortage. In particular, if a weak market is constrained by government regulations, it will put less pressure on decision-makers to approve proposals for new refineries. If capacity increases could, to a large extent, be achieved by debottlenecking and expansion of existing refineries, building new refineries would be the second best choice. However, the problem remains that China needs foreign capital for almost any significant expansion in refinery capacity, and this creates its own difficulties.

While domestic supplies have been inadequate to meet demand for the last five years, the consequence of a delay in building new deep-processing plants should lead to a prolonged period of China's importing more refined products. Our projection shows that the petroleum product market is expected to be balanced when more new refineries come onstream. A closer analysis reveals that if the completion year for each joint venture refinery is further postponed for one year beyond our projection, China will add 100,000–120,000 b/d of products to its deficit in 1999 and 2000, and reduce 100,00–200,000 b/d of refined products in its surplus from 2001 to 2005. On the other hand, if the completion year for each joint venture refinery is moved a year ahead of our projection, China will turn from product importer to product exporter one year earlier, and could gain a product surplus of 100,000–200,000 b/d beyond the projected level during the years 2001–2005.

Many of the proposed new and debottlenecking refineries are in the south — Guangdong and Guangxi provinces — with three new refineries totaling 700,000 b/d, more than doubling the current regional capacity. However, current proposals will also have a significant impact on eastern China — Shandong, Shanghai, and Zhejiang provinces — with three proposed refineries and with a total capacity of 370,000 b/d, which equals 60 percent of existing capacity in the region. There have been few proposed refinery projects in the crude producing areas of the north and west, and only one in the northeast. In crude producing areas, such as Daqing and Shengli, officials are particularly eager to diversify and add value to crude production as their aging oilfields approach a period of decline. However, northern and northeastern provinces already have enough refineries, and the emerging domestic markets are in the south and east, particularly the coastal provinces. Foreign investors have shown they prefer to invest in the consuming regions to avoid extra transportation costs and uncertainties created by China's infrastructure. Locating in a southern coastal province gives access to growing domestic markets as well as international crude and product markets.

Due to cash shortages in both Chinese currency and foreign exchange in all level of government and Sinopec, foreign capital will be needed for new refineries for the foreseeable future. However, Sinopec's financial resources are significant and should be perfectly adequate to meet their commitments in joint-venture projects. Joint-venture refineries have more operational and financial flexibility than domestic refineries, such as being allowed to import crude and export refined products without permits. If they like, part of the output can be channeled into China's domestic market, but limited currency convertibility is an obstacle to these sales. Coastal refinery construction is premised on export markets that would pay for the capital and foreign exchange costs of the refinery, as well as their crude supplies.

4.2.2 Assessment of capacity increases

The profit incentive, plus zooming oil demand growth in China, has caused a rush of refinery construction and planning. While most assessments of the level of refinery investment point to a lingering deficit in capacity, these projections usually present a static picture; that is, they compare 10 years or so of demand with all the refinery projects planned for the next 10 years. The problem with this approach is that refinery projects are less and less certain as we peer farther and farther into the future, making it very difficult to talk about refining capacity 10 years down the road.

In order to cope with the anticipated surge in petroleum product demand, China has drawn up future refinery expansion plans. Table 4.8 presents Sinopec's forecast refining capacity forecast for China. FCC units, the most popular and widespread upgrading step, are projected to be built mostly in coastal joint ventures, giving total planned capacity of 960,000 b/d in 1995 and 1.3 million b/d in 2000. Hydrocrackers are also popular among the expansion plans, with aggregate capacity reaching 190,000 b/d in 1995 and 220,000 b/d in 2000. China also has expansion plans for additional coker units, but it is not clear how many desulfurizing units for kerosene and diesel gas oil components production will be built.

While Table 4.8 is a good reference for seeing how Sinopec sees future refining capacity, we did not use their projection for two reasons. First, this forecast, by our judgment, reflects the desires of Chinese planners rather than actual projects — wishful thinking that may not materialize.

Table 4.8 Additions of refining distillation capacity and projection

	1994	1995	1996	1997	1998	1999	2000	2001	2002	2003	2004	2005	2006	2007	2008	2009	2010
Capacity at the beginning of year	3935	3965	3985	4160	4320	4445	4681	4861	5031	5111	5281	5551	5671	5771	5871	6071	6231
Debottlenecking/expansion																	
Anqing Petrochemical		20															
Cangzhou					50												
Dalian						50				50							
Fujian (Amoco?)				30					80								
Guangzhou				50				70									
Jingmen General Petrochemical Works						46											
Jiujiang Petrochemical					30												
Lanzhou											90						
Maoming	30			80			75										
Qilu						40											
Zhenhai			50				80				80						
Tianjin Petrochemical			20														
Other refineries			50		25		25										
Sub-total	30	20	75	160	105	136	180	70	80	50	170						

Continued below

Table 4.8 *Continued*

	1994	1995	1996	1997	1998	1999	2000	2001	2002	2003	2004	2005	2006	2007	2008	2009	2010
Capacity at the beginning of year	3935	3965	3985	4160	4320	4445	4681	4861	5031	5111	5281	5551	5671	5771	5871	6071	6231
New refineries																	
Korla (CNPC/provincial government)					20												
Dalian (Total/Sinochem)			100														
Hainan (Hohbond, UK)						100											
Ningbo (Concor)								100									
Beihai (Kuok Kerry/ provincial government)										120							
Pengzhou (CNPC and local government)										100			100				
Shanghai (?/Sinopec)												120					
Shenzhen (Yukong/CNPC)													100				
Huangdo (Aramco/ Ssan gyong/Sinochem)															200		
Huizhou (Shell/MSN/Sino/ CNOOC)																160	
Sub-total			100		20	100		100		120	100	120	100	100	200	160	
Annual Additions	30	20	175	160	125	236	180	170	80	170	270	120	100	100	200	160	
Total capacity at the end of year	3965	3985	4160	4320	4445	4681	4861	5031	5111	5281	5551	5671	5771	5871	6071	6231	6231

Note: The author made an independent evaluation of the completion date of each refinery, rather using the "published completion" date. Again, this table reflects the author's efforts in collecting Chinese refinery information and forecasting their completed date. This information could change periodically.

Second, the 1995 projection seems too optimistic to come true, as we are already in the late 1990s.

Based on the proposed refinery lists, our view is that there still will be some shortage of refinery capacity in China over the next 10 years. Our forecast, as Table 4.7 shows, predicts capacity will be about 4.9 million b/d in 2000, around 5.7 million b/d by 2005, and up to 6.2 million by 2010, while total throughput will be about 3.7 million b/d, 4.4 million b/d and 4.9 million b/d, respectively. In fact, excess capacity is projected to be built in Asia, making it a significantly more competitive market. Based on construction plans, we predict that cracking capacity will increase to approximately 1.5 million b/d by 2000 (and same in 2005 due to lack of information beyond 2000).

4.3 Refinery Yields and Output

While refined petroleum product output has grown by approximately 5 percent annually over past 10 years, domestic demand has grown by 6–8 percent each year. In order to cope with the demand increase, refinery output becomes an important issue.

4.3.1 Current refinery yields and output

We see the distillation capacity utilization around the world was around 80 percent — the level that would require new investment in upgrade capacity. China follows other regions, having a small but steady increase in capacity utilization during recent years. This trend is expected to continue into the future (see Fig. 4.3).

Over the past decade, gasoline production has grown faster than total petroleum production, averaging 7 percent per year, compared with 5 percent growth for refinery throughput. Table 4.9 shows that gasoline production in 1997 totaled approximately 810,000 b/d, for a yield of 25 percent of crude throughput, substantially higher than the 20 percent obtained in 1985. However, over the past few years, diesel oil output increasingly exceeded gasoline production, reflecting the refinery configuration changes of recent years. Total middle distillate production in 1997 averaged about 1.1 million b/d, and its yield on crude throughput reached 33 percent, the highest yield in Chinese refining history. Refinery yields for light products have also experienced substantial increases in recent years. With the efforts to increase distillate yields, FCC capacity reached 1 million b/d in 1996. In 1997, the average yield of gasoline and middle distillates (excluding LPG) at Chinese refineries was 58 percent, up by 10 percent from 1990.

Among middle distillates, kerosene production has had low growth

Table 4.9 Refinery throughput, utilization rate, product output, and refinery yield

	1985	1986	1987	1988	1989	1990	1991	1992	1993	1994	1995	1996	1997
Refinery runs (000 b/d)	1718	1852	1955	2173	2269	2321	2433	2529	2651	2639	2881	3000	3324
Utilization rate (%)	68	73	74	79	76	70	69	66	68	67	72	72	77
Refinery output (000 b/d)													
Refinery gas*	38	42	44	45	46	49	48	55	56	66	78	87	101
Naphtha*	153	117	156	191	205	213	236	238	263	287	299	323	345
LPG*	52	65	69	74	81	83	88	84	131	141	169	217	257
Motor gasoline	343	392	405	442	459	506	560	628	736	665	709	753	811
Aviation gasoline*	2	2	2	2	2	2	2	2	2	2	3	3	3
Jet fuel	12	12	15	15	16	20	29	31	39	44	48	56	64
Kerosene	75	78	76	68	70	65	59	58	45	45	94	116	116
Diesel	414	456	484	503	528	533	583	622	724	711	794	883	993
Fuel oil	517	545	571	587	604	596	592	604	583	556	509	425	405
Lubricants	27	29	31	35	35	41	42	45	46	33	44	49	68
Bitumen*	33	34	35	35	34	35	48	53	51	57	55	56	58
Petroleum coke*	15	15	15	15	16	17	24	25	26	27	28	30	31
Other products*	64	93	89	61	55	41	21	16	17	51	108	63	124
Refinery yields on throughput (%)													
Motor gasoline	20.0	21.2	20.7	20.3	20.2	21.8	23.0	24.8	27.8	25.2	24.6	25.1	24.4
Kerosene	4.4	4.2	3.9	3.1	3.1	2.8	2.4	2.3	1.7	1.7	3.3	3.9	3.5
Diesel	24.1	24.6	24.7	23.1	23.3	23.0	24.0	24.6	27.3	26.9	27.6	29.4	29.9
Fuel oil	30.1	29.4	29.2	27.0	26.6	25.7	24.3	23.9	22.0	21.1	17.7	14.2	12.2
Lubricants	1.6	1.6	1.6	1.6	1.6	1.8	1.7	1.8	1.7	1.3	1.5	1.6	2.0
Others*	19.9	19.0	19.9	24.7	25.3	25.0	24.5	22.6	19.5	23.8	25.4	25.8	28.1

*1996 and 1997 numbers are estimated.

Sources: International Energy Agency, OECD and State Statistical Bureau.

Fig. 4.3 Refinery capacity, runs and utilization

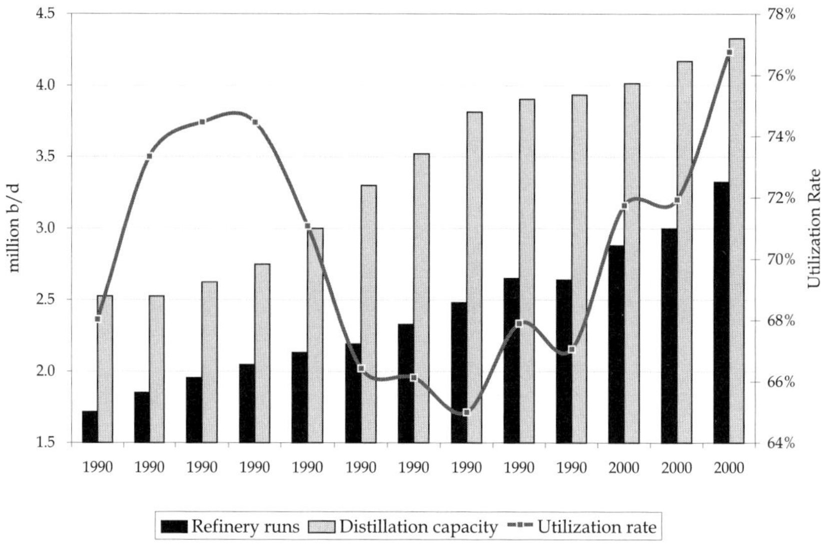

rates since 1985, particularly with a small drop in the second half of the 1980s and the early 1990s. Refinery yields of kerosene, however, fell from 4.4 percent in 1985 to 3.5 percent in 1997. In contrast, diesel oil production has increased in that time at an average rate of 7 percent, from 414,000 b/d in 1985 to near 1 million b/d in 1997. Refinery yields of diesel oil also rose significantly, reaching 30 percent in 1997.

Although fuel oil production kept a slight increase in the late 1980s and was flat in the early 1990s, its production has fallen dramatically during the past 5 years. Fuel oil's yield on crude throughput has declined significantly, from 30 percent in 1985 to only 12 percent in 1997.

As Table 4.10 shows, the production of light distillate oil comes primarily from the northeast and east, which together account for more than half of all crude processed in China. The three provinces of Guangxi, Yunnan and Guizhou, rely on refineries in Guangdong province for most of their refined products supply. In 1995, Guangdong province produced 10 percent of total light products, while its total refined product production made up 9 percent of the national total. If and when proposed refinery capacity of 340,000 b/d is built in the next several years, Guangdong province should have a greater product surplus for supplying neighboring provinces.

Table 4.10 Principal petroleum product production (000 b/d) by region

	Gasoline			Diesel oil			Fuel oil		
	1985	1990	1995	1985	1990	1995	1985	1990	1995
National total	342.8	506.1	708.6	413.5	533.2	794.4	517.4	596.3	509.2
North	42.7	70.8	80.9	56.7	52.3	74.8	97.7	79.0	63.3
Beijing	20.4	33.0	23.1	35.3	19.3	16.1	52.4	43.8	26.2
Tianjin	12.7	16.3	20.9	13.6	17.1	22.8	29.8	26.8	25.8
Hebei	9.6	21.4	31.5	7.8	16.0	31.9	15.5	8.4	7.8
Inner Mongolia			5.4			4.0			3.4
Northeast	104.1	154.5	189.3	149.0	190.2	233.8	244.5	209.5	156.4
Liaoning	64.4	92.7	93.8	103.6	119.4	137.1	169.6	142.6	108.5
Jilin	13.9	26.9	34.6	8.2	16.3	20.4	24.2	22.0	15.5
Heilongjiang	25.9	34.9	61.0	37.2	54.5	76.4	50.7	44.9	32.4
East	80.4	121.3	180.0	107.1	139.4	225.1	122.5	188.1	177.6
Shanghai	14.3	25.1	26.0	19.1	21.1	26.0	52.0	53.1	52.6
Jiangsu	15.8	21.9	25.7	26.9	32.2	42.9	36.8	50.7	44.7
Zhejiang	9.6	13.9	29.0	9.3	14.6	40.9	15.3	14.4	12.9
Anhui	8.9	13.6	17.3	11.1	16.5	20.2	18.2	11.1	8.8
Fujian			18.8				16.9		
Jiangxi	5.7	11.2	15.1	4.0	9.4	17.1		7.8	7.9
Shandong	26.0	35.6	48.1	36.7	45.5	61.0	0.1	51.0	50.6
South-central	74.8	106.6	153.3	69.1	97.6	156.2	29.0	84.1	70.3
Henan	6.7	19.0	32.0	5.4	14.1	24.1	5.5	2.6	3.2
Hubei	26.4	28.1	29.5	20.7	25.3	30.9	0.1	25.0	17.3
Hunan	15.1	18.9	21.1	14.3	18.1	21.3	5.5	11.1	10.1
Guangdong	26.5	40.4	67.1	28.6	39.8	76.9	2.1	45.2	39.1
Guangxi	0.1	0.3	3.5	0.2	0.3	3.0	15.7	0.3	0.6
Hainan									
Southwest	0.4	1.0	4.1	0.2	0.5	3.3	0.4	0.4	0.8
Sichuan	0.4	1.0	1.2	0.2	0.5	1.0	0.4	0.4	0.8
Guizhou									
Yunnan			3.0			2.3			
Tibet									
West	40.4	52.0	100.9	31.5	53.3	101.1	23.3	36.3	40.9
Shaanxi	1.0	4.1	9.0	0.8	4.0	8.0	1.5	2.7	3.4
Gansu	18.3	15.6	38.2	12.5	18.2	38.0	13.1	16.0	13.9
Qinghai	0.5	0.7	6.8	0.8	0.7	4.2	2.5	1.2	2.5
Ningxia	0.3	0.8	5.5	0.5	0.7	5.0	0.6	0.3	1.8
Xinjiang	20.2	30.7	41.4	16.8	29.6	45.9	5.6	16.1	19.3

Source: State Statistical Bureau.

Table 4.11 "Producibility" forecasts for principal products and total product

	Production (000 b/d)						Annual growth (%)			
	1990	1995	1997	2000	2005	2010	1990–95	1995–00	2000–05	2005–10
Total product	2208	2939	3390	3765	4477	5001	6.3	3.6	3.5	2.2
Gasoline/naphtha	719	1007	1156	1266	1488	1687	7.0	12.0	3.3	2.5
Middle distillate	598	889	1108	1239	1502	1726	8.3	15.7	3.9	2.8
Fuel oil	596	509	405	432	477	495	(3.1)	(6.2)	2.0	0.8
Others	295	533	722	828	1,010	1,091	12.6	22.9	4.0	1.6

The quality of China's petroleum products is low compared to international standards. For instance, higher volumes of 87-octane gasoline will be produced, but will be blended with other, lower quality blendstocks for domestic use. There is no sign that China will eliminate lead in gasoline in the foreseeable future, but the use of nonlead additives in gasoline will gradually increase over the next decade. Refineries have been running domestic low-sulfur crude produced in China, and therefore producing relatively low-sulfur middle distillates, but as increasing volumes of sour crude are processed in Chinese refineries, the average sulfur content of distillates may increase.

The petrochemical industry relies heavily on petroleum refining for feedstocks, and petrochemical operations are typically integrated into refineries. So far, natural gas has not been extensively exploited in China, though the potential is promising. Most of the discovered resources are "dry gas" with few LPGs. Development of China's petrochemical industry will require large amounts of naphtha and other light hydrocarbons from refineries. China's policy goal of petrochemical self-sufficiency by 2000, combined with the fact that economic development is increasing petrochemical demand rapidly, means there will be dramatic growth in petrochemicals in China.

4.3.4 Third-party processing

Due to slow growth of domestic crude production, the share of imported crude in Chinese refinery throughput has risen from 3 percent in 1990 to 12 percent in 1995, and 21 percent in 1997. With continuing robust demand, this share is expected to stay over 15 percent, even more for the next decade. China's petroleum net imports will not reverse its course, and oil deficit will remain when we move into the next century.

China has spare refining capacity for third-party processing because, in the past, every refinery lobbied central planners for more capacity than necessary in order to get a larger quota of crude from the central government. Consequently, most refineries have operated below capacity. Third-party processing was actually initiated domestically in 1988 when certain provinces were given permission by the central government to buy a certain amount of crude, with no restrictions on origin. After that, a number of refineries along the coast were approved to accept foreign clients for third-party processing. Current Chinese policy is more open and any refinery with excess capacity is encouraged to do such business so long as it has suitable facilities to receive crude and handle the products.

After 1990, China started to reduce expensive product imports by allowing third-party processing in its coastal refineries. Only six of 40 refineries in China are located along the eastern and southern coast, where they have the opportunity to take third-party processing, because they are geographically closer to international markets. They are in Dalian, Zhenhai, Maoming, Nanjing, Fujian and Shanghai. The bulk of imported sweet crude is sourced from Indonesia, Malaysia, and, more recently, Vietnam, though some crude is still exported to Singapore for refining and then reimported. Not all products refined by third-party processing are re-exported — many are sold into China. For most third-party processing, China buys some of the products, such as kerosene, gasoil, and fuel oil, that are in short supply in the domestic market. China's open policy towards foreign processing is the result of underutilized refining capacity and slow increases in crude production. Its primary advantage stems from short shipping distances between China and other East Asian countries. The short shipping distance offsets higher processing fees in China relative to Singapore. For example, processing in Maoming in Guangdong province costs over $2.00 per barrel, while the term fees for processing in Singapore vary between $1.50 and $1.80 per barrel.

Contracts for third-party processing usually run for one year. Several varieties of crude have been involved, but most are low-sulfur grades such as Southeast Asian and Omani crude. Maoming is the only refinery that can handle sour crude. In 1993, the Zhenghai refinery had the largest volume of third-party processing at 24,000 b/d. Other coastal refineries such as Maoming, Fujian, and Dalian had the balance. At present, Zhenghai refinery on the eastern coast takes nearly half of all third-party processing, and Dalian on the northeastern coast has about a quarter of the national total. Some other refineries are taking only small volumes (tens of thousand tons) for trial.

Despite the absence of third-party processing data from 1988 to 1991, we can infer that the volume of third-party processing was very small, because crude import volumes were low over that period. The volume of third-party processing has increased significantly in recent years. The data in Table 4.12 show that third-party processing grew from 38,000 b/d in 1992 to 60,000 b/d in 1993, and slid after then (excluding processing for domestic clients), while third-party processing as a percentage of total refinery runs remained small.

Third-party processing recently lost some of its appeal. The ability to sell some of the products made third-party processing very profitable, and that ability was reduced when the 1994 import restrictions went into place. Most products refined via third-party processing must be re-exported, and only a limited volume may be sold into Chinese markets. Significant third-party processing volumes should persist for some time, as long as there is spare capacity in some of the larger coastal refineries. However, the third-party processing must contend with: (1)

Table 4.12 Third-party processing in China

	1988	1989	1990	1991	1992	1993	1994	1995	1996
Total refinery runs (000 b/d)	2048	2132	2192	2328	2480	2651	2639	2881	3000
Crude oil imports									
Volume (000 b/d)	17	66	59	120	229	315	247	341	452
As a % of refinery runs	0.8	3.1	2.7	5.2	9.2	11.9	9.4	11.8	15.1
Third-party processing									
Volume (000 b/d)	n.a	n.a.	n.a.	n.a.	38	60	24	20	12
As a % of refinery runs	n.a.	n.a.	n.a.	n.a.	1.5	2.3	0.9	0.7	0.4
As a % of crude imports	n.a.	n.a.	n.a.	n.a.	16.6	19.0	9.7	5.9	2.7

Source: Sinopec.

high product prices and low crude prices on the international market; and (2) competition from Singapore and other refining centers in Asia.

At present, third-party processing clients are from various countries and regions, but most of them are Hong Kong-based firms. As long as there is no refinery in Hong Kong, their demand for third-party processing will continue. But if labor and utility costs rise significantly in China, in addition to relatively high refinery energy consumption and lower efficiency, the profitability of third-party processing will fall. The total volume in the next few years, therefore, is expected to be not much larger than 50,000 b/d. As joint-venture refineries enter service, they may displace some third-party processing because of their freedom to sell into Chinese markets (observing the valid regulations on prices and distribution system set by central government) and higher efficiency.

4.4 Oil Transportation

For a long time, transportation and distribution systems were controlled and administrated by the central government. Transportation and distribution are the link between the supply and demand, and their efficiency is critical to overall efficiency of Chinese oil markets. As uneven economic development continues around the country, transportation and distribution sectors will play more critical roles in efficiently meeting demand in all provinces (Tables 4.13 and 4.14).

With crude production flat, imports rising, and refinery capacity expanding, China is facing severe transportation problems. As demand has grown, the need to rehabilitate and increase the capacity of some pipelines and to remove bottlenecks on some rail lines and ports has become urgent. Transportation logistics, including availability of ships and port restrictions, are limiting factors in crude and product distribution in China. Bottlenecks in the transportation system frequently lead to delays in crude and product deliveries.

In China, some crude oil is transported by pipeline from oilfield to refinery, but some is also carried by rail in areas where pipeline construction is too difficult, or where the refineries are too far away from oilfields. For instance, about 70 percent of crude produced in 1995 was moved by pipeline. Because product pipelines are generally uneconomic (most refineries are scattered and relatively small), petroleum products are shipped on railroads and waterways, with some help from tank trucks. Only about 1700 kilometers of petroleum product pipeline are in operation in China[4], and while new pipelines are planned in congested

[4] Of which the Golmud-Lhasa pipeline accounts for 1080 kilometers, about 64 percent of total petroleum products pipeline.

Table 4.13 Crude oil transportation, 1986–1996

	Pipeline		Railway		Ship	
	Amount transported (mmt)	Proportion (%)	Amount transported (mmt)	Proportion (%)	Amount transported (mmt)	Proportion (%)
1986	118.34	61.9	15.74	8.3	56.99	29.8
1987	121.04	62.4	15.72	8.1	57.08	29.5
1988	123.77	63.9	14.24	7.4	55.67	28.7
1989	125.35	65.5	13.6	7.1	52.46	27.4
1990	123.39	64.6	13.05	6.8	54.5	28.6
1991	122.62	65.9	13.38	7.1	50.15	27.0
1992	121.91	66.6	13.20	7.2	48.00	26.2
1993	122.08	66.7	16.64	9.1	44.37	24.2
1994	119.58	65.8	15.28	8.4	46.94	25.8
1995	117.66	67.2	17.11	9.8	40.30	23.0
1996	118.12	69.0	16.93	9.9	36.22	21.1

Source: *China Petroleum Industry, 1996*, the China Petroleum Information Institute, CNPC, 1997, p. 16.

Table 4.14 Oil and gas pipelines under construction and planned in the near future

Project location	Length (km)	Pipeline diameter (mm)	Service type	Project status
Korla to Shanshan	480	720	Transportation, oil	Under construction
Shaanxi to Beijing	900	660	Transportation, gas	Under construction
Shanbei to Xi'an	280	426	Transportation, gas	Under construction
Tazhong 4 to Luannan	2 × 310	426	Transportation, oil and gas	Under construction
Shanshan to Urumqi	310	457	Transportation, gas	Under construction
Luannan to Korla	190	377	Transportation, oil	Under construction
Jingbian to Yinchuan	330	426	Transportation, gas	Planned
Dongying to Linyi	130	720	Transportation, oil	Prepared
Beihai to Nanning	230	426	Transmission, oil products	Planned

Source: *China Petroleum Industry, 1995*, the China Petroleum Information Institute, CNPC, 1996, p. 39.

corridors, completion timetables have not been established (see Table 4.14). A few product pipelines are under consideration, but a comprehensive network of crude and products pipelines is not expected to be completed over the next 10 years.

Total crude trunk pipeline length is about 9272 kilometers, mainly covering the northeastern and northern parts of China, running from oilfields to seaports and river ports on the Yangtze River. Refineries located along the trunk pipelines receive crude from branch lines, and those along the coast or major rivers receive crude from harbors, sometimes with the help of a short pipeline. For example, crude from the Daqing oilfield is transported through pipelines to the Dalian refinery (140,000 b/d), and is also discharged to the refineries at Shanghai and Maoming using 20,000-ton vessels. Only inland refineries have to depend upon railroads for crude transportation.

China lacks deep-water harbors, so the single buoy mooring systems (SBMs)[5] have to be built together with sub-sea pipelines. Most Chinese ports suffer from shallow draft, inadequate infrastructure and low storage capacity. Storage is expected to expand rapidly through joint-venture projects. Vitol helped build storage facilities for products (300,000 barrels) and crude (3 million barrels) at the Maoming Refinery in Guangdong province, enabling its port to receive very large cargo carriers (VLCCs). Maoming completed a crude dock in Bohe Bay that is able to anchor 250,000-ton oil tankers. About 2 million barrels of Oman crude can discharge from a VLCC at the SBM in Maoming. Elf plans to build a crude and products terminal in support of its 120,000 b/d refinery project in Shanghai on nearby Cezishan Island.

Although China lacks the facilities to accommodate large oil tankers, the port at Dalian is able to handle ships of 150,000 tons and is ideally situated, being only a day's voyage from South Korea and a few days from Hong Kong and Japan. The port at Zhenhai, at the door of the Zhenhai refinery, has fewer freight restrictions than Port Dalian. The Zhenghai refinery in Zhejiang Province, about 200 kilometers to the south of Shanghai, has a harbor good for 150,000-ton tankers. Fujian refinery can handle 100,000-ton tankers and will be suitable for 150,000-ton after it is expanded in a few years. Maoming is building its own SBM suitable for 250,000-ton VLCCs. The Shenzhen refinery will have a deepwater terminal in Dapeng Gulf, several kilometers away from Hong Kong that could handle oil cargoes of as much as 1.46 million b/d. A few refineries at the lower reaches of the Yangtze River such as Jingling may also handle smaller tankers.

4.5 Refining Policy

With the rapid pace of economic growth and industrialization, China's refining industry faces a constantly increasing demand for refined petroleum products and petrochemical feedstocks. In order of cope with the

[5] SBM will allow shippers to discharge crude directly from VLCCs.

anticipated surge in petroleum product demand, China has drawn up future refinery expansion plans, primarily relying on debottlenecking and revamping its existing refineries. However, the tight economic environment and strict oil market regulation may delay foreign investors' participation.

4.5.1 Tight economic environment: bad moment for foreign investment in refining

The next few years will see crucial developments in the Chinese refining industry. Significant attention should be paid to China's macroeconomic environment, upon which the health of its oil market depends. If we remain bullish regarding China's economic liberalization and decentralization, then we can be bullish on the future of China's oil market. The opposite is also true.

In both 1996 and 1997, the central government continued its tight monetary policy in order to cool down the overheated economy. However, preliminary statistics indicate that total investment in new projects was still higher than in 1995. The number of state-owned enterprises (SOEs) with losses did not decline. The Chinese official statistics indicates that the average asset–liability ratio (debt vs. assets) in SOEs as a whole has increased from 1994's 74 percent to more than 80 percent. This means that the majority of assets in state enterprises are being transformed into debts owed to state banks. Continuing losses in state enterprises could result in an increase of "dead debt" within the state banking system. As of the end 1997, the rate of unemployment was close to 10 percent (the official government figure only acknowledges 2.8 percent unemployment in urban areas, while some estimates present a range of between 12 percent and 20 percent), indicating that unemployment has replaced inflation as a primary government concern. Other than a limited number of foreign bank branches in a few cities, banks in China are completely under government control. Bank loans to SOEs for the expansion of operating capacity have been cut significantly, resulting in an increasing amount of debt. One positive development is that the inflation rate has dramatically fallen in both 1996 and 1997, compared to 1995. In conclusion, the macroeconomic environment in China has shown some degree of improvement, but not significant.

On the one hand, the inflation threat could come back again. Attracted by high interest rates, foreign money poured into China in 1994.[6] With

[6] Under the pressure of soaring inflation in 1994, the interest rates on Chinese savings in 1995 reached a record high: 24 percent for a three year deposit, 26 percent for 5 years, and 29 percent for 8 years. Short-term interest rates are also much higher than in the US.

all these dollars, the Central Bank's money supply grew rapidly. In 1994, the Central Bank had to print 300 billion yuan more than it planned, in order to buy dollars in the market. Consequently, inflation reached an all time high in 1994. For existing and projected joint-venture refiners in China, high domestic inflation coupled with an appreciation of the Chinese yuan not only increases their costs and risk for domestic supplies, but also damages the competitiveness of their product exports in the world market.

On the other hand, investors in the Chinese refining sector should be cautious about the currently over-valued yuan. In particular, depreciation of yuan in the next two years will devalue all the investments made now. Moreover, a surge in prices of labor and land use in the coastal areas will cut the rate of return on foreign investment.

How will China's macroeconomic situation affect foreign investment in the oil industry? There is a direct and an indirect impact. The direct impact is more obvious, resulting from China's oil policy and regulation of the oil market. The indirect impact, while not directly connected to the oil industry and market, is a factor resulting from the overall economic situation, and has an important role in determining the fate of an investment project. For instance, when the government is trying to sort out its priorities among the many multi-billion-dollar infrastructure projects that are under consideration, its previous promises regarding particular oil projects may be put on hold. When the central government attempts to rein in inflation, bank credit will be tightened, thus cutting the money supply to Chinese counterparts in joint ventures. As a result, project construction could slow down markedly in these circumstances.

4.5.2 Protected refining industry may lose its competitiveness

In its released ninth five-year plan (1996–2000) approved by the National People's Congress earlier in 1996, the Chinese government has made a priority of restructuring SOEs. If significant changes are made, there may be important repercussions for the oil industry and market in China. Regulated oil prices are protecting Chinese oil companies. While guaranteed refinery margins allow refiners to continue to operate inefficiently, the increasing cost of crude exploration and development (E&D) may squeeze the profitability of crude producers. In this regulated market, foreign investment will be welcome upstream, but will be limited downstream.

As many economists acknowledge, China's economic reform has been limited by an impediment: its many large, unprofitable SOEs, the legacy of decades of central planning. The dominance of these enterprises has caused three serious problems: (1) excessive investment in fixed assets, which are not supported by existing material and energy supplies, and could consequently exacerbate inflation; (2) operating losses (two-thirds of SOEs are running in the red and depend on state bank loans to meet basic operating expenses); and (3) a chain of debt, which exacerbates the credit crisis in China. While some oil companies in China do not face operating losses, they all have other problems in common with SOEs.

In the upstream, the Chinese government recontrolled prices in May 1994, and increased the crude prices of both regulated low price and regulated high price (also see Section 2.2 in Chapter 2, Oil Pricing). All of these regulated crude prices are significantly higher than the crude price level prior to May 1994.

As a result, 1994 is the first time in several years that CNPC, a state-owned upstream company, has shown a profit. To avoid squeezing refinery margins, higher crude prices have led to higher retail product prices. More importantly, a high retail product prices squeeze on wholesale and retail margins. Sinopec, a state-owned downstream company, remained profitable through 1994's changes. However, CNPC is greatly concerned about its future. As its costs for E&D grow, CNPC's profit margin will shrink year by year. Regulated crude price averages approximately US$ 14.50 per barrel, with Daqing crude, the highest priced Chinese crude on the world market, averaging $16.30 per barrel (f.o.b.) in September 1995. The closing spread between regulated and world market crude prices leaves the Chinese government little room to adjust domestic crude prices higher unless world market prices rise in the future.

In the downstream sector, Sinopec has enjoyed profitability for a long time, benefiting not only from government-guaranteed refinery margins, but also from a monopoly on the sale of basic petrochemical products. In addition, when product inventories have been high, the government has been able to temporarily suspend product imports in order to protect domestic refinery production. For example, China cut product imports in the second quarter of 1995 so as to maintain the profitability of domestic refinery operations — consequently refinery runs surged from June to August. However, high product prices have a dual impact. On the one hand, the high retail prices of petroleum products are a constraint on demand. On the other hand, high prices have also led to product smuggling, which competes for market share with domestic refineries.

China's macroeconomic policy ensures that the oil industry remains protected by the government. Chinese officials have been discussing the possibility of reorganizing the national oil industry into a system of regional, integrated, oil companies as a first step toward introducing a measure of competition and reshaping China's oil market. However, no action has been taken as yet. Product demand and production trends in China remain open to intervention from nonmarket forces.

4.5.3 Holding rein on foreign participation in the domestic market

In the middle of 1995, the Chinese government announced its new policy on foreign investment, which divided industrial sectors into three investment categories: those sectors where foreign investment would be encouraged, where it would not be encouraged, and where it would be prohibited outright. Foreign investment in the construction of new refineries is not encouraged by current Chinese government policy, which is a dramatic shift from the government's stance in 1993 and 1994. However, foreign participation in refinery expansion and revamping is encouraged. This reflects the fact that, on the one hand, China has placed a high priority on expanding its relatively small existing refineries to gain economies of scale. Joint ventures in expansion of existing refinery capacity could also preserve China's initial capital input. On the other hand, expanding its existing refineries may not necessarily build complete infrastructures, such as harbor, railroads, telecommunication facilities, water and power supply, etc. Projects for refinery expansion can be achieved relatively easily compared to the plans to construct grassroot refineries, largely due to cost differentials. This is also acceptable for foreign investors. In addition, revamping the existing plants could also avoid the creation of competition for Sinopec from joint-venture refineries. Several coastal refineries have been listed on the agenda for expansion in the next few years. The Chinese government has invited foreign oil companies to participate in these projects and has begun discussions on various proposals.

This latest government policy regarding refinery investment will add great uncertainty to the fate of previously proposed new refineries under the former joint-venture structure. In 1993 and early 1994, due to the booming oil market, foreign oil companies were rushing to plan joint-venture refineries in China. Following the central government's "re-regulation" in the middle of 1994, negotiations were halted and investors stepped back to wait for China's oil market to adjust.

It is important to keep in mind that China's economy is still dominated by a central command system, especially in the energy sector. The central government requires several levels of approval for all investment projects. A project with an initial investment of less than $6 million (RMB 50 million) can be approved by the local government. However, construction projects with an initial capital investment greater than $6 million are to be handed over to the SPC for approval, and projects with an investment greater than $25 million (RMB 200 million) have to be approved by the State Council. Obviously, the anticipated amount of refinery investment will be far beyond these limits, necessitating government approval at the highest levels. Therefore, the central government's policy will be the final determinant of the success of any proposed foreign investment, and may conflict with the desires of local governments, the Chinese counterparts in the joint ventures.

There is an additional procedure to be considered regarding proposed construction projects: the annual planning report prepared at the end of each year by the SPC, which lists all the proposed projects for the next year. These projects are divided into two categories: projects that have received construction approval and projects yet to be considered. This annual planning report is drafted in a nationwide planning meeting by December of each year, and is then submitted to the National People's Congress for "final approval" in March. Projects which are promoted for construction in the next year need to be on the decisive list in this annual planning report, especially in the category of approved construction. Foreign investment or joint-venture projects are not excluded from this requirement.

Thus far China is willing to share its petroleum resources with foreign investors as a means of capital exchange (such as offshore and Tarim oil exploration and development), but China is not ready to open its oil market to foreigners due to the limited market competitiveness of the Chinese refining industry. Opening its oil market to foreign investors would probably lead to severe losses in the refining industry, as the majority of Chinese refineries are small, overstaffed, and relatively inefficient. Except for some individual gas stations set in the SEZs, foreign oil companies are not allowed to market their petroleum at the retail level. In an effort to maintain unified prices, the central government designated Sinopec as the exclusive agent to run the country's wholesale business. Crude oil allocated to Sinopec and its affiliated refineries for import and refined products for export are handled jointly by three licensed state-owned international oil-trading companies. It is an immense disincentive to foreign investment, since the ultimate target in investing in refining is to create a retail market network.

Although China favors foreign investment in refinery expansion and infrastructure (port, pipeline and storage facilities), uncertainty remains. In addition, to evaluate political risk, market risk, economic risk, currency risk, etc., a foreign company should also consider the relevant regulations and policies that affect its potential investment, such as the hurdles of distribution regulations, import/export controls, and frequent central government interference. Moreover, Chinese regulations prescribe that foreign entities participating in existing refinery projects can take a stake of no more than 49 percent, reflecting a policy that the Chinese must maintain control of equity investments. The tightening of oil policy has lost its luster for refining investment in China. If China opens its retailing business for foreign investors in the near future, current oil distribution and transportation networks in China are still not sophisticated enough to allow foreign companies to sell refined products beyond the markets around coastal metropolitan areas.

When will this situation change? China's oil policy depends heavily on its macroeconomic policy. The slow evolution of economic liberalization will have a consequential effect of retarding the opening of the Chinese oil market to outside market forces. However, the full set of oil regulation and policies affecting refining industry and oil market has inevitably conflicted with China's desire to pursue admission to membership of the World Trade Organization (WTO). If China enters the WTO, which requires "national treatment" of foreign companies, refining policy may reluctantly be altered by the external forces.

4.5.4 Policy constraint of refinery expansion

According to the ninth five-year plan (1996–2000), China planned to increase its crude distillation capacity by 1 million b/d during the period from 1996 to 2000. However, there is no sign that any significant amount of capacity will be added in the near future, other than the long-awaited Dalian refinery (100,000 b/d) which was completed in October 1996. If the five-year plan is to be taken seriously, how can China reach its refining capacity target in such a short period of time?

As we have mentioned before, refining policy in China has favored debottlenecking and revamping existing facilities rather than building new grassroots refineries. Evidence of this is the fact that the SPC has delayed approval of several proposed joint-venture refineries. Indefinitely shelving approval for these projects can be interpreted as a rejection.

Meanwhile, Sinopec has planned to revamp and expand its seven existing refineries that are either situated in coastal areas or are inland and have access to river transportation. Sinopec expects financial help from foreign sources in order to achieve this goal. The refineries that were reported to be in the process of expansion or improvement are: Jiujiang in Jiangxi province (adding 30,000 b/d); Dalian (this is an existing refinery, not the new joint venture) in Liaoning province (100,000 b/d); Fujian in Fujian province (160,000 b/d); Guangzhou in Guangdong province (120,000 b/d); Cangzhou in Hebei province (50,000 b/d); and Lanzhou in Gansu province (90,000 b/d). Adding it all up, there is a total of 550,000 b/d of expansion capacity in the works. However, it is unlikely that all of this capacity will come onstream by 2000. For example, expanding Fujian refinery's capacity by 160,000 b/d would make it three times the size of its current distillation facility (50,000 b/d), making it difficult to complete by 2000. However, a number of other refinery expansion projects by CNPC and Sinopec are feasible, making additions of 550,000 b/d by debottlenecking and revamping before 2000 a more realistic goal.

The planned expansion of 1 million b/d from 1996 to 2000, minus 550,000 of debottlenecking, leaves about 450,000 b/d for new refineries. This could mean that China would approve at least four new refinery projects. Where should these new refineries be? Three joint-venture refineries have submitted feasibility studies to the SPC. They are: Shenzhen refinery, in partnership with Yukong (100,000 b/d); Juizhou refinery, in cooperation with Shell; and Huangdo refinery, a joint venture with Saudi Aramco and Ssangyong Refinery. The first two refineries are located in southern China, where there are high consumption levels and insufficient refining capacity. But the proposed location for Huangdo refinery is near other Chinese refineries, and in a region that has a surplus in refined products.

In addition, there are five refineries totaling 750,000 b/d proposed for Hainan province. Hainan province (an island south of Guangdong), is treated as an SEZ. The central government has given it significant autonomy and the benefit of preferential policies. These include granting the provincial government authorization to approve refinery projects if investment funding is provided by foreign sources. Thus, refineries wholly owned by foreign companies could channel their products into the local market in Hainan province. However, the proposed addition to capacity is far more than this region can absorb. It is unlikely that all five of these proposed refineries will come onstream together. However, one or two refineries (perhaps the 100,000 b/d Lingao refinery and one of the others) are likely to be completed by 2000.

To add 1 million b/d of refining capacity, China *must* build some new joint-venture refineries. Over the past five years, while petroleum product exports fell, imports almost doubled. Further delay in approving several planned joint venture refineries will cause China to import more products both before and after 2000.

These projections indicate that if the setting-up of joint-venture refineries is delayed, revamping and expansion of existing refineries may prevent the situation from worsening, but cannot solve the problem of capacity shortage.

What are the prospects for changes in China's refining sector? Two important factors stand out. First, Sinopec and the provincial governments are directly aware of the gaps between supply and demand in certain regions, and have more motivation then the central government to expand existing refineries and build new ones. Unfortunately, however, under a centrally planned economy, they have no ability to influence the macroeconomic and investment environment. A tough economic environment only weakens their bargaining power in joint-venture negotiation. In the post-Deng era, if provincial governments grasp more decision-making power from the central government, the provinces could enjoy more freedom to decide on needed refinery projects and begin construction quickly.

Second, self-sufficiency objectives in refining are very important to the government, and this concern will force China to adopt a more flexible policy toward foreign investment when the central government has inadequate capital. If China's refinery expansion is delayed, the Chinese economy will require more product imports to meet demand increase. But if this were to occur, there still would be stringent limitations on product imports. The government will place great priority on minimizing product imports, especially if the world oil price is high and China suffers a large trade deficit. As product imports creep up, however, policy makers will realize that self-sufficiency is a goal that China will need foreign assistance to reach. This could change the government's attitude and lessen the restrictions and red tape that foreign refiners now face.

While domestic supplies have been inadequate to meet demand, the consequence of a delay in building new deep-processing plants should lead to a prolonged period of China's importing more refined products. An ongoing question for China's decision-makers will be the precise trade-off between importing crude for domestic processing capacity and importing refined products for direct marketing. It is estimated that one ton of crude creates at least 10 times more value in China if it is used to produce petroleum or petrochemical products. It can also be expected that revenues from growing exports of value-added petrochemicals will exceed outlays for rising crude and feed imports over time, reducing the drain on government hard currency coffers. From an economic point of view, therefore, it makes more sense for China to build adequate refining capacities to import crude and produce its own refined products.

4.6 Conclusion

Almost half of the country's refining capacity is situated in the northeast and north, but there is a big deficit in the east and south. However, most new refining capacity will be in China's booming eastern and southern coastal regions, in order to locate product supplies close to demand and simplify logistics for domestic and export sales.

There will be sufficient crude distillation capacity available, while the upgrading unit capacity is relatively insufficient. Secondary units will continue to be insufficient to meet the increasing demand for middle distillates.

The success of refineries funded completely or partially by non-Chinese companies will depend in part on Chinese government policy. There will be demand for products from third-party processing facilities, but hard currency is necessary for the investors to repatriate profits and for

Sinopec to bid on the products from such facilities. These refineries may need yuan to pay workers, taxes, and the like, but crude purchases will be in hard currency so most sales must also be in hard currency. The limited convertability of Chinese currency constitutes a major central control over the whole economy, one which is controllable by limiting who can participate in the exchange and the volumes to be traded. This limitation will limit non-Chinese access to China's markets, and is not likely to be eliminated in the near future.

We do envision that some targeted investment in refining could produce improvements in reducing anticipated refined product deficits. If our calculations of demand and supply for the next few years are correct, then foreign firms who intend to build new refineries might want to consider which configuration they will need in order to beat Chinese regulated product prices.

China is willing to share its petroleum resources with foreign investors as a means of capital exchange, but China is not ready to open its oil market to foreign investors. In addition, the regulated oil prices and distribution system have limited foreign investors' access to China's market. The slow evolution of economic liberalization will have a consequential effect of retarding the opening of the Chinese oil market to outside market forces. As China has given priority to debottlenecking and revamping its existing refineries, we will see limited additions in new refinery capacity during the next few years.

Chapter 5
Petroleum Product Demand

For a long time, "oil demand" in a centrally planned economy such as China's has meant the amount of petroleum products allocated by the central government for distribution and consumption. By and large, it has not reflected the reality of market need for petroleum products. The central government in China has played an active and direct role at all stages of demand management through the annual plan.

In the case of petroleum products, the annual plan controlled by the central government provides for most of the refined products used by state enterprises and fixes allocations for commercial departments, which make retail sales to residential and commercial consumers. The planned allocation system resulted in the establishment of quotas for regions and enterprises. The central government assigned quotas to the various ministries or provinces, which then allocated them to various units or subdivisions under their control. The consumer at the lowest level is the final assignee in such supply administration. As a result, demand has been primarily supply driven and has operated under the constraints imposed by the country's self-sufficiency policy.

Since economic reform began in the early 1980s, however, market forces have gradually begun to play a role in petroleum product distribution, sales, and pricing. The central allocation system — which regulates the quantities of fuel which companies and individuals can produce, sell, and consume — has changed in three aspects. First, refiners have been allowed to sell above-quota volumes directly to end-users or marketers. Second, refiners have more freedom to sell their products domestically, if they purchased crude oil abroad, and to sell their product surplus, thus allowing them to operate as third-party processors. All these sales can be handled directly by refineries to consumers at the refiners own set prices without interference by state-owned distributors. Third, the central government has permitted provincial governments and consumers with their own foreign exchange earnings to import products through licensed, state-owned international trading agents (i.e. Sinochem and newcomers Unipec and ChinaOil).

As a result of these market reforms, more and more of China's product distribution and sales have moved out from under the control of the central government, and the ability of the state to collect national petroleum product consumption data has deteriorated significantly. This is evidenced by official publications of petroleum statistics that report only the consumption of petroleum products in the industrial, construction, and transportation sectors, but exclude the agricultural, commercial, and residential sectors. The petroleum product consumption figures that appear in this book are estimated by using refinery output minus exports plus imports. Therefore, all consumption figures refer to *notional demand or apparent consumption*. In addition, due to the lack of stock figures, we have ignored changes in inventories in recent years. On an annual basis, however, changes in inventories are very small in relative terms and make little difference in the analysis of the overall Chinese market.[1]

This chapter addresses a number of issues relating to demand for the various petroleum products, and it raises some of the key questions which lie at the heart of the debate over China's oil future: Will economic liberalization continue to unleash strong demand for petroleum products, as in recent years? At what rate and for which products? Will China mirror Asia's thirst for distillate, or will motorization of society boost gasoline consumption? What is fuel oil's role as an industrial fuel?

This chapter presents a variety of demand forecasts — in the aggregate, by principal product, by sector, by region — in order to look at demand forecast from a number of perspectives. Inevitably, this leads to a certain amount of repetition. The purpose of the analysis is to identify possible ranges for future petroleum product demand, analyze the key factors determining demand trends, and discuss some of the issues raised and the options which the demand trends create. We start with a look at current oil consumption in the larger context of total energy consumption in China. Then, we focus on the principal petroleum products: gasoline, diesel oil, and fuel oil, and also touch on kerosene, jet fuel, naphtha, and LPG. And finally, we aggregate these products to give a picture of total petroleum product demand.

The long-run oil demand forecasts in this book were generated in two ways: from the "bottom up", looking at the demand situation for each petroleum product; and from the "top down", looking at total petroleum product demand as a part of energy demand trends. For readers interested in the details of our modeling, Appendix 2 (Methodology of

---

[1] We have included stock changes of gasoline and diesel oil in 1993 and 1994 when they are confirmed by Chinese sources. These assessments takes into account stockbuilds caused by a surge of product imports at the end of 1993 and a stockdraw of gasoline and diesel oil during the first half of 1994.

Demand Forecasts) presents a more extensive description of the demand model and analytical process.

5.1 Oil in Total Energy Consumption

In China, ideology, economics, and geology have combined to provide an interesting history of oil consumption (see Fig. 5.1). After falling and then stagnating in the early 1960s, owing to the after-effects of the Great Leap Forward and the chaos of the early stages of the Cultural Revolution, energy consumption rebounded as heavy industry proliferated in the late 1960s. During this period as well, the country was exploiting major oil finds at Daqing and Shengli. The net result was that, for the first time since the revolution, China became self-sufficient in its energy supply. The 1960s and 1970s marked the heyday of the policy of self-reliance, a period when China cut itself off from the rapid technical progress occurring elsewhere in the world. Policies before the late 1970s focused primarily on expanding production, not on how efficiently energy was used in the country. By 1979, however, growth in energy production had slowed, and in 1980 and 1981 production actually declined. From then on, strategies to improve the efficiency with which China consumed energy became as critical as those to expand domestic energy production. Until the mid-1980s, China held to the fundamental tenets of energy self-reliance, but by that time it had become clear that maintaining this policy came at the expense of suppressed domestic demand, especially for oil.

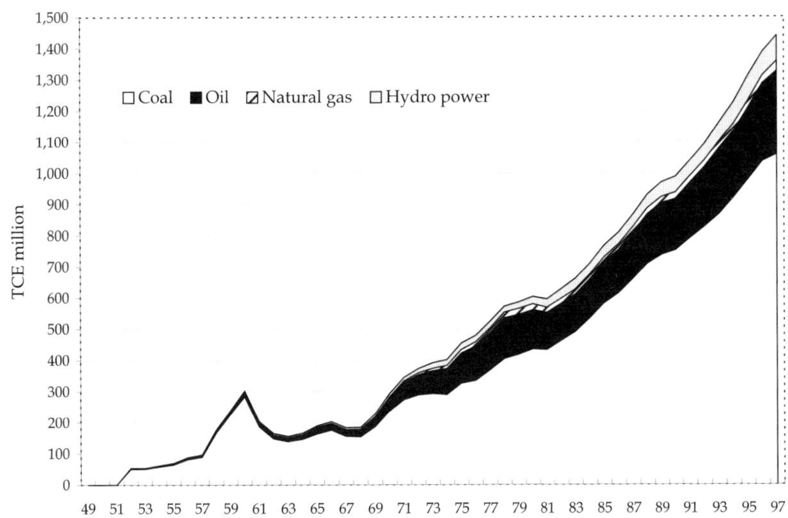

Fig. 5.1 Energy consumption by type of primary energy
Sources: *China Statistical Yearbook, 1997*, and *A Statistical Survey of China, 1998*, China Statistical Publishing House.

Table 5.1 Total energy consumption

	Energy consumption (million tons of TCE*)	Percentage of total energy consumption			
		Coal	Oil	Natural gas	Hydro power
1960	301.88	93.9	4.1	0.5	1.5
1961	203.90	91.3	5.5	0.9	2.3
1962	165.40	89.2	6.6	0.9	3.2
1963	155.67	88.9	7.2	0.8	3.1
1964	166.37	88.0	8.0	0.7	3.3
1965	189.01	86.5	10.3	0.9	2.7
1966	202.69	86.2	10.2	0.7	2.9
1967	183.28	84.8	10.9	0.8	3.5
1968	184.05	83.8	12.1	0.8	3.4
1969	227.30	81.9	13.8	0.8	3.5
1970	292.91	80.9	14.7	0.9	3.5
1971	344.96	79.2	16.0	1.4	3.4
1972	372.73	77.5	17.2	1.7	3.6
1973	391.09	74.8	18.6	2.0	4.6
1974	401.44	72.1	20.7	2.5	4.7
1975	454.25	71.9	21.1	2.5	4.6
1976	478.31	69.9	23.0	2.8	4.3
1977	523.54	70.3	22.6	3.1	4.1
1978	571.44	70.7	22.7	3.2	3.4
1979	585.88	71.3	21.8	3.3	3.6
1980	602.75	72.2	20.7	3.1	4.0
1981	594.47	72.7	20.0	2.8	4.5
1982	620.67	73.7	18.9	2.5	4.9
1983	660.40	74.2	18.1	2.4	5.3
1984	709.04	75.3	17.4	2.4	4.9
1985	766.82	75.8	17.1	2.2	4.9
1986	808.50	75.8	17.2	2.3	4.7
1987	866.32	76.2	17.0	2.1	4.7
1988	929.97	76.2	17.0	2.1	4.7
1989	969.34	76.0	17.1	2.0	4.9
1990	987.03	76.2	16.6	2.1	5.1
1991	1307.83	76.1	17.1	2.0	4.8
1992	1091.72	75.7	17.5	1.9	4.9
1993	1159.93	74.7	18.2	1.9	5.2
1994	1227.37	75.0	17.4	1.9	5.7
1995	1311.76	74.6	17.5	1.8	6.1
1996	1389.48	74.6	18.0	1.8	5.5
1997	1440.00	73.5	18.6	2.2	5.7

*TCE = tons coal equivalent. For explanation, see footnote 1 in Chapter 1.
Sources: *China Statistical Yearbooks, 1985–1997,* and *A Statistical Survey of China, 1998,* China Statistical Publishing House.

A unique feature of energy consumption in China is the high share of coal in the energy mix, as both Fig. 5.1 and Table 5.1 illustrate. Coal has remained the primary energy source in China: accounting for over 95 percent of primary energy consumption in 1960, it declined to a low of some 70 percent in 1973 before rebounding to its current level of 76 percent. Much of the rise since the late 1970s was due to a program of substituting coal for oil in power plants and other under-boiler uses. The fundamental reason for the heavy reliance on coal is the existence of giant, low-cost (and low-quality) reserves in the country. For instance, in northern China, there are coal reserves of 2–3 billion tons, selling for less than the international market prices, keeping coal more competitive with other energy. We believe that the other principal forms of energy — oil, natural gas, and nuclear — are highly unlikely to take market share away from coal in the foreseeable future.

As a percentage of total demand, oil consumption has increased markedly over the past three decades, but the rise has been very unsteady, as Table 5.1 reveals. With government's efforts to replace oil demand with other fuels, in 1997, China's oil dependence was less than 20 percent in primary commercial energy consumption, compared to about 50 percent in the rest of the Asia-Pacific region and 40 percent in the global total. China uses oil much less intensively than the rest of the world.

In China, petroleum product demand is met primarily through domestic refinery production, but some crude is also burned directly. The use of crude oil as a direct fuel has fallen sharply over the past 10 years, but our estimates show that in 1997 more than 100,000 b/d of crude oil were still being burned in the power and industrial sectors. The reduction of direct burn has resulted largely from improving product transport infrastructure and expanding refining in certain locations. However, the decline in direct burn should slow in the future, since the majority of China's power plants have converted from oil to coal use. In the economy as a whole, there have been few other areas in which the government has been able to squeeze oil consumption. Thus, the rapid economic expansion and modernization that have occurred over the past few years have led to a sustained rise in product demand, especially for gasoline, diesel, and petrochemical feedstocks.

An important point concerning Chinese oil demand merits mention here: even with a closed energy market, China's oil demand has turned out to be somewhat sensitive to high or low world oil prices. Such sensitivity arose mainly from the ability to generate hard currency earnings in the 1970s and early 1980s through crude exports. The reduced quantities of crude available to the domestic market and the anti-import bias resulting from the country's self-sufficiency policy thus combined to constrain growth in oil demand. Following the price collapse of 1986, Chinese planners developed a new strategy to build refineries in order to export value-added petroleum products instead of crude.

Leaving aside oil for the moment, we expect that coal will resist challenges from other energy sources and remain the backbone of China's energy economy. China has depended on coal for more than 70 percent of its energy needs, compared to 29 percent in the rest of Asia-Pacific and 28 percent in global total. Coal is and will continue to be a substitution fuel for oil in China's industrial and electric power sectors.

Even though China has enormous hydroelectric potential, the growth of hydro output in China will be moderate in the years to come, because most of the major, viable undeveloped resources are located far from the major cities, and the construction cost of hydroelectric plants has increased enormously. China should achieve its target of about 10 gigawatts of nuclear power by the turn of the century. While the nuclear capacity to be added is large by comparison with most other countries, nuclear power will still supply only a small percentage (about 3 percent) of China's energy by the early 2000s, compared with about 70 percent in France, 20 percent in the US, and 30 percent in Japan.

Available supplies of natural gas are expected to be limited, except in Sichuan or Guangdong, where natural gas could play a significant role in the energy future. Discoveries of significant new quantities of natural gas could have a large impact on petroleum used as feedstock in the chemical industry and on fuel oil applications in industrial processes, because natural gas has a relatively high value per unit of energy compared with distillate and fuel oil. However, the future for natural gas consumption is constrained predominantly by the locational mismatch between supply and demand.

5.2 Gasoline

China's gasoline market is just balanced, and gasoline demand has been growing at a slower pace than diesel oil. The general quality of China's gasoline still lags behind the world standard, and gasoline exports comprise the highest quality gasoline, leaving the domestic market with leaded, but nevertheless low-octane gasoline.

China's gasoline demand grew at an average rate of 40,000 b/d each year during the last decade. At 770,000 b/d, gasoline demand in 1997 was more than half a million barrels higher than in 1980. With the number of vehicles doubling every seven years, gasoline consumption also grew accordingly. Compared with the 5.6 percent annual increase in total petroleum product consumption during the past decade, there has been an 7.4 percent annual average hike in gasoline consumption. As a result, China's percentage of total Asia-Pacific gasoline consumption has risen from 20 percent in 1990 to 25 percent in 1997. Nevertheless, when viewed in an international context, it is apparent that the Chinese market has plenty of room to boom: Chinese per capita consumption of gasoline was only 8.5 gallons per year per person in 1995, even less than one-tenth of the global average. And while sales growth is even higher in other countries in the region, Chinese demand for gasoline is growing rapidly in absolute terms.

5.2.1 Gasoline demand overlook

As Table 5.2 shows, the average growth rate of gasoline consumption has been heavily influenced by the substantial economic achievements in several major provinces on the southern and eastern coasts.[2] Gasoline demand in these provinces grew at the following annual rates: 12.1 percent in Shandong (the highest growth rate of any area the country), 9.4 percent in Jiangsu, 10.3 percent in Shanghai, 9.2 percent in Henan, and 9.4 percent in Guangdong. In 1995, the five biggest consuming provinces — out of a total of 30 regions — were Liaoning (40,000 b/d), Heilongjiang (45,000 b/d), Guangdong (42,000 b/d), Shandong (50,000 b/d), and Sichuan (47,000 b/d). Total consumption in the "Big Five" thus accounts for one-third of national gasoline demand. Sichuan province, with the largest population at over 100 million, has seen its gasoline demand skyrocket. The 47,000 b/d consumed in 1995 was near double the 1985 consumption of 25,000 b/d.

The consumption (in liters) of gasoline per vehicle per day slightly declined in China during the last eight years, slipping from 14.8 in 1985 to 9.7 in 1995. This is not surprising since it follows the pattern set by other countries, in which gasoline consumption per vehicle per day falls as the economy develops. However, gasoline consumption per vehicle per day in China is still significantly higher than that in other Asian developing countries, which was 4.2 in Bangladesh, 4.5 in Hong Kong, 5.5 in India, 11.3 in Indonesia, 5.3 in South Korea, 5.4 in Malaysia, 6.0 in Pakistan, 12.5 in Philippines, 5.5 in Sri Lanka, 6.6 in Taiwan, and 8.8 in Thailand in 1987. Within China, gasoline consumption per vehicle per day varies immensely, ranging from a low of 3.1 in Beijing to a high of 23.2 in Gansu. After sluggish sales during 1989 and 1990, the coastal regions resumed their upward climb in gasoline demand. The central government has, however, constrained gasoline demand in other regions, vividly demonstrating its ability to influence consumption. For example, the quota of gasoline in Beijing for each institute-owned car is as low as

[2] Estimates of consumption by province are based on the partial consumption data by province which are received from the China's Statistical Bureau.

Table 5.2 Estimated gasoline consumption by region (000 b/d)

	Gasoline consumption (000 b/d)			Gasoline demand (liters/vehicle per day)			Gasoline demand (gallon/person per year)		
	1985	1990	1995	1985	1990	1995	1985	1990	1995
National total	**325.9**	**438.8**	**669.0**	**14.8**	**11.2**	**26.5**	**5.0**	**5.9**	**8.5**
North									
Beijing	8.9	12.3	12.3	8.4	6.4	8.6	14.2	17.3	14.9
Tianjin	7.2	9.1	7.8	13.3	10.9	10.5	13.6	15.8	12.6
Hebei	13.1	19.3	28.0	9.5	6.8	17.5	3.6	4.9	6.6
Shanxi	10.9	14.0	17.2	10.7	7.6	19.5	6.4	7.6	8.5
Inner Mongolia	9.8	13.7	14.7	16.4	14.4	30.5	7.5	10.0	9.8
Northeast									
Liaoning	18.8	25.8	39.3	12.5	9.3	26.9	7.8	9.9	14.6
Jilin	9.9	13.3	21.3	14.3	10.8	34.7	6.6	8.1	12.5
Heilongjiang	21.0	28.8	45.3	19.9	14.9	49.2	9.7	12.7	18.6
East									
Shanghai	10.1	13.8	27.0	15.8	13.8	34.3	12.7	15.7	29.2
Jiangsu	15.3	23.1	37.7	15.2	11.9	34.9	3.8	5.2	8.1
Zhejiang	8.8	13.9	20.5	15.4	11.8	23.9	3.4	5.2	7.2
Anhui	9.3	11.7	14.5	13.0	9.5	22.5	2.8	3.2	3.7
Fujian	7.1	7.7	7.0	15.1	8.4	15.2	4.0	3.9	3.3
Jiangxi	8.6	10.0	14.8	14.9	12.1	37.2	3.8	4.0	5.5
Shandong	16.1	25.6	50.3	10.8	9.6	26.4	3.2	4.7	8.8
Sourth-central									
Henan	13.6	19.0	32.7	9.6	7.5	27.7	2.7	3.4	5.5
Hubei	15.4	21.4	29.5	15.5	14.8	33.8	4.8	6.0	7.8
Hunan	11.5	14.6	18.5	13.5	9.8	21.5	3.1	3.7	4.4
Guangdong	17.1	18.8	42.1	10.4	6.0	15.8	4.2	4.6	9.3
Guangxi	6.3	9.2	11.2	11.0	9.4	20.5	2.5	3.3	3.7
Hainan	n.a.	2.3	8.5	n.a.	7.6	37.6	n.a.	5.6	17.8
Southwest									
Sichuan	24.6	34.1	46.9	16.7	14.4	32.7	3.7	4.9	6.3
Guizhou	7.0	7.7	15.9	15.3	10.5	41.8	3.6	3.6	6.9
Yunnan	14.7	19.6	23.9	20.8	16.5	26.8	6.6	8.2	9.1
Tibet	3.2	n.a.	n.a.	24.5	n.a.	n.a.	24.6	n.a.	n.a.
West									
Shaanxi	9.2	11.6	16.6	14.4	10.7	27.2	4.7	5.5	7.2
Gansu	8.6	11.1	24.1	17.6	14.3	63.7	6.4	7.4	15.0
Qinghai	3.8	4.2	6.0	17.5	13.2	39.5	14.4	14.7	18.7
Ningxia	2.2	2.9	4.5	14.9	11.8	33.8	7.9	9.4	13.1
Xinjiang	14.0	20.1	30.9	21.4	20.9	52.0	15.8	20.1	28.1

Sources: *China Statistical Yearbooks, 1988–1997* and author's calculation.

2.5 to 3.1 liters per car per day (i.e. 20 to 25 gallons per car per month). The actual gasoline allocation resulting from the inadequate supply has caused a large number of motor vehicles and tractors to lie idle.

The gasoline consumption (gallons) per person per year also varies significantly within the country, from 3.3 in Fujian to 29.2 in Shanghai. The metropolitan regions, such as Beijing, Shanghai, and the northeast, generally had the highest gasoline consumption per capita — over 15 gallons per person per year — because of relatively developed transportation system in these areas. In industrial and remote regions, people also used more gasoline than other regions — between 10 and 30 gallons per person per year — because of heavy industrial activities and low population density.

5.2.2 Motor vehicles

While the global automobile industry is challenged by operational losses, employment cutbacks, and market weakness, China's automobile industry is booming and the market for new cars is strengthening. After two consecutive years of growth, the output of automobiles in 1997 reached 1.6 million vehicles, up more than three times from 1990's level.[3] It was estimates that production of automobiles could top 2 million vehicles by the turn of the century.

China is one of the biggest market for the world's carmakers, and almost all major automobile companies have either created joint ventures with China or opened an office in China. An agreement on car production opened the way for Audi, Chrysler, Toyota, and others to establish joint-venture production facilities in China a number of years ago. Other foreign carmakers, such as Ford, Hyundai, and Daewoo entered the market somewhat later and as a result must initially produce automotive parts with Chinese partners before being permitted to assemble vehicles. Foreign carmakers are finding that the Chinese market yields extremely high profits, averaging 16 percent in China, compared to about a 4–6 percent profit margin in Japan. A joint venture between Shanghai and Germany Santana earned a 22.4 percent return on investment.[4]

Chinese versions of foreign cars are on the rise, but an enormous number of foreign-made cars also have poured into China. China has seen two significant increases in imports during recent years. During the first, in 1985 and 1986, the import of passenger cars, primarily Japanese,

---

[3] See *A Statistical survey of China, 1998*, China Statistical Publishing House, May 1998, p. 107.

[4] Kang Ning, "Could China Block the Wave of Automobile Imports?" *Developing Zone Herald*, January 1994, p. 15.

peaked at 60,000 vehicles per year. After those two years, car imports cooled down for a while, but in 1992 and 1993, the fever of nationwide economic development brought another tide of automobile imports. In 1992, for example, imported passenger cars soared by 90 percent from 1991 levels. Despite superficial similarities, there were two critical differences between these import booms. First, in 1985–1986, imports were limited in the southern regions; but in 1992 and later, the imports flowed into all parts of the country, both north and south. Most importantly, perhaps, imports also penetrated into the internal provinces. Second, in 1985–1986, the imported passenger cars were of low or medium quality; but in 1992 and later, luxury cars — Cadillac, Lincoln, Mercedes Benz, Lexus LS400, and so on — made up a noticeable portion of total imports. How many cars were imported during recent years? No one knows exactly, because many additional cars were smuggled into the country. For example, from January to April, 1993, South Korean publications claimed that 26,000 passenger cars were exported to China. But Chinese Customs listed only 166 cars. In all, it is estimated that there are at least 80,000 cars of South Korean origin — Hundai, Kia, and Daewoo — on the road in China. However, Chinese customs official statistics claim that the total number is only about 1000 cars.[5]

China levies a 200 percent tariff for passenger car imports. This compares with only a 10–15 percent tariff in the developed countries, and below 100 percent in most developing countries. China has maintained this high tariff for passenger cars for a long time. Its current tariff even represents an increase from the mid-1980s' figure, but this still has done little to prevent increasing car imports. Why has the high tariff been ineffective? The reasons are two-fold: first, the prices of passenger cars in China are approximately three times the world market prices. Even with a 200 percent tariff, it is still possible for foreign car dealers to earn attractive profits. Second, the majority of consumers for these imported cars are government agencies. They do not care about the high prices, due to the fact that none of the money is coming out of their pockets.

What is the status of privately owned cars in China? In China, the number of privately owned cars was officially 3.6 million at the end of 1997[6], of which some are used as taxis. Considering China's population of 1.2 billion, this number is next to nothing. There can be little doubt that people aspire to own motorized vehicles, and that as their per capita income rises, they desire to own automobiles in particular. According to

---

[5] See Kang Ning, *op. cit.*, 1994, p. 18.

[6] See *A Statistical Survey of China, 1998*, China Statistical Publishing House, May 1998, p. 119.

research conducted by the Chinese Academy of Social Sciences, the consumer goods most desired in a household with annual income over 15,000 yuan (about US$ 1760 per year equivalent, at the exchange rate of 8.5 yuan per US dollar) are a house or decent apartment and a passenger car. Such households in the cities account for about 11 percent of total households. According to a survey conducted by the State Statistical Bureau, there are 5.3 million households with an income over 50,000 yuan per year (about $5880 per year.) These households already have the financial capability to purchase passenger cars.

To own a car is still only a dream for most Chinese, however, because to buy an ordinary, domestic-made passenger car would cost them 25 to 50 years' worth of income based on current prices and wages. If China can expand the scale of its carmaking and lower the cost to the level of 5 to 10 years' worth of income, this Chinese dream could become a reality. In addition, if China enters into the World Trade Organization (WTO) in the near future, its car prices are expected to decline further to satisfy the WTO requirement. At that moment, China would become a second South Korea, where mostly domestically produced cars are running in the street, but the tariffs for imported cars are moderate.

Chinese infrastructure would also need to undergo major changes to accommodate this growth, however. The number of privately owned cars is still negligible, and does not place a tremendous burden on the infrastructure. But transportation conditions would not be sufficient to support an increase in private cars in China. China's roads are barely adequate for a country a fraction of its size. According to China's Statistical Yearbook, the total length of roads was 1.2 million kilometers in 1996, of which about 20 percent were unpaved roads, and less than 30 percent were rated above "second class", according to international standards. Road density is about 195 kilometers per million persons in China (using the above "second class" road number), while it is 229 in Pakistan, 236 in South Korea, 242 in the Philippines, and 513 in Thailand.[7] Average speed on the main roads is between 19 and 25 mph. Transportation facilities have already encountered real challenges as the economy booms in the coastal areas, and only recently has China started to build highways linking north and south — thanks to financial aid from foreign funds and World Bank loans. China's road situation is likely to get worse before it gets better.

[7] The road densities for Asian developing countries are 1988's numbers, and taken from *World Development Report 1994*, the World Bank.

5.2.3 Gasoline demand forecasts

We have used two approaches to forecast gasoline demand in order to capture the potential range of demand fluctuation in the next 10 years: country comparison analysis and econometrics.

5.2.3.1 Comparative analysis

It is commonly believed that developing countries have some similarities in their stages of economic development.[8] Comparative analysis draws a picture of future development in a specific country by making a comparison between this and others with comparable economic indicators. For the purpose of this book, we present comparisons of three corresponding data sets — driving-age person per passenger car, GNP per capita, and gasoline consumption per passenger car — which taken together, offer insight into the future of the Chinese gasoline market.

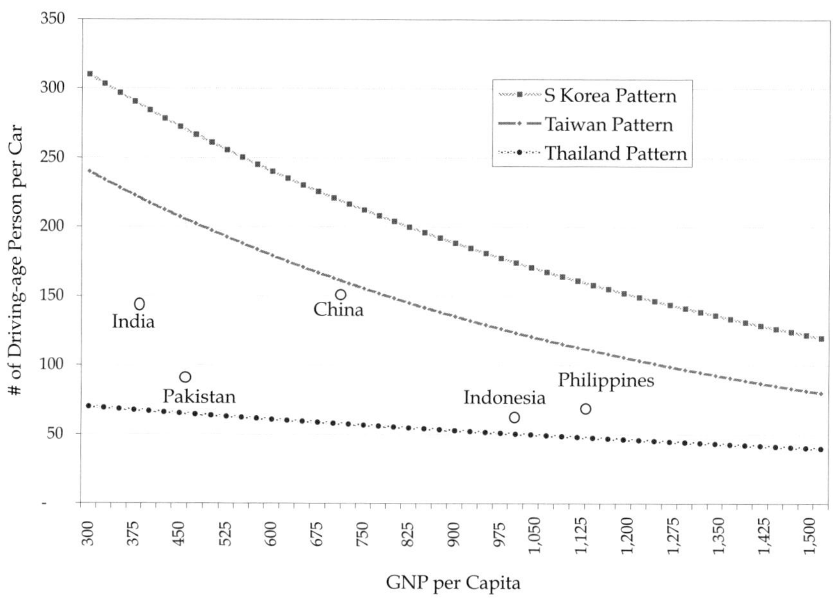

Fig. 5.2 Driving-age person per passenger car in selected Asian countries

[8] The data used for this analysis are taken from (1) *World Motor Vehicle Data, 1996*, Motor Vehicle Manufacturers Association of the US, Inc.; (2) *The Sex and Age Distribution of the World Populations*, the 1996 Revision, UN; (3) *World Table, 1997*, the World Bank; and (4) *World Development Report, 1998*, the World Bank. These data may be different from other data provided by Chinese sources.

First, the likelihood of gasoline demand "exploding" in Asian developing countries can be assessed in terms of the number of passenger cars, indicating that demand for gasoline "explodes" in a country primarily because the number of cars increases rapidly. Put another way, gasoline demand in Asian developing countries may increase marginally if the number of gallons per vehicle increases — such increases are at the heart of demand increases in mature motorized societies. But demand in the Asian developing countries can explode only if the number of cars increases significantly.

Such an increase in the number of cars in service tends to occur only in countries where: (1) GNP per capita increases beyond a threshold level, and (2) cars are manufactured locally. The relationship between personal income and car ownership varies from country to country, but the presence of domestic car companies seems to be an important variable. Fig. 5.2 shows the relationship between people per car and GNP per capita in selected low income Asian countries. The historic pattern in South Korea and Taiwan saw increased car ownership as income rose during the 1980s. The Taiwanese pattern has been more "automobile intense" (i.e. fewer people per car at a given income per capita level), and whether countries proceed along a "Taiwan" or "South Korea" pattern is a matter of some interest. In Fig. 5.2, we see that China, India, and Indonesia all seem to be on the "Taiwan" pattern, but they were on different points in the evolution of their automobile societies. In China, there is a tremendous number of people below the age of 20. As they grow older, if the same number of people-per-car ratio were to be maintained, the number of cars will have to increase significantly. Thus, even remaining on the "Taiwan" pattern (rather than the even more auto-intensive "Thailand pattern") the number of vehicles in these countries will increase significantly.

Second, as Fig. 5.3 shows, when we compare the percent increase in the crucial 20–65 year age group in a number of Asian developing countries, China is one of those countries which have uniformly younger people than others. The process of industrialization and the accompanying increase in per capita income creates a growing number of consumers who see automobile ownership as a vital symbol of their success. China certainly has the demographic profile for significant increases in motor vehicles ownership and use.

Fig. 5.4 exhibits the historical relationship between GNP per capita and driving-age person per passenger car in China, Taiwan, and South Korea, respectively. As of 1996, China had 180 driving-age persons (ranging from age 20 to 65) per passenger car, much higher than the Asian average of 35. However, the total registrations of passenger car has increased three times between 1990 and 1997. China may follow the gasoline use per car will continue to decline to about 3 gallons per car for the year 2005.

Finally, by varying assumptions about economic growth, population increase, and fuel efficiency improvement, we find that the range for Chinese gasoline demand in 2005 will be from 800,000 b/d to 1.4 million b/d, again falling between the Taiwan and South Korea patterns (see Fig. 5.6). A sensitivity analysis with a low economic growth rate of 6 percent per year suggests that gasoline consumption could still be as high as 1.0 million b/d in 2005.

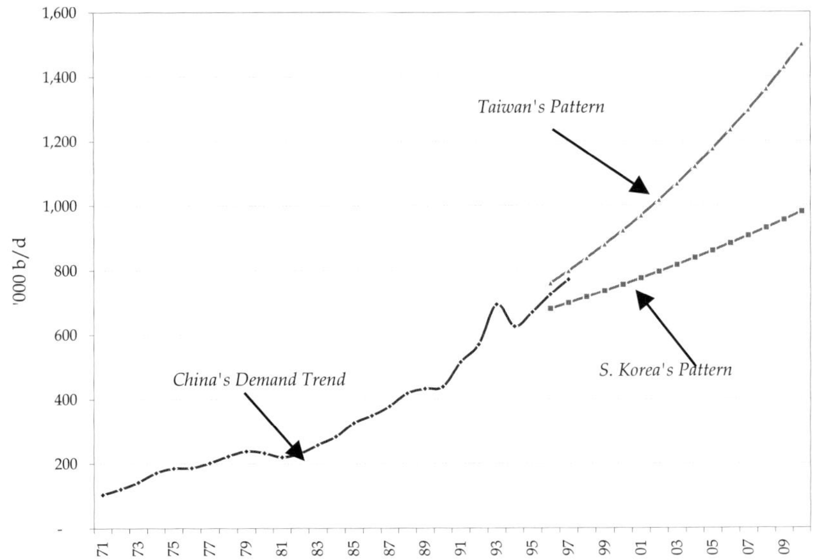

Fig. 5.6 Projected gasoline consumption in China

5.2.3.2 *Econometric analysis*

In conventional econometric models, gasoline demand can be related to the level of economic activity, population, the relative price of gasoline, and the number of vehicles.[9] We use national income per capita as the variable of economic activity, total registrations of passenger cars, and implied petroleum product prices as a proxy of gasoline price due to lack of price information in China (for detail, see Appendix 2). In addition, despite the fact that gasoline prices have been controlled by a central

---

[9] The data used for this analysis are taken from various *China Statistical Yearbooks, 1987–1997*, and the State Statistical Bureau.

administration, price elasticity of gasoline demand may exist, because prices are increasingly determined by supply and demand. As shown in the first half of 1994, the central government may periodically move to assert its control over petroleum demand in China, but the continued development of real markets is gradually ushering in real internal market prices. We believe that gasoline prices will play a more and more important role in affecting demand in the Chinese domestic market in the years to come.

The assumptions adopted are: (1) economic growth rates during the late 1990s and the 20th century at least as high as those of the last two decades — about 12 percent per year in GDP in coastal regions, and 6–7 percent in internal regions; (2) relative high growth in car registration in the late 1990s and early 2000s, followed by a slowdown; and (3) a moderate crude price path in the next 10 years. Dramatic increases in the number of passenger cars are not expected in China during the next 10 years, but their use as a major transportation vehicle is expected to increase rapidly, particularly in rural areas. Unlike the situation in India, Thailand, and Taiwan, motorcycles, which could become a factor in gasoline consumption, are expected to remain regulated by high taxes due to undeveloped roads, traffic congestion in big cities, and a high population of young people.

Given the above assumption, Chinese demand for gasoline will grow substantially over the next 10 years. In our "base case" forecast (Table 5.3), we project a rather paltry 3 percent annual growth rate for gasoline demand over the next 10 years, which results in a growth from the present rate of 770,000 b/d in 1997 to 950,000 b/d in 2005, and 1.1 million b/d in 2010. Growth will average 3.0 percent before 2000 and 2.7 percent between 2000 and 2010. This econometric forecast falls at the low end of the range in the comparative country analysis. We realize this is a moderate forecast, but even if economic growth exceeds 9 percent annually, the central government is likely to try to contain gasoline demand by taxing car and gasoline sales, as part of a broader effort to prevent further vehicle congestion in metropolitan areas. In addition, the projected increase in gasoline demand can just barely be met by the new refining projects on the drawing board. If "excessive demand" for products requires imports, diesel oil will be the first priority with limited foreign exchange because diesel oil could meet industrial demand, which would then generate more economic output. We will put these demand forecasts on a more realistic scenario, when we look at the gasoline supply/demand balance in Chapter 7, Supply/Demand Balance: Implications for Trade.

Could our demand estimates be too high? It is estimated that 50 million potential new drivers will emerge in China in the next 10 years, more production expands during the next 10 years. Fuel oil used as bunker fuel will remain the smallest part of total demand, due to fuel efficiency gains in the marine bunkers fleet. Even though fuel oil demand for bunkers will still account for only 8 percent of total demand by 2010, it will have the highest annual growth rate — 2.2 percent per year from 1995 to 2005. The share of total fuel oil consumption accounted for by refinery use *as fuel* is expected to remain 12 percent from 1995 to 2010, as refinery activities expand. Out of the total projected increase in volumes, both industrial and electric power demand will account for the most of the increase, following by refinery and bunker demand. Fuel oil demand in other sectors is expected to fall.

Table 5.8 Fuel oil consumption forecast by sector

	Consumption (000 b/d)					Average annual growth (%)			
	1990	1995	2000	2005	2010	1990–95	1995–00	2000–05	2005–10
National total	**605.9**	**615.2**	**640.1**	**659.6**	**674.2**	**0.3**	**0.8**	**0.6**	**0.4**
Industry	227.2	253.3	271.6	281.2	288.9	2.2	1.4	0.7	0.5
Electric power	178.3	199.4	218.4	226.4	232.8	2.3	1.8	0.7	0.6
Refinery	79.2	73.9	77.9	81.1	83.7	(1.4)	1.1	0.8	0.6
Bunker	38.0	43.4	50.5	54.1	57.6	2.7	3.1	1.4	1.2
Others	83.2	45.2	21.7	16.7	11.1	(11.5)	(13.6)	(5.1)	(7.8)

Note: For forecast model specification, see Appendix 2.

Fuel oil sulfur content will also change, suggesting a declining share of low-sulfur fuel oil (less than 0.5 percent) in total demand. Over the next decade, we expect demand for high-sulfur fuel oil (greater than 1 percent) to increase in absolute terms and as a percentage of total product consumed. At 630,000 b/d of fuel oil demand in 1997, low-sulfur fuel oil demand accounts for roughly 75 percent of total consumption. We expect low-sulfur fuel oil demand to fall to 450,000 b/d by 2005 and to account for about 69 percent of total demand. By 2010, low-sulfur fuel oil demand should reduce to 440,000 b/d and account for around 65 percent of total consumption.

5.5 Other Refined Products

5.5.1 Kerosene

Almost a hundred years ago, John D. Rockefeller had a dream that, some day, "every Chinaman will own a kerosene lamp". While kerosene consumption has been so vital to expansion of the economies of other Asian developing countries, the kerosene lamp in China is not really a symbol of modernization. In fact, kerosene has become an insignificant fuel in China (it averages only 1.8 kilograms per person per year), because electricity has more advantages than kerosene for improving the living conditions of hundreds of millions of peasants. Nor is China likely to emulate Japan and South Korea in their heavy use of kerosene-based home heating. Instead, the vast regions of China will continue heating and cooking with traditional fuels — coal, charcoal and renewable energy. As rural electrification goes on, kerosene demand for lighting will shrink; to an extent, kerosene will be displaced by electricity in modernizing regions of the country.

Table 5.9 illustrates that kerosene is used primarily in the transportation and commercial / residential sectors, with these sectors consuming about 50 percent and 40 percent of total consumption, respectively. The major source of kerosene demand growth has been the expansion of the aviation network and the military jet fleets. Kerosene consumption in transportation has experienced a rapid increase in the last five years because commercial jet fuel demand more than doubled between 1990 and 1995. However, the kerosene demand in the commercial / residential sector, which consumes another large amount of kerosene, fell dramatically, along with a significant market share drop. The military also consumes large amounts of kerosene — about 10,000–20,000 b / d — which accounts for the unexplained gap between the calculated apparent demand and officially published consumption figure. Finally, the agriculture sector consumes only 800 b / d of kerosene, accounting for only 0.7 percent of total demand.

Therefore, kerosene demand in China is unlikely to take off. We estimate an increase of about 20,000 b / d to 30,000 b / d by 2010 based on the

Table 5.9 Kerosene consumption by sector

	1985		1990		1992		1995	
	000 b / d	%	000 b / d	%	000 b / d	%	000 b / d	%
Total consumption	83.2	100.0	75.8	100.0	90.2	100.0	110.6	100.0
Industry	4.3	5.2	4.4	5.9	5.4	5.9	9.7	8.8
Agriculture	0.7	0.9	0.7	0.9	0.8	0.9	0.8	0.7
Construction	0.3	0.3	0.3	0.4	0.3	0.3	0.8	0.7
Transportation	12.1	14.6	20.2	26.6	35.6	39.5	54.0	48.8
Commercial / residential	65.8	79.0	50.2	66.3	48.1	53.4	45.4	41.0

Sources: *China Energy Statistical Yearbook, 1991, China Statistical Yearbook, 1997,* and State Statistical Bureau.

In the late 1970s, Chinese leaders overestimated potential Chinese oil resources, which lead to the widespread direct burn of unrefined oil as fuel in many sectors, and to the construction of a number of oil-fired power plants. Table 5.11 presents the estimated volume of crude burn, based on total crude production minus refinery throughput, minus net

Table 5.11 Estimated crude direct burn

	Crude oil Production (000)	Refinery Runs (000 b/d)	Crude Exports (000 b/d)	Crude Imports (000 b/d)	Losses	Crude Direct Burn* (000 b/d)	Direct Burn as % of of output
1971	788	650	5	1	20	114	14.5
1972	913	734	13	–	23	144	15.8
1973	1072	840	37	–	27	169	15.8
1974	1297	990	101	–	32	173	13.4
1975	1541	1050	198	–	38	255	16.6
1976	1743	1180	170	-	43	350	20.1
1977	1873	1260	182	10	46	394	21.0
1978	2081	1344	226	14	52	473	22.7
1979	2123	1400	269	5	53	407	19.2
1980	2119	1574	266	7	53	249	11.8
1981	2024	1495	275	1	50	206	10.2
1982	2042	1507	304	13	49	200	9.8
1983	2121	1608	304	7	53	167	7.9
1984	2292	1650	446	5	48	162	7.1
1985	2498	1718	601	5	46	133	5.3
1986	2614	1852	570	9	46	164	6.3
1987	2683	1955	545	10	45	153	5.7
1988	2733	2048	521	17	53	125	4.6
1989	2754	2132	488	66	50	158	5.7
1990	2766	2192	480	59	51	93	3.4
1991	2820	2328	452	120	51	109	3.9
1992	2828	2480	430	227	52	93	3.3
1993	2872	2651	389	314	50	95	3.3
1994	2913	2639	371	247	47	103	3.5
1995	2968	2881	377	342	32	75	2.5
1996	3137	3000	431	452	31	127	4.0
1997	3198	3324	397	710	31	112	3.5
Annual growth rates (%)							
1971–80	11.6	10.3	54.7	24.8	11.6	9.0	–2.3
1981–90	2.7	3.4	6.1	23.0	–0.4	–9.4	–11.8
1991–97	2.1	6.1	–2.7	42.9	–6.7	2.7	0.6

*The number shown here may include crude stock changes.

Sources: State Statistical Bureau, International Energy Agency, Chinese Customs, and Author's calculation.

crude exports, and minus losses. In 1978, the fraction of crude burned is estimated to have been as high as one-quarter of total crude production, reaching 470,000 b/d.

In the late 1980s, oil shortages forced the oil-burning power plants to be retrofitted to burn coal. Almost all new power plants built since then have been based on coal or gas. As a result, use of crude oil as a direct fuel fell sharply, to about 200,000 b/d in the late 1980s. In 1997 only slightly more than 110,000 b/d of crude oil were estimated to be burned as a direct fuel in the power and industrial sectors. If we consider that a surge in crude imports at the end of 1997 may have caused a crude stockbuild in 1997, the actual 1997 direct burn may have been the same as that in 1992 or 1993, below 100,000 b/d. This is the lowest level for crude burn since China became a giant in crude production. The reduction of direct burn of crude oil resulted largely from the improving transport infrastructure and the expansion of refining in certain locations.

As China becomes a net crude importer, crude shortages will further curtail direct burn to utilize every possible barrel for refining. Crude direct burn levels will not be dramatically altered, but may slightly reduce the levels as they have been in the past few years. We estimate about 50,000 b/d by 2010.

5.7 Total Petroleum Product Demand

In 1997, China ranked fourth among all countries in total petroleum product consumption, following the US, Japan, and Russia. China's product consumption accounted for about 15 percent of the Asia-Pacific total, and 4 percent of the global total. However, comparing total and per capita oil product consumption of the 30 largest petroleum product consumers in the world in 1997, China's rank descends from fourth to about 30th in terms of per capita usage. Petroleum product consumption per capita in China averages 0.4 liter per day, compared with 1.1 liters in Asia overall, 8.2 liters in Japan, 7.7 in South Korea, 5.1 in Taiwan, 2.2 in Thailand, and 3.4 in Malaysia.

5.7.1 Total product demand overview

Table 5.12 and Fig. 5.7 present in the starkest form what has happened to China's petroleum product market over the last 18 years. During the 1970s, petroleum product demand kept pace with the economic growth. In early 1980s, product demand growth slowed down due to low oil production growth. During this period government policies focused on fuel oil, and there was a big drop in fuel oil demand, but there was almost no deterioration in the consumption of other petroleum products.

After that, total petroleum product demand grew 6.7 percent annually during 1985–1988 and was up another 5.4 percent in 1989. After registering negative growth of 1 percent in 1990, product demand surged at a fairly fast tempo, averaging a 6 percent growth annually between 1991 and 1993. However, 1994 reversed the trend of the past three years, showing sluggish growth in both oil demand and refinery runs. After then, the rise in product consumption outpaced the growth of crude throughput, which averaged 7 percent a year to progress from 2.9 million b/d in 1994 to 3.7 million b/d in 1997.

Table 5.12 Consumption of petroleum products

	Total	Gasoline	Diesel	Kerosene	Fuel Oil	Jet Fuel	Naphtha	LPG	Lube	Asphalt	Others
1980	1750	233	340	70	565	7	48	39	38	25	386
1981	1670	220	336	64	520	8	54	35	28	27	378
1982	1660	231	324	65	515	9	63	39	26	27	361
1983	1700	258	337	70	527	9	72	48	22	27	331
1984	1760	284	358	69	526	11	75	49	24	32	332
1985	1870	326	397	69	520	12	115	51	25	33	323
1986	1990	349	432	72	550	12	89	64	26	34	362
1987	2120	378	478	67	569	15	129	68	30	35	349
1988	2250	418	528	65	594	15	163	74	34	35	324
1989	2350	432	567	64	609	16	184	80	34	34	329
1990	2320	439	550	56	606	20	196	82	39	35	297
1991	2500	515	612	54	603	29	224	96	42	47	279
1992	2614	570	655	54	651	31	230	93	45	52	234
1993	2908	693	828	57	681	39	268	154	46	51	93
1994	2943	625	762	53	623	44	295	170	33	57	281
1995	3077	669	893	103	615	46	298	195	41	61	156
1996	3269	725	945	118	575	48	301	224	50	65	218
1997	3663	771	1,086	124	628	50	303	258	70	71	303
Annual growth rates (%)											
1980–90	2.9	6.5	4.9	−2.3	0.7	11.5	15.2	7.8	0.4	3.4	−2.6
1990–97	6.7	8.4	10.2	12.2	0.5	13.7	6.4	17.8	8.5	10.7	0.3

Note: Others include refinery gas, ethane, paraffin waxes, petroleum coke, and other petroleum products.
Sources: *Energy Statistics and Balances, Non-OECD (1971–1995)*, International Energy Agency, OECD, and State Statistical Bureau.

To elaborate a bit on our statistical analysis, gasoline's share in total petroleum consumption during the last decade jumped from 17 percent in 1985 to 21 percent in 1997, and diesel oil rose from 21 percent in 1985 to 30 percent in 1997. During the period 1985–1997, gasoline demand increased by 7.4 percent per year, and diesel oil demand by 8.7 percent per year. The "slow" growth occurred in kerosene, which increased at 5.0 percent per year, and fuel oil, which managed just 1.6 percent annual growth. Before 1990, fuel oil consumption had the largest volume in all petroleum product demand. In 1991, the actual volume of diesel oil used exceeded fuel oil demand for the first time, and gasoline accomplished the same feat in 1993, evidence of their driving role in China's total petroleum product consumption.

1994 demand growth was slow compared to the record high growth of 1993 primarily because the Chinese government put the brakes on what had, in 1993, been run away demand. The barriers set by government policy on product trade between provinces and regions have severely retarded demand growth, since regions which are short of oil products have a reduced ability to find sources of supply both domestically and internationally. Consequently, China had one of those years that bewilder Western economists — a year of 10 year economic growth but no oil demand growth. In 1994, total product demand averaged 2.9 million b/d, only marginally higher than 1993. According to Chinese oil sources, there was a huge stockbuild (500,000 b/d of diesel oil and 400,000 b/d of gasoline) at the end of 1993, then stocks were drawn down

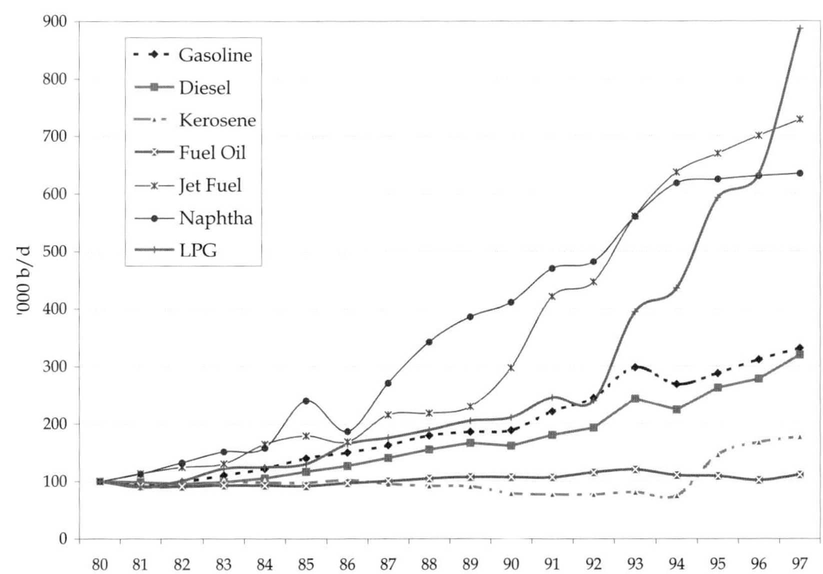

Fig. 5.7 Petroleum product demand growth (1980=100)

during the year of 1994. The drawdown squeezed the country's requirement for product imports, especially given that demand was weakened by high regulated prices and tight central control of oil distribution.

Both 1996 and 1997 were high-growth years similar to 1993, a year when the Chinese oil market experienced a sizeable demand surge. The robust growth thus far in 1996 and 1997 is different from that in a true "go-year" such as 1993. In 1993, the oil market enjoyed a short-lived period of freedom: less government restriction of oil prices, retail business, domestic exchanges, and international trade. This year's strong demand growth in the face of a tight macroeconomic environment indicates a steady level of demand persisting despite higher levels of government control. Current demand levels may be more sustainable in the long-term if government regulation loosens somewhat.

Will this strong demand force the Chinese government to again implement a tighter policy? We do not believe so. There were significant changes in China's macroeconomic environment between 1993 and 1997. In 1993, inflation skyrocketed and the economy was overheated by excessive investment and a large domestic money supply. But in 1997 and 1998, inflation was under control and the economy appeared much more stable. A cooling-down economy will not necessitate governmental tightening of oil policy. China's economy and oil market each continue to follow a boom–bust cycle, but their respective cycles are not necessarily concurrent. As long as the central government is not struggling with an overheated economy while fighting inflation (or sliding into a bust cycle), the bullish oil market will last.

Each of the major petroleum products, in theory, has close ties to its own determinate variables. Gasoline demand is closely correlated with income and geographic factors, while fuel oil demand is more sensitive to displacement by other energy sources. As a result, total petroleum product demand is a function of the assorted variables behind the different products. This makes it extremely difficult to forecast total petroleum product on the basis of past relationships, especially given the two most important variables governing product use in China: government allocation and economic growth.

China's petroleum product demand is and will remain very sensitive to government policy. Inflation's negative effect on social stability remains a primary concern of the central government. A tight economic policy to curb soaring inflation would, at the same time, control expansion in the oil market. Although the Chinese economy is still suffering year-on-year losses from its state-owned enterprises, the impetus of economic growth has steadily shifted to its nonstate-owned industries and sectors.

Energy policy in China is likely to place great priority on minimizing oil demand. The most costly mistakes in analyzing the demand outlook for China are likely to arise from assuming that government has little or no power to shape petroleum demand. There are limits to what government can do; each product has its own economics, and indeed, its own politics. Nevertheless, petroleum product pricing, distribution, and taxation policies are — after the price of crude oil itself — the most important variables affecting China's oil future.

The effects of such policies will manifest themselves differently in each petroleum product. Middle distillates are the "fuel for the engines of economic growth", and thus are the least-taxed fuels. Gasoline, by contrast, is seen as a luxury item, to be rationed and/or taxed heavily. Fuel oil is a purely industrial product whose demand was depressed by coal substitution; in the future, if China imports more fuel oil, as industry expands, fuel oil may make a comeback, recapturing parts of its market share in China.

Nevertheless, there are two policy objectives that may motivate a turn away from oil. One is fiscal restraint — limiting expensive imports. China's moratorium on oil imports during the first quarter of 1994 may be a sign of things to come. The other goal is continued government regulation of oil distribution — reducing end-use consumption through the administrated allocation system in uneasy conjunction with the price system. The Chinese government has demonstrated a willingness to adopt command-and-control policies that directly cut oil consumption in the name of fiscal austerity. These phenomena are likely to be repeated time and again until a change in the economic system. However, as China moves toward a market economy, the impact of such allocation regulation will gradually become marginal. Such powers to allocate, however, will continue to have a significant impact well into the next century, keeping a lid on long-term (10–20 year) oil demand growth.

Having said this, the forecast of total petroleum product demand presented here is primarily based on sector models which make explicit projections (see Appendix 2). We assume that: (1) the annual rate of real economic growth will continue to fluctuate between 6 and 12 percent, averaging about 9.0 percent; (2) China's energy intensity relative to economic output will be decreased without compromising the goals of improving standards of living; and (3) the industrial share in total GDP/national income will continue to increase over the next 10 years.

Economic growth is unlikely to escape from this trend because: (1) monetary and fiscal policies will be deployed against inflation — which puts a ceiling on how fast the economy is allowed to grow; and (2) the movement towards a market economy and privatization provides a significantly better framework for sustainable economic growth than was previously present. Therefore, for the next 10 years, during recessions, the economic growth may be below 7 percent, but only for a year or two.

Table 5.13 Demand of forecasts for principal products and total product

	Consumption (000 b/d)					Average annual growth (%)			
	1990	1996	2000	2005	2010	1990–96	1996–00	2000–05	2005–10
Total product	2320	3269	3824	4608	5500	5.9	4.0	3.8	3.6
Gasoline	439	725	816	923	1070	8.7	3.0	2.5	3.0
Diesel oil	550	945	1194	1523	1898	9.5	6.0	5.0	4.5
Kerosene	56	118	122	135	142	13.3	1.0	2.0	1.0
Fuel oil	606	575	584	591	597	–0.9	0.4	0.2	0.2
Jet fuel	20	48	68	100	140	15.3	9.0	8.0	7.0
Naphtha	196	301	395	528	674	7.4	7.0	6.0	5.0
LPG	82	224	305	428	541	18.2	8.0	7.0	4.8
Lubricant	39	50	61	74	86	4.1	5.0	4.0	3.0
Asphalt	35	65	95	140	187	11.1	10.0	8.0	6.0
Other	297	218	185	166	164	–5.0	–4.0	–2.2	–0.2

Note: For forecasting model specification, see Appendix 2.

Similarly growth may average well above 12 percent for a few years, but this is unlikely to persist.

Table 5.13 and Fig. 5.8 offer the current and incremental demand for principal petroleum products. Putting it all together, we project that by 2010, product-by-product demand will be as follows:

- Gasoline will rise vigorously, climbing an average of 2.8 percent annually to reach a total 1.1 million b/d. This represents an increase of 330,000 b/d by 2010. A resurgence in personal income and passenger car production will be responsible for this strength in demand.
- Diesel oil will rise by 5.0 percent per year to 2.1 million b/d by 2010. Price liberalization for diesel oil may further boost diesel demand prospects, but we project an increase of about 75,000 b/d per year. Of all the fuels, diesel oil demand has the greatest potential, but that potential depends on whether demand could "takes off" as part of deliberate efforts of further industrialization.
- Fuel oil will rise by 0.6 percent per year to 670,00 b/d, an increase of about 50,000 b/d over 10 years, albeit at the slowest rate among the other products.
- Kerosene will have an average annual increment of 1.4 percent — the low growth rate among the principal petroleum products — or move up by 25,000 b/d over the next decade.
- Jet fuel will increase by 7.8 percent per annum, gaining a total of about 85,000 b/d, the highest growth rate among the refined products.

- Naphtha's annual rise will be 5.8 percent, yielding a total increase of about 3300,000 b/d.
- The total petroleum product demand will thus increase by 3.8 percent per year from 1997 to 2010 to reach 5.9 million b/d, or, put another way, will increase by about 170,000 b/d per year over the next decade.

Some 80 percent of the total incremental demand volume will consist of the light products. The largest gains in terms of petroleum product consumption are in gasoline, diesel oil, naphtha, and LPG. The shares of gasoline and diesel oil in total petroleum product demand will rise from 21 and 30 percent in 1997 to 19 and 35 percent in 2010, respectively. Conversely, fuel oil will remain marginal as coal and hydropower remain the key sources of energy for many areas of possible fuel oil use.

5.7.2 Demand forecasts: are they plausible?

The major uncertainty in our demand forecasts is how long the current Asian economic and financial turmoil will last? Up to the publication of this book, there is no sign of economic turnaround in Asia. Moreover, Asian financial turmoil may spread into a global economic crisis. The economic and financial crisis in Asia has undermined China's growth formula in 1998. When China has had to absorb some of the impact of the regional downturn, the further deep crisis and slow recovery may force us to scale down demand projections.

We want to use three measures to test how our forecasts of individual product demand stack up against forecasts of total product demand. Such "top down" forecasts typically examine GDP:oil consumption elasticities, oil intensity, and the relation between GNP per capita and oil consumption.

5.7.2.1 GDP: oil ratio

The GDP:oil ratio is the most critical variable affecting the long-term oil demand outlook. Many developing countries are thirsty for oil, to fuel rapidly expanding or industrializing economies (indeed, the GDP:oil ratio in LDCs has averaged close to 1:1 over the last 20 years). Some of these countries, however, are concerned with the balance of payments consequences of rising imports, and others have had to adopt measures to limit consumption, resulting in a decreased GDP:oil ratio. This is especially true in potentially high-growth, oil-importing countries.

Given the preceding economic growth outlook and projected total petroleum product demand, we calculate that China's GDP:oil demand elasticity for the next decade will average 0.5, which is low compared with other Asian developing countries, but which resembles the average ratio in China from 1983 to 1997 (see Fig. 5.8). That ratio means that for very 1.0 percent increase in economic growth there is a corresponding 0.5 percent increase in oil demand growth. Some forecasters assessing the future of the Chinese oil market are promoting the view that the GDP:oil ratio in China may be as high as 1:1, as in other LDCs.

Supporting our projection, there are three important factors operating in China to create a relatively low elasticity of petroleum product demand. First, the projected change in economic structure will be a continuing expansion in the industrial, commercial, and service sectors, with a decline in output share of the agricultural sector. Commercial and service sectors are considerably less energy intensive than the industrial sector, thus, the faster the expansion in commercial and service sectors in relation to the industrial sector, the lower the energy intensity. Second, the industry sector generates the highest demand for petroleum products, but a natural process of capital turnover, improving management, technological progress, and energy efficiency in the production of various commodities will greatly reduce energy intensity, even in the absence of a larger shift away from heavy industry. In this regard, we expect China to make hefty gains through the increased importation of foreign technology and machinery. Steady progress in the formulation and implementation of an energy demand management program, including price

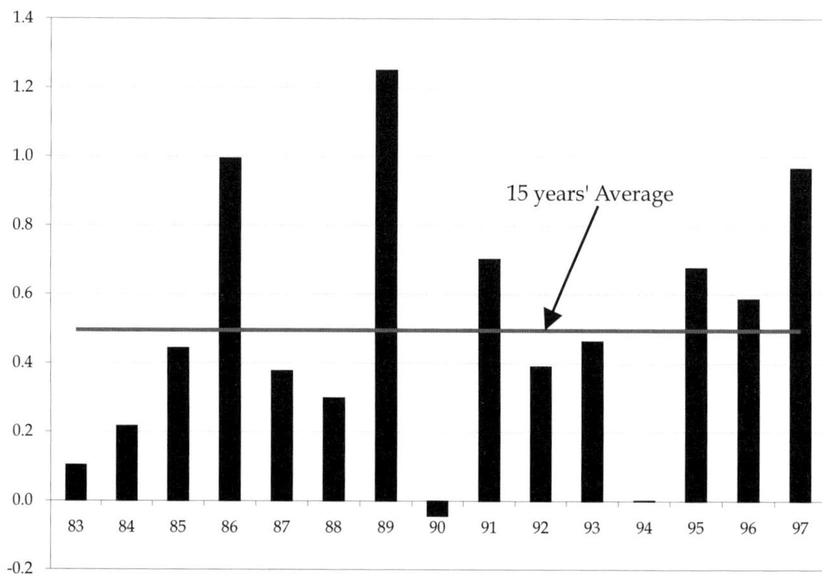

Fig. 5.8 GDP:Oil consumption elasticities

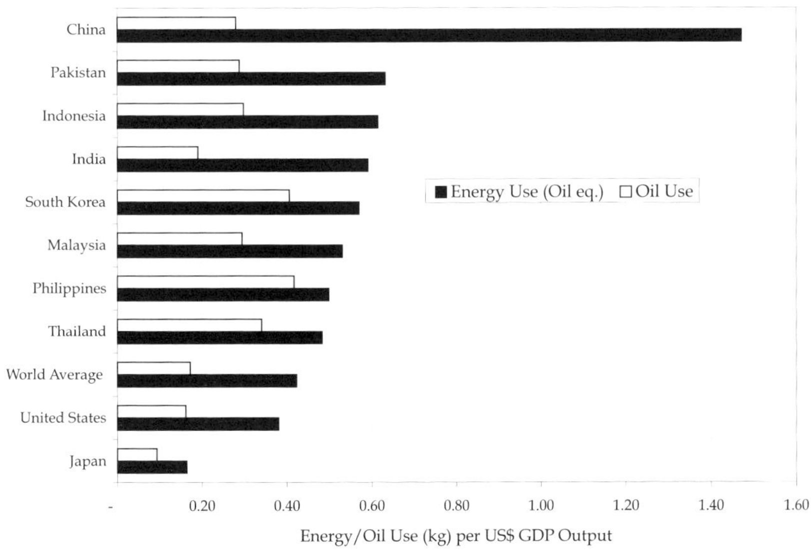

Fig. 5.9 Energy and oil intensities in selected countries
Sources: *The World Development Report, 1998,* World Bank, and author's calculation.

reform, will be critical in realizing future energy efficiency improvements commensurate with the potential for oil savings. The final factors lowering the GDP:oil ratio are balance of trade and currency concerns. As the crude deficit boosts imports and the government tries to limit its import bill, oil demand may be constrained, resulting in a lower growth rate.

5.7.2.2 Energy and oil intensities

Energy and oil intensities measures energy and oil use per unit GDP value. Cross-country comparisons of *energy* intensity reveal an exceptionally high energy use in the Chinese economy.[15] As Fig. 5.9 exhibits, energy consumption per GDP output in China is well above that of any other selected developing countries in Asia. Energy intensities in China are more than four times higher than the US, and about nine times higher than Japan. One factor contributing to China's high level of energy use relative to GDP is that most industries still use technology from the 1950s and 1960s. Only about a small percent of Chinese industries use technology from the 1970s. Few Chinese industries apply modern technology.

[15] Due to different methods of calculating output value, relative prices, and problems resulting from use of official exchange rates to convert monetary statistics to comparable units, cross-country comparisons provide only a rough picture of energy intensity.

the energy intensity per unit of value created in the economy was several times higher in China than in developed countries, the oil shortage in China could be partially solved by promoting energy conservation. On the other hand, since per capita oil consumption in China is so low, even if energy conservation is successfully implemented, a rapid increase of oil demand is almost inevitable.

The relationship between economic growth and oil demand in China will bear no similarity to that experienced in any other countries. There is less scope to decrease oil intensity in China as living standards are raised and country is industrialized. Our demand forecasts envisions that oil consumption per capita per day in China will rise to 0.5 liter by 2005 from 0.4 liter in 1994, as GNP per capita increases from $750 in 1996 to above $1200 in 2005. As Fig. 5.10 exhibits, if we draw lines for high and low oil consumption per capita given GNP per capita among the selected developing countries, China will remain near the lower boundary in this picture even in 2005. However, an increase of 0.1 liter in oil consumption per capita per day represents an astonishing increase in a country with such a huge population. This might be a small step for most LDCs, but it is a great leap forward for China.

5.8 Conclusion

For the past 40 years, China's oil demand has been driven by domestic supply. Although this pattern has changed slightly at times when a surge in petroleum product imports caused the government to temporarily slowdown scheduled refinery output, domestic demand continues to rely primarily on domestic refinery production capability.

While market forces are playing an increasing role in petroleum product distribution, sales and pricing, the central government continues to play a significant role in demand management. Oil plays a secondary but growing role in total energy consumption, which continues to be dominated by coal.

In fact, the fate of the Chinese oil market is strongly influenced by macroeconomic policy. Inflation's negative effect on social stability remains a primary concern of the central government. A tight economic policy to curb soaring inflation would, at the same time, control expansion in the oil market. So far, the good news for the oil industry is that the inflation rate fell under 10 percent in 1996, even below 5 percent in 1997, after hovering around 20 percent for more than two years. If pressures within the macroeconomic environment could be eased, the government's tight control over the oil industry might be relaxed somewhat. Oil statistics have tended to quickly reflect any shift of government regulation, so a loosening could lead to sustained growth in consumption.

We project China's GDP growth to average just about 9.0 percent over the next 10 years, although we believe growth to be cyclical, rising well over 12 percent in some years and under 7 percent in other years. We estimate the GDP:oil demand growth ratios to be in the range from 1:0.4 to 1:0.6. For example, if GDP is growing at 1 percent, then oil demand is growing at 0.4 or 0.6 percent.

We expect oil demand to grow by roughly 2.2 million b/d between 1997 and 2010, or about 170,000 b/d per year. This rate of growth depends on an annual growth rate of 5.0 percent in diesel oil and 3.8 percent in gasoline during this period. If we differ dramatically from other forecasts of petroleum product demand, it is most likely due to different assessments of: (1) economic growth; (2) government policy on oil allocation, and regulation on sales of automobile and gasoline; and (3) import tariffs and quotas.

Chapter 6
Oil Quality and its Environmental Impact

Crude oil quality has a large and lasting impact on the economics of the oil industry. It affects the quantities of each petroleum product that the refining industry can create, given its current configuration, and thus affects the profitability of various kinds of refineries. Crude oil quality also affects the relative costs of producing the various products, and thereby influences product prices and the demand for each product over time. This chapter examines changes in crude production and consumption quality based on our long-term crude projection, and addresses a number of issues, such as how demand will vary across regions, how China's environmental policies will evolve and to what degree oil quality specifications will be introduced or expanded.

6.1 Crude Oil Quality

6.1.1 Crude production quality

For the purposes of this chapter, we have divided crude oil into three grades of sulfur content: low-sulfur is defined as less than 0.5 percent, medium-sulfur as 0.5 to 1 percent; and high-sulfur as more than 1.0 percent. We also divide crude into two categories of gravity: light crude is more than 30 degrees API, and heavy crude is less than 30 degrees API (see Table 6.1). The current and future quality of Chinese crude oil is based primarily on data from the China National Petroleum Corporation (CNPC), China's upstream oil company, on the basis of each oilfield (see Table 6.2). Our forecasts of production by type of crude rely on crude production projections.

Our research into the quality of crude reveals that, the Chinese crude oil barrel became somewhat lighter in gravity in the early 1980s, and since then the proportion of light crude (more than 30 degrees API) has remained fairly constant at about 66 to 69 percent of total

Table 6.1 Crude quality classification

Heavy high-sulfur (heavy sour):	API < 30	S > 1%
Heavy medium-sulfur (Heavy medium):	API < 30	0.5% ≤ S ≤ 1%
Heavy low-sulfur (Heavy sweet):	API < 30	S < 0.5%
Light high-sulfur (Light sour):	API ≥ 30	S > 1%
Light medium-sulfur (Light medium):	API ≥ 30	0.5% ≤ S ≤ 1%
Light low-sulfur (Light sweet):	API ≥ 30	S < 0.5%

Table 6.2 Crude gravity and sulfur content by oilfield

	API gravity	Sulfur content (%)
Daqing	32.4	0.076
Shengli	26.8	0.860
Gudao (Shengli)	21.0	1.620
Huabei	32.6	0.230
Renqiuyan Dazhan (Huabei)	29.4	0.360
Dagang	25.2	0.250
Erlian (Inner Mongolia)	25.9	0.160
Fuyu (Jilin)	28.5	0.110
Shuangyang (Jilin)	38.0	0.060
Xinglongtai (Liaohe)	33.9	0.160
Huanxiling (Liaohe)	19.1	0.260
Jidong	30.0	0.160
Jiangsu	35.1	0.180
Chuanzhong Mixer (Sichuan)	35.8	0.025
Taindong-1 (Guangxi)	30.2	0.180
Zhongyuan Mixer	35.9	0.450
Laojunmiao-1 (Yumen)	33.8	0.180
Ya'erxia-1 (Yumen)	33.3	0.240
Lenghuhua Tugou (Qinghai)	36.3	0.480
Majiatan (Changqing)	35.2	0.040
Ganguyi (Yanchang)	40.1	0.039
Nanyang Mixer (Henan)	33.0	0.130
Jianghan Mixer	31.5	1.270
Karamay	28.5	0.080
Turpan-Hami Mixer	40.0	0.030
Tabei Mixer (Tarim)	32.1	0.640
Huizhou 21-1 (South China Sea)	44.6	0.030
Weizhou (South China Sea)	40.6	0.060
Bozhong 34 (Bohai Sea)	35.2	0.050
Chengbei (Bohia Sea)	23.0	0.640
Shale Oil	26.8	0.622
Tarim	36.0–39.0	1.000

Source: CNPC.

crude production. The proportion of heavy crude produced has stayed at 31 to 34 percent. Looking ahead, and based on our projections of future production developments, heavy crude will comprise about 29 to 31 percent of oil production, primarily as a result of the decrease in heavy crude production in Eastern China, while light crude production will slightly increase to stay in the range of 69 to 71 percent of total output (see Table 6.3 and Fig. 6.1). In terms of sulfur content, the proportion of low-sulfur crude increased from roughly 70 percent of the total output in 1970 to about 78 percent in 1997. The proportion of medium-sulfur crude has fallen from about 30 percent in 1970 to 21 percent in 1997.

During the next 10 years, crude production is projected to remain predominately sweet, but the low-sulfur share will have declined from

Table 6.3 Crude production quality and projection

	1980	1985	1990	1994	1995	1996	1997	2000	2002	2005
Crude oil production by quality (000 b/d)										
Heavy medium-sulfur	365	541	670	616	598	581	559	542	545	548
Heavy low-sulfur	171	215	284	308	311	326	340	380	410	448
Light high-sulfur	21	20	17	17	17	17	16	15	15	13
Light medium-sulfur	0	12	20	51	62	79	113	175	227	352
Light low-sulfur	1603	1710	1776	1905	1963	2115	2161	2260	2221	2175
Total	2160	2498	2766	2898	2951	3118	3189	3372	3418	3536
Share in total production (%)										
Heavy medium-sulfur	16.9	21.6	24.2	21.3	20.3	18.6	17.5	16.1	15.9	15.5
Heavy low-sulfur	7.9	8.6	10.3	10.6	10.5	10.5	10.7	11.3	12.0	12.7
Light high-sulfur	1.0	0.8	0.6	0.6	0.6	0.6	0.5	0.5	0.4	0.4
Light medium-sulfur	0.0	0.5	0.7	1.8	2.1	2.5	3.5	5.2	6.7	9.9
Light low-sulfur	74.2	68.4	64.2	65.7	66.5	67.8	67.8	67.0	65.0	61.5

Sources: CNPC, State Statistical Bureau, *China Statistical Yearbooks, 1989–1996*, China Statistical Publishing House, and author's prediction.

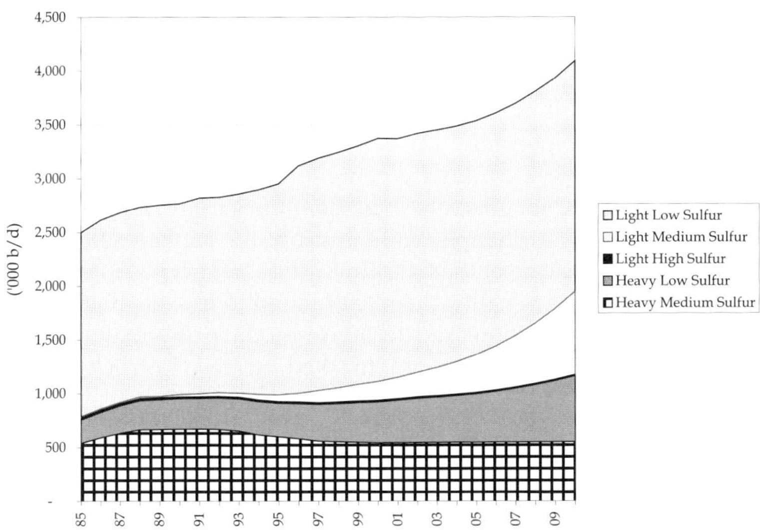

Fig. 6.1 Crude oil production quality and forecasts, 1985–2010

78 percent in 1997 to less than 74 percent by 2005. The decline in the production of Daqing oil, mostly light low-sulfur, was compensated for by increased production by other onshore and offshore oilfields with light but more sour crude. A significant change in crude production will be an increase in medium-sulfur crude output, which will yield more sour barrels. The crude slate between light and heavy output will remain primarily at 1997's level over the next 10 years. Within this time frame, production of light sweet crude will have the lowest growth, while production of light medium-sulfur crude will have the most rapid growth, largely because of an increase in the Tarim Basin's output. By 2005, China's capacity to produce light crude is forecast to grow by 260,000 b/d, while heavy crude production is expected to increase by only about 110,000 b/d. This trend marks the reversal of the 1970–1990 trend towards refining a heavier crude. As crude output increase over the next decade, crude production for all grades will also rise.

Onshore production is projected to become lighter and sweeter by 2005, but its share in total production will decline by 7 percent. Production in Daqing is expected to drop slightly after 2000, but this will be partially offset by an increase in light medium-sulfur crudes in other onshore fields. The effect will be to reduce onshore production of light sweet crude by about 30,000 b/d per year after 2000, largely due to declines in Daqing's production. Output of light medium-sulfur crude is expected to rise by about 240,000 b/d during the period 1997 to 2005, and again most of the increase will be in the Tarim Basin.

Crude output in offshore exploration is forecast to peak to about 400,000 b/d around the year 2000, and become lighter and sweeter. Most of the increase is expected in the South China Sea and Bohai Sea. Heavy medium-sulfur crude is expected to make up about 10 percent of the rise, while light low-sulfur crude and light medium-sulfur crude will account for an additional 90 percent.

In addition, Chinese crudes are generally deficient in asphalt, which is at present obtained in relatively small quantities from Tarim and Liaohe crude and some of the Shengli crude. Most of the asphalt sold in China is the carbonized residue of the paraffinic crude, which has inferior properties as a bonding and sealing agent compared to true asphalt. The supply would need to be increased to accommodate any major highway construction program in China or a marked increase in truck traffic.

6.1.2 Crude consumption quality

This section looks at changes in crude consumption quality in the light of our assessments of the changes in crude production quality and the possible sources of crude imports. Significant crude imports will reshape

Chinese crude consumption quality, which has been determined by indigenous production. Changes in crude consumption quality will affect the petroleum product yields that refineries can produce, which we considered as a basis for the product output projection.

The sulfur content of crude oil imports is presently restricted by the limited capabilities of Chinese refineries. The refinery in Maoming, for instance, is equipped with a hydrocracker and is capable of accepting 2 percent sulfur crude. However, Maoming is designing a 38,000 b/d desulfurization unit to facilitate increasing sour-grade from the Middle East. Maoming is also optimizing lubricant production by shifting to high-sulfur, intermediate crude oils with a high content of heavy minerals.[1] Zhenhai refinery, situated about 100 miles north of Shanghai, will be able to raise sour crude processing capacity to 260,000 b/d over the next five to 10 years, with no released details on desulfurization capacities. Ningpo refinery, not having a desulfurization unit, is said to have a maximum acceptable level of 1 percent sulfur content.

With net crude oil imports rising to about 400,000 b/d in 2000, and then to 900,000 b/d by 2005, questions about crude quality become very important. Will China buy medium- and high-sulfur crude to feed new coastal refineries? And will it buy sweet crude to run in domestically designed refineries? Will China need more light crude to satisfy demand for light products?

In 1997, about 40 percent of China's crude imports was light sweet crude (30–45 degrees gravity and less than 0.5 percent sulfur content) from Indonesia, Yemen, Angola, Papua New Guinea, and Vietnam, and about 35 percent was light sour crude (greater than 0.5 percent sulfur content) from Oman, Iran, and Saudi Arabia. The balance came from Malaysia. Pakistan, and Russia.

As traditional suppliers (such as Indonesia) become net oil importers themselves, China is looking to Middle East countries for crude supplies in the years to come. Over the next decade, the possible sources for supply from the Middle East are Kuwait, Oman, Saudi Arabia, Iran, and Iraq, where most of the crudes are light and sour. Based on our crude projection, we assume that crude imports from the Middle East will rise, and the pattern of China's crude consumption quality will change from the pattern of crude production quality.

Thus, China's crude consumption will shift to be more sour and light. As Table 6.4 and Fig. 6.2 illustrates, the trend toward a sour barrel will receive a big boost if, as expected, there is a tremendous increase in Middle Eastern crude imports. Consumption of light high-sulfur and light medium-sulfur crude will increase dramatically, as crude net imports

[1] At present, one-third of the crude oil Maoming refines is from Middle East oil producers.

Table 6.4 Crude consumption quality and projection

	1980	1985	1990	1994	1995	1996	1997	2000	2002	2005
Crude oil consumption by quality (000 b/d)										
Heavy medium-sulfur	288	362	544	573	586	588	559	542	545	548
Heavy low-sulfur	171	215	284	308	311	326	340	380	410	448
Light high-sulfur	21	20	17	17	17	17	137	331	452	706
Light medium-sulfur	0	12	20	51	62	79	126	196	257	398
Light low-sulfur	1421	1293	1481	1825	1940	2129	2296	2344	2338	2360
Total	1901	1903	2346	2774	2916	3139	3458	3793	4001	4460
Share in total consumption (%)										
Heavy medium-sulfur	15.1	19.0	23.2	20.7	20.1	18.7	16.2	14.3	13.6	12.3
Heavy low-sulfur	9.0	11.3	12.1	11.1	10.7	10.4	9.8	10.0	10.2	10.0
Light high-sulfur	1.1	1.1	0.7	0.6	0.6	0.5	4.0	8.7	11.3	15.8
Light medium-sulfur	0.0	0.6	0.9	1.8	2.1	2.5	3.6	5.2	6.4	8.9
Light low-sulfur	74.8	68.0	63.1	65.8	66.5	67.8	66.4	61.8	58.4	52.9

Sources: CNPC, State Statistical Bureau, *China Statistical Yearbooks, 1989–1996*, China Statistical Publishing House, and author's prediction.

jump to 400,000 b/d in 2000. The shifting quality of crude consumed means that new refineries will have to be designed to process sour crude. Our research shows that many of the planned refineries take account of this fact and will indeed be able to process relatively high-sulfur crudes. The overall result of the increasing share of sour crude is that the market share of light, low-sulfur crude (which characterizes the region) will decline in China from 67 percent in 1995 to about 54 percent in 2005.

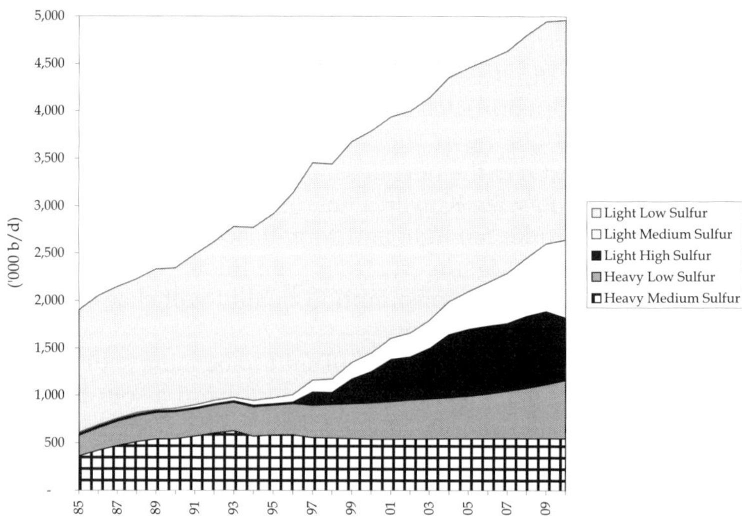

Fig. 6.2 Crude oil consumption quality and forecasts, 1985–2010

In addition to growing sourer, the refinery-input barrel in China will become slightly lighter over the next 10 years. Although Chinese crude production will retain the current profile with minor variations, Chinese crude consumption will become lighter, as imported light barrels from the Middle East become an increasingly important component of domestic consumption. We project that light crude consumption (both sweet and sour) will rise from 74 percent of total crude consumption in 1997 to 77 percent in 2005, increasing by 900,000 b/d between 1997 and 2005. Of course, this has important implications for the types, quality, and amounts of products that refineries will produce in the years to come, and on the quantity and types of investments that China will make in cracking units.

6.2 Environmental Impact

The environment in China is fragile and deteriorating. Air pollution caused by the burning of coal in industry and for household cooking and heating is a very serious problem in urban China. The needs to minimize the potential adverse environmental impacts associated with petroleum product consumption are not as severe or urgent as with coal. Environmental problems created by coal result from the ways in which it is consumed, while problems with petroleum product consumption result primarily from the ways in which it is supplied. Both coal and oil contribute to the emission of sulfur dioxide, but coal consumers (the larger plants and power stations) discharge sulfur dioxide on a greatly magnified scale compared to oil consumers. Moreover, while most oil consumers in the industrial sector discharge sulfur through relatively tall stacks, household coal consumers discharge at or near ground level, with a more devastating effect on the local environment. For instance, in northern China, the coal burned has a sulfur content averaging about 1 percent, while fuel oil and diesel have a lower sulfur content. In addition, China's industry is more than nine times the size it was in 1965[2], and emissions of industrial and toxic pollutants have increased exponentially. The majority of these incremental emissions have come from coal.

6.2.1 Environmental policy

According to a new Environmental Law, which was passed in the National People's Congress in 1994 and was in force by the end of 1996, all businesses should pay for any discharging of waste into the environment,

[2] Source: *China Statistical Yearbook, 1996*, China Statistical Publishing House.

Table 6.5 Environmental regulations on sulfur content for petroleum products

Petroleum product	Product sulfur content (%)
Gasoline	
#97	0.15 or less
#93	0.15 or less
#90	0.15 or less
#70	0.15 or less
Light diesel	
Super-grade	0.2 or less
First-grade	0.5 or less
Normal-grade	1.0
Fuel oil	
#20	1.0 or less
#60	1.5 or less
#100	2.0 or less
#200	3.0 or less

Source: *Petrochemical Product Standard*, China Petrochemical Publishing House.

instead of only being charged when their waste exceeds set environmental limits. This new law is being brought in because the existing penalty cost is so small that some enterprises prefer to be charged rather than investing in the new pollution-control devices. The new Environmental Law could force those plants, which emit an excessive amount of pollutants to shut down or to bring criminal charges against violators.

While environmental regulations are playing an increasingly important role in influencing the petroleum product market worldwide, they have not yet been a crucial determinant of petroleum supply and demand trend in China. China has set some environmental regulations for petroleum product consumption (see Table 6.5), but in comparison with other countries in East Asia they are lax. The current limit on sulfur content in diesel in China is more than 0.5 percent by weight. The limit on fuel oil is between 1.0 and 3.0 percent. It was reported that these limits were reduced in some regions, forced down by local legislation in areas where local authorities can set up higher standards than the state regulations. There is no sign, however, that China will tighten its national limits further. At present, these regulations are far from restrictive in terms of oil product demand, at least according to Western standards.

Another critical realization is that China has not restricted sulfur levels for *end-users*, even when the government has mandated the allowable sulfur content for the various petroleum products. The regulations affect only refineries, and not consumers, who simply use whatever the government allocates to them under the centrally planned system — and desperately acquire whatever they can buy in an infant market system.

It seems reasonable that Chinese environmental authorities should pay more attention to dirty polluters instead of relatively clean petroleum products. In principle, environmental regulators have the power to order polluters to reduce emissions or to cease operations if the emissions pose a serious threat to public health or ecological integrity.[3] Nevertheless, a practical remedial measure to limit the output of pollutants in China would require polluters to pay for pollution, instead of forcing them to suspend their operation.[4] Thus, for a new refinery, power plant, or industrial enterprise, meeting environmental standards could be a part of the requirements necessary to gain governmental approval for construction.[5] It has also been suggested by those concerned with improving the Chinese environment that punishment of violations should be more severe, including stoppage of construction and/or shutting of the relevant operation or plant. However, there has been no vigorous enforcement of legislation or public disclosure requirements, which might embarrass polluters and prod them into action. Due to booming demand and high costs, refineries have not been willing and able to pay the price for cleaner products.

Another possibility for reducing pollution caused by small-scale use of coal is for the government to promote oil. However, if the country is a net importer, this is unlikely to occur. Very little oil or gas is used in home heating in China; coal briquettes and similar products dominate the market. If natural gas discoveries were made in China, the present natural gas systems would be converted and expanded to bring gas to the household sector for heating and cooking purposes, as well as for

[3] Article 16 of the Environmental Protection Law of 1989 indicates that local governments at all levels shall be responsible for the environmental quality within their jurisdictions and shall take measures to improve environmental quality. Environmental Protection Law of the People's Republic of China (1989) was published in *China Environmental News*, January 15, 1990.

[4] According to Article 3 of Environmental Protection Assessment of Industrial Enterprises (for trial implementation) (1985), pollution emissions from an enterprise must be assessed in 10 major pollution categories: sulfur dioxide, smoke and dust, carbon dioxide, cyanide, phenols, oil, mercury, chromium, waste residues of smelting operations, and waste residues of chemical plants. See *Selections of Documents of Environmental Protection Commission of the State Council*, China Environmental Science Publishing House, Beijing 1988, pp. 47–48.

[5] A nationwide investigation of pollution in China shows that industries contribute some 70 percent of the nations' pollution. See Hou, Daijun "Industry, Chief Polluter", *China Environmental News*, January 1991.

Table 6.6 Gasoline production by quality (%) in Sinopec

	1989	1990	1991	1992	1993	1994	1995	1996
70 RON	64.34	62.00	54.19	52.03	54.28	48.65	43.13	33.59
90 RON	34.40	35.35	42.58	44.63	42.31	43.60	47.07	56.89
93 RON	0.96	2.51	2.61	2.47	3.07	6.07	7.30	7.78
95 RON	0.00	0.07	0.56	0.70	0.09	1.26	1.12	0.52
97 RON	0.30	0.06	0.06	0.18	0.24	0.43	0.74	1.22
Total	100	100	100	100	100	100	100	100

Source: *Sinopec Yearbook, 1997*, Sinopec Publishing House, April 1998, p. 138.

Table 6.7 Percentage of unleaded gasoline production and the contents of leaded gasoline in Sinopec

	1992	1993	1994	1995	1996
The percentage of unleaded gasoline in total gasoline production in Sinopec (%)	39.2	54.9	55.0	56.6	52.6
The content of leaded gasoline (g/l)	0.037	0.035	0.031	0.031	0.028

Source: *Sinopec Yearbook, 1997*, Sinopec Publishing House, April 1998, p. 138.

industrial use.[6] This would tend to ameliorate some of the problems of air pollution and of coal transportation and distribution in China.

6.2.2 Product quality and environmental impact

Because China cannot "afford" a cleaner environment, environmental regulations and taxes are likely to play a limited role in shaping the petroleum product market over the next decade. Oil, to a great extent, is more environmentally friendly than coal. Nevertheless, China needs to improve the quality of its oil products and improve the efficiency with which they are used.

Two primary factors will determine petroleum product specifications in China over the next decade: (1) the quality of crude oil processed in domestic refineries; and (2) the government's policy concerning petroleum product quality for both domestic output and imports.

China has shown concern about harmful automotive emissions. China claims that it taxes gasoline consumption not only for fiscal reasons, but

[6] It is reported that due to environmental concerns, after 1995, coal used in industrial and residential sectors in Shanghai, the largest city in China, will be gradually replaced by natural gas, which will be produced from the East China Sea near Shanghai.

also for environmental and health reasons, including both harmful automotive emissions and traffic congestion.

Up to 1998, the gasoline produced in China was primarily 70 octane and 90 octane. (A very small amount of 93 octane, 95 octane and 97 octane is produced as well.) In 1996, approximately 40 percent of the gasoline output was 70-octane gasoline, while the other 60 percent was 90 octane or above (see Table 6.6). China plans to increase production of high-octane gasoline during the next few years. China Petrochemical Corporation (Sinopec), the state-owned downstream company, anticipates ending the production of 70-octane gasoline by 1998, and producing primarily 90-octane gasoline, as well as producing 30 percent more of 93-octane gasoline by 2000.

While other Asian developing countries are starting their reformulated and alternative fuels programs, China is only initiating its policy towards unleaded gasoline.[7] China has increased its output of unleaded gasoline substantially since the early 1990s (Table 6.7). In 1993, China issued a regulation establishing a standard for unleaded gasoline, which stimulated additional production. In 1996, unleaded gasoline accounted for about 50 percent of total gasoline production. The percentage of unleaded gasoline in 90 octane rose from 30 percent in 1990 to more than 50 percent in 1996. However, the quality of high-grade unleaded gasoline in China is still far behind that of reformulated gasoline in the US. In mid 1998, China announced an aggressive program, which indicated to stop using leaded gasoline nationwide within two years. China planned to increase sales tax of leaded gasoline after January 1, 1999, and will force the refineries to terminate leaded gasoline production after January 1, 2000. Therefore, there will be no more consumption of leaded gasoline by July 2000.[8]

When unleaded gasoline is produced, it is usually blended with leaded gasoline, because almost no Chinese car engines require the use of unleaded gasoline, and some made-in-China vehicles consume *only* leaded gasoline. To reduce pollution in metropolitan areas, China is planning to eliminate the use of leaded gasoline in four major cities: Beijing,

[7] Thailand is presently undertaking vigorous measures aimed at increasing unleaded gasoline use. At 21 percent of the gasoline pool, unleaded gasoline sales are at an all time high, encouraged by tax breaks and subsidies to promote its use. Singapore introduced two grades of unleaded in 1991, while India proposed to use unleaded gasoline in 1993. Taiwan introduced unleaded gasoline in 1987, which currently accounts for about 30 percent of gasoline consumption. South Korea has comprised roughly 60 percent of unleaded gasoline in the total gasoline market. In addition, South Korea uses MTBE as an octane-booster in about 40 percent of the total gasoline market, while Taiwan produces all the MTBE needed domestically and also uses it as an octane-booster.

[8] See *People's Daily* (Overseas Edition), September 21, 1998, p 1.

Shanghai, Tianjin and Guangzhou, and to encourage the use of lower-lead gasoline (having a lead content less than 0.15 gram per liter) in other areas in 1998.

China is self-sufficient in gasoline. This situation could continue into the early 2000s. Therefore, gasoline production specifications will determine the quality of the gasoline consumed. In 1997, China consumed approximately 770,000 b/d of gasoline, of which about 50 percent was unleaded. It was reported that China expects to increase its consumption of unleaded gasoline to 80 percent of the total by 2000.[9] At that time, total gasoline consumption will reach 840,000 b/d, a slower pace of growth

Table 6.8 Diesel oil production by quality (%) in Sinopec (the first type of classification)

	1989	1990	1991	1992	1993	1994	1995	1996
#0	82.40	82.23	82.20	81.89	82.90	85.04	85.66	81.83
–#10	8.75	9.55	8.63	7.71	7.29	7.79	6.20	6.18
–#20	2.94	3.19	3.09	3.06	2.29	3.01	1.93	2.10
–#30	0.16	0.10	0.15	0.21	0.29	0.19	0.09	0.19
–#35	2.06	2.04	1.93	1.86	1.74	1.22	1.13	1.08
+#5	1.22	0.62	1.63	2.12	1.97	0.21	2.84	5.99
+#10	0.11	0.11	0.05	0.01	0.47	0.06	0.42	0.59
Diesel for farm use	0.09	0.01	0.01	0.02	0.01	–	–	–
Special diesel oil	1.60	1.44	1.32	1.21	1.12	1.24	1.02	0.81
Other diesel oil	0.66	0.71	1.01	1.91	1.91	1.24	0.71	1.24

Source: *Sinopec Yearbook, 1997*, Sinopec Publishing House, April 1998, p. 139.

than diesel oil consumption, which may help China to achieve its target of 80 percent of unleaded gasoline in consumption. By 2000, leaded gasoline (20 percent of total consumption) will have further restrictions on its lead content, reducing the amount to less than 0.1 gram per liter.

China produces about 40,000 b/d methyl tertiary butyl ether (MTBE), or 5 percent of gasoline production by volume. China's MTBE output is relatively high compared with other countries in the region. China has exported its MTBE to Japan, and it uses the rest as an octane-booster in its 90-octane gasoline. If global demand for MTBE rises, however, China may constrain its domestic use and export MTBE instead.

Over the next five years, China will continue to lag behind other countries in terms of quality of gasoline produced and consumed. Therefore,

[9] See: (1) Wen, Guifang "Challenges to China's Oil Market and Necessary Countermeasures", *International Petroleum Economics*, (in Chinese), 1995, Vol. 3, No. 6, p. 25, (2) Tong, Yongrui, "Problems Facing the Petroleum Processing Industry at the Turn of the Century", *International Petroleum Economics*, (in Chinese), 1995, Vol. 3, No. 6, p. 6, and (3) Han, Jingcheng and Gai, Hongyu, "Auto Fuel Demand and Supply in China", *International Petroleum Economics*, (in Chinese), 1996, Vol. 4, No. 5, p. 17.

Table 6.9 Diesel oil production by quality (%) in Sinopec (the second type of classification)

	1994	1995	1996
Super-grade (0.2% sulfur content or less)	22.2	18.8	14.2
First-grade (sulfur content between 0.2% and 0.5%)	37.4	39.2	42.1
Normal-grade (1% sulfur content)	40.4	42.0	43.7

Source: *Sinopec Yearbook, 1997*, Sinopec Publishing House, April 1998, p. 139.

if gasoline is still in surplus in the north, exports will need to be China's best unleaded gasoline, leaving low-grade leaded gasoline for the domestic market.

Diesel oil in China is used in three sectors: bus/truck transport (23–25 percent of total consumption); farming (37–39 percent); and others (including power generation, railway transportation, and water transportation, etc.). However, diesel quality specifications are not divided by sector usage. There are two divisions in diesel oil classification. The first classification depends on a pour point: for example, #0 diesel is a light diesel fuel with a pour point not higher than zero degree centigrade, and #10 diesel is a heavy diesel fuel with a pour point not higher than 10 degrees centigrade. In China, the largest volume produced is #0 diesel oil, which makes up more than 80 percent of total diesel oil output. The #0 diesel is primarily used for the electric power and transport sector, while heavy diesel oil is used mainly in industry (Table 6.8).

The second classification separates diesel oil into three grades: super-grade (0.2 percent sulfur content or less); first-grade (sulfur content between 0.2 percent and 0.5 percent); and normal-grade (sulfur content approximately 1 percent). All these grades are 45 cetane, which is lower than the world average of 48–50. In addition, about 20 to 60 percent of diesel oil is produced by catalytic cracking, a higher share from this type of facility than in most other countries. Currently, only 16 percent of total diesel oil output is super-grade diesel, with the rest being evenly divided between first-grade and normal-grade. However, it is important to note that most other industrialized countries have introduced lower than 0.1 percent sulfur diesel oil, while even the highest grade diesel produced in China has higher sulfur levels.

China has proposed increasing production of super-grade diesel oil to 50 percent of its total by 2000, with the remaining 50 percent being made up of first-grade diesel oil.[10] To reach this target, China will increase its diesel oil output from hydro-treating and hydro-refining facilities.

[10] See Han, Jingcheng and Gai, Hongyu, "Auto Fuel Demand and Supply in China", *International Petroleum Economics*, (in Chinese), 1996, Vol. 4, No. 5, p. 18.

Consequently, China needs to add more hydro-treating and hydro-refining capacity to meet its environmental requirement by 2000. However, in any circumstance, this proposal seems unrealistic.

At present, about 75 percent of fuel oil produced and consumed in China is low-sulfur fuel oil. However, if China has to import more Middle Eastern sour crude, the sulfur content of its fuel oil will naturally increase. The penetration of low-sulfur fuel oil will be constrained by the high cost of desulfurization versus other compliance options. With limited desulfurization capacities and more sour crude processing, it will be difficult for China to implement stricter fuel oil standards. Moreover, as long as China has a fuel oil deficit, low-priced high-sulfur fuel may appear more attractive to Chinese importers, adding more high-sulfur fuel oil to total consumption levels.

Neither has the environmental revolution come to China's diesel and fuel oil markets, with the accompanying segmentation between high- and low-sulfur fuels. In other countries, there are high prices for low-sulfur and low prices for high-sulfur products — consumers pay an environmental premium. In China, however, the prices are the same. In 1992 and 1993, attracted by low prices, China imported the bulk of its high-sulfur diesel and fuel oil from Japan and South Korea, through direct and indirect trade (via Hong Kong and other countries), and sold them into the domestic market as the same price as other low-sulfur fuel oil.

As China's oil market starts to merge with the world oil market and a large deficit in its fuel oil balance, China may import more high-sulfur petroleum products, encouraged by the margin between low-sulfur and high-sulfur products. Moreover, when China imports more Middle East sour crude for coastal refineries, the sulfur content of its diesel oil and fuel oil will dramatically increase. Nevertheless, we expect that product standards in the country — such as lower sulfur content in diesel oil and fuel oil, and the introduction of unleaded gasoline — will gradually become stricter, which could significantly improve the standards presented in Table 6.5, even though this will result in further increases in the costs of refineries.

Chapter 7
Supply/demand Balance: Implications for Trade

The Chinese oil market in both 1996 and 1997 surprised the world again, and showed striking signs of a boom year, recovering from its 1994 downturn. Even with higher levels of production of crude oil and refined products, strong demand kept fundamentals very tight, resulting in more oil imports. While the Chinese government acted to keep inflation in check in 1996 and 1997, the oil market enjoyed an environment with slightly loosened control and less interference from the central authorities. In the foreseeable future, crude output will have limited increases, while oil consumption continues to grow. Meanwhile, we will see only a slight degree of government flexibility in its regulation of the oil industry.

In this chapter, we will use the crude production analysis and demand analysis for refined products presented in the previous chapters, as an input to formulate our crude oil balance and petroleum product balances. The final objective is to arrive at a characterization of the overall oil market outlook. We begin with an overview of the import/export picture. Then, changes in crude supply and crude demand (which is composed largely of refinery throughput, but which also includes direct burn and annual losses) are compared to derive China's crude import requirements over the next decade. Changes in petroleum product supply and demand are analyzed in a similar way to project China's need to import or ability to export value-added refined products. In reviewing this picture, we also discuss China's ability to support the balance we predicate.

Seen in the widest perspective, four major changes have characterized China's oil imports and exports during the last two decades. First, like many oil exporters (e.g. Indonesia), China has made vigorous efforts to diversify its total exports away from oil-based goods to non-oil items. The rapid increase in Chinese trade volume in recent years has seen the country's hard currency earnings shift from crude oil to a broader range of goods — electronic goods, textiles, shoes, toys, and non-oil raw materials. In other words, China is trying to produce highly

Table 7.1 Crude oil exports and imports, 1971–1997

	Exports (000 b/d)	Imports (000 b/d)	Net exports (000 b/d)	Net exports as a % of oil output	Net imports as a % of oil output
1971	5	1	4	0.5	–
1972	13	0	13	1.4	–
1973	37	0	37	3.4	–
1974	101	0	101	7.8	–
1975	198	0	198	12.8	–
1976	170	0	170	9.7	–
1977	182	10	173	9.2	–
1978	226	14	212	10.2	–
1979	269	5	263	12.4	–
1980	266	7	259	12.2	–
1981	275	1	274	13.5	–
1982	304	13	291	14.3	–
1983	304	7	296	14.0	–
1984	446	5	441	19.2	–
1985	601	5	596	23.8	–
1986	570	9	561	21.5	–
1987	545	10	534	19.9	–
1988	521	17	504	18.4	–
1989	488	66	423	15.3	–
1990	480	58	421	15.2	–
1991	452	119	333	11.8	–
1992	430	227	203	7.2	–
1993	389	314	75	2.6	–
1994	371	247	124	4.3	–
1995	377	342	35	1.2	–
1996	431	451	–20	–	0.7
1997	397	710	–314	–	9.8

Sources: International Energy Agency, OECD and Chinese Customs.

southern provinces may make sense economically, and could ease the problems caused by the inadequate transportation infrastructure for delivering domestic crude. However, the rising Chinese yuan caused the c.i.f. price for imported crude in Guangdong province to rise much higher than the f.o.b. price for Daqing crude in the north. Despite this imbalance, China's crude imports into the south did not decline, because the cost of shipping crude domestically is still higher than the spread between the c.i.f. price in the north and f.o.b. price in the south. However, this pattern of oil flow may change, if China devalues its currency under the pressure of the Asian financial crisis.

7.1.2 Destinations of exports and sources of imports

With China's crude production increasing at a rate of only 2 percent during the last 10 years, and its oil consumption growing at 6 percent a year, the resource surplus has dwindled rapidly. A few statistics suffice to tell the story. From only 10,000 b/d as recently as 1987, crude imports rose by 310,000 b/d in 1993 and another 710,000 b/d in 1997. At the same time, crude exports plummeted from a peak of more than 600,000 b/d in 1985 to only 400,000 b/d in 1997. This surge in crude imports reflects the need of coastal refineries, particularly to feed the sharp increase in third-party processing activities there.

Table 7.2 shows the volume of China's crude exports to various markets for the last nine years. On the export side, the numbers reveal a steadiness in terms of trading partners, while the import picture shows a number of changes. Japan, South Korea, and the United State were the three countries to which the bulk of Chinese crude barrels moved. Exports to Japan have remained at about 200,000 b/d during the last five years, representing more than half of China's total crude oil exports, while the share taken by South Korea and the US has stayed at around 10–20 percent, respectively. Looking at shipments of Chinese crude from the other side of the dock reveals a different story. According to Japanese and US statistics, China ranks as the fifth largest source of Japanese crude imports, but its contribution to total US crude imports is negligible. China has also relied significantly (and continues to rely) on the US and Japanese markets for its non-oil trade.[1]

China's crude exports to its major crude client, Japan, fell gradually during the past years. In September 1995, China and Japan negotiated the terms of a crude export deal covering the next five years. The amount China will export has been cut approximately 50,000 to 100,000 b/d for the period between 1996 and 2000. (The price formula for this transaction consists of 50 percent of the combined prices of Minas and Cinta crude plus 50 percent of the market price. Minas and Cinta are both low-sulfur crude, but Minas is light and Cinta is heavy. This formula essentially simulates the quality of Daqing crude — light, low sulfur — and Shengli crude — heavy, low sulfur — which makes up the bulk of crude exported to Japan for power generation.) The reduction of crude exports reflects the emerging crude shortage in China and the consequent decline of the amount of crude available for export.

[1] According to both Japanese and US statistics, since 1980 China has climbed from the fifth to second in Japan's imports market, while its position in US import market has been upgraded from 13th to fifth place.

Table 7.2 Crude exports (000 b/d) by destination and imports by source

	1990	1991	1992	1993	1994	1995	1996	1997	1997's share in total
Crude exports by destinations:									
Japan	273	241	256	254	239	241	233	209	52.8
Singapore	66	59	45	15	12	16	8	8	2.0
US	80	93	72	56	61	50	53	46	11.6
N. Korea	21	22	20	21	17	20	19	10	2.6
S. Korea	20	23	33	32	21	38	63	91	22.8
Indonesia		–	–	–	1	11	23	22	5.6
Others	19	14	4	12	19	0	32	10	2.6
Total	480	452	430	389	371	377	431	397	100.0
Crude imports by sources:									
Indonesia	25	55	95	80	94	106	126	132	18.6
Pakistan	2	2	0	4	5	1	–	2	0.3
Papua New Genia	–	–	10	16	16	7	9	7	0.9
Vietnam	–	–	6	6	12	15	20	30	4.2
Malaysia	7	4	9	10	6	12	4	5	0.6
Australia	1	-	11	8	1	1	4	7	0.9
Others	1	0	2	7	2	0	1	7	1.0
Sub-total from Asia-Pacific	35	62	134	131	137	142	164	188	26.5
Oman	17	53	61	82	67	73	113	181	25.4
Yemen	–	–	9	33	25	50	76	82	11.5
UAE	–	1	5	11	1	7	–	1	0.1
Saudi Arabia	–	2	4	4	3	7	5	10	1.4
Iran	6	1	2	1	1	19	46	55	7.8
Iraq	–	–	–	–	–	–	–	5	0.7
Kuwait	–	–	–	–	–	–	–	1	0.2
Qatar	–	–	–	–	–	–	–	2	0.2
Sub-total from Middle East	23	57	81	132	98	156	240	336	47.3
Angola	–	–	4	24	7	20	33	77	10.8
Libya	–	–	6	14	–	4	3	1	0.2
Nigeria	–	–	–	–	–	8	–	–	–
Congo	–	–	–	–	–	1	3	20	2.8
Egypt	–	–	–	–	–	–	–	6	0.8
Gabon	–	–	–	3	2	2	–	8	1.1
Algeria	–	–	–	–	–	3	–	–	–
Others	–	–	–	1	–	–	–	7	1.0
Sub-total from Africa	–	–	10	43	10	37	39	118	16.6
Russia	–	–	–	–	–	–	6	10	1.4
Other regions	–	1	2	8	2	8	2	58	8.1
Total	58	119	227	314	247	342	451	710	100.0

Source: Chinese Customs.

China has been a major energy supplier to North Korea through its program of crude exports. This deal was made years ago between the two governments as part of a long-standing political agreement. North Korea receives at least 10,000 b/d of Daqing crude in exchange for steel and iron ore. A former agreement between North Korea and Iran for the supply of up to 30,000 b/d of crude oil blend to North Korea was terminated early in 1993 because of North Korea's inability to make payment in hard currency. After losing the support of the former Soviet Union (FSU), North Korea has come to rely heavily on China for its energy supply. Crude exports to North Korea are quite stable, at 3–5 percent of total crude exports from China. However, monthly exports have been erratic, with high exports one month followed by lower amount in the following month. Chinese statistics show that, for example, China exported 28,000 b/d of crude to North Korea in August 1995, then cut exports to less than 10,000 b/d in September. Until ties between the two countries are loosened at a high political level, China will maintain its obligation to export crude to North Korea despite in increase in its own oil shortage.

Before 1991, crude exports to South Korea averaged about 20,000 b/d. After diplomatic relations were established between the two countries in 1992, crude exports jumped to 40,000 b/d in 1995, 60,000 b/d in 1996, and more than 90,000 b/d in 1997. A considerable amount of Chinese crude used to go to Singapore for processing, but this flow has slowed to a trickle, falling from 74,000 b/d in 1989 to only about 8000 b/d in both 1996 and 1997. This shift primarily reflects an increase in refining capacities in China's own coastal regions.

The sources of crude imports to China show greater variety. Before 1992, crude imports primarily came from Asia-Pacific nations. Indonesian, Malaysian, and Australian crude accounted for more than 50 percent of total imports. But this situation has begun to shift, and after 1995, the volume of crude imports from the Middle East exceeded the crude flow from Asia for the first time. Since 1992, China has also imported crude from African countries, particularly from Angola and Libya. In 1992 crude imports from Africa accounted for only 4 percent of total imports, but jumped dramatically in 1997 to reach 16 percent. In absolute terms, the volumes increased from 10,000 b/d in 1992 to 120,000 b/d in 1997. These import increases from Africa have largely displaced crudes from the Asia-Pacific region.

Although crude imports from other Asian countries increased, Middle Eastern crude poured into China at a much higher rate (Fig. 7.1). All the Middle Eastern countries exported more barrels into China in 1995, which marked the first year that China's imports from the Middle East, 160,000 b/d, surpassed imports from Asian countries, 140,000 b/d. This signals that, in China, the pattern of crude importing sources has shifted, and the Chinese market will grow increasingly dependent on crude imports from the Middle East.

Oman and Yemen are the two major Middle Eastern suppliers of crude to China. Oman was the first Middle East country with which China signed a term contract to import crude, back in the mid-1980s. Crude imports from Oman have since grown to more than 100,000 b/d. Yemen began exporting crude to China in 1992, but the amount of crude it exported in 1997 has more than doubled from 1994 levels, reaching 80,000 b/d.

In March 1995, just at the time when the US announced a ban on Iranian oil, China began monthly imports of crude from Iran, a significant increase over the initial one-month crude deal set up between China and Iran in 1994. China signed a term contract for 20,000 b/d of crude imports from Iran prior to 1995. Crude imports from Iran peaked at 55,000 b/d in November 1997, a record. However, the lack of desulfurization capacity in refineries will prevent China from increasing its order for high-sulfur Iranian crude, since only the few refineries possessing desulfurization capacity (in the southern and eastern coastal provinces) can process Iranian crude. Perhaps as a reward for China's purchases,

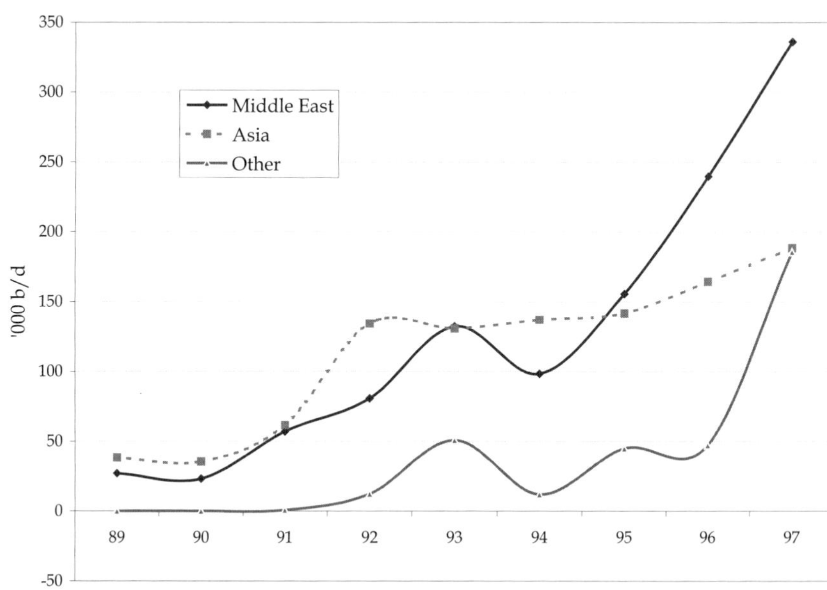

Fig. 7.1 Sources of crude oil imports
Source: Chinese Customs.

Iran has altered its policy regarding its destination ban, which requires that Iranian crude buyers not resell its oil. This has given China the flexibility to ship Iranian crude to Singapore or other countries for third-party processing when necessary.

China has dramatically increased its crude imports from Iran since 1995, while China started to import crude from Iraq in 1997. China also has a good relationship with Saudi Arabia, which has mainteind oil exports to China for a long time. Despite that the divisions between Saudi Arabia, Iran and Iraq are far too deep to be healed, China holds a balance between the three and avoids alienation of any one politically and economically.

Since the Middle East crude with its high-sulfur content cannot be refined in most Chinese refineries, this fact explains why Oman and Yemen (both have low-sulfur crude) are the two major countries in the Middle East exporting to China. For the first time in 1993, imports of Omani crude exceeded Indonesian crude imports. These increasing levels of exports of Middle Eastern crude are making the Chinese a more important force in the global oil market — and perhaps less of a force in the regional market.

Up to 1998, Indonesia was still the largest supplier of crude imported to China in the Asian market. In 1997, the annual amount of crude imported from Indonesia averaged approximately 130,000 b/d, varying monthly from a high of 220,000 b/d to a low of 35,000 b/d. In turn, China exported a small amount of crude to Indonesia, probably for third-party processing and/or for price differential profits.

China increasingly imported West African crudes that could replace Indonesian Minas as the availability of crude from Indonesia's production shrunk, or its crude became very expensive. When steadily falling Brent prices tightened the spread to Minas to less than $2.00 per barrel, it made Brent-linked West African grades more attractive.

While China will continue to maintain its obligation to export crude to Japan and North Korea, we will see China's crude imports rising annually until a significant number of barrels are provided by its offshore and Tarim oil exploitation and development projects.

7.1.3 Crude balance outlook

China's crude supply will not be adequate to handle its rising crude demand. In 1994, China consumed 3.5 million b/d of crude (including refinery throughput, crude direct burn and losses). While remaining a net refined product importer, China also showed a crude oil deficit in 1997. Overall, China's crude production was about 460,000 b/d below its total product demand during 1997, without considering its direct burn of

Table 7.3 Crude oil net exports projection

	Crude Oil Production (000 b/d)	Refinery Runs (000 b/d)	Losses (000 b/d)	Direct Burn (000 b/d)	Crude Net Import and Forecasts (000 b/d)	Net Exports as a % of Crude Production	Net Imports As a % of Refinery Runs
1995	2968	2881	32	75	–35	1.2	n.a.
1996	3137	3000	31	127	20	n.a.	0.7
1997	3198	3324	31	112	269	n.a.	8.1
1998	3264	3329	31	105	201	n.a.	6.0
1999	3326	3571	31	99	375	n.a.	10.5
2000	3395	3692	31	93	421	n.a.	11.4
2001	3392	3848	30	85	572	n.a.	14.9
2002	3441	3915	30	79	584	n.a.	14.9
2003	3474	4062	30	74	692	n.a.	17.0
2004	3508	4282	30	68	872	n.a.	20.4
2005	3560	4390	29	64	924	n.a.	21.0
2006	3631	4582	29	60	940	n.a.	21.0
2007	3720	4575	30	57	940	n.a.	20.6
2008	3830	4746	30	54	999	n.a.	21.1
2009	3954	4887	30	51	1015	n.a.	20.8
2010	4115	4903	31	49	868	n.a.	17.7

Notes:
Numbers after 1998 are forecasted.
This projection for crude net imports does not consider a significant increase in crude supply from the Tarim Basin.

crude, which was about 140,000 b/d. Such a delicate balance will remain, when product consumption rises along with economic growth.

This crude consumption is projected to increase at a rate of 100,000–150,000 b/d per year in the near term. As Table 7.3 shows, based on the distillation capacity projections made in Chapter 4, and on the assumption of operating rates of about 90 percent for new refineries, China will be forced to halt oil exports much earlier than many China analysts expect.[2] We do not expect a great increase in crude supply from the Tarim Basin during the next 10 years, so the prospect is for relatively flat crude exports in the north, overwhelmed by increasing crude imports in the south. The developments we foresee would result in a deficit amounting to as much as 500,000 b/d by the turn of the century, and about 1 million b/d by 2010 (Fig. 7.2).[3] That would soak up a major stream of oil production in world markets, and would also entail China's spending

[2]Both the World Bank and the Central Intelligence Agency of the US projected that China would suspend oil exports by the end of the decade.

[3]According to Xin Hua News Agency, China will consume an average 3.6 million b/d of oil during the years from 1996 to 2000, but will face an annual shortfall of some 560,000 b/d per year. See *Wen Hui Daily*, March 21, 1996.

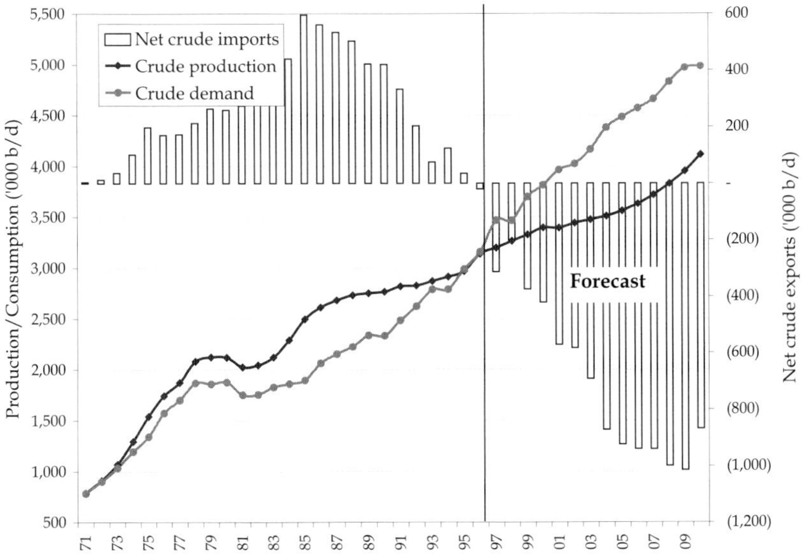

Fig. 7.2 Oil production and consumption (including crude direct burn and losses)

US$ 5–8 billion per year for crude imports, a phenomenon that could hinder Chinese domestic investment.

Only two variables could reverse this projection: if major new crude resources are discovered *and brought to market*, or if refinery capacity changes in a manner different from what is outlined in the refining chapter of this book. Based on our supply/demand projection and evidence on crude direct burn, the absolute domestic crude supply shortfall will continue a crude oil deficit that will grow larger unless and until Tarim Basin oil comes to market. In this scenario, China could import as much as one-fifth of its total refinery throughput.

Despite the coming domestic supply shortfall, China will continue to export crude oil from extra barrels in the north to fulfill contracts with Japan and North Korea. But it will be difficult to keep crude exports at the 300,000 to 400,000 b/d levels for the remainder of the decade, let alone to increase exports by significant margins. This assessment is supported by two factors. First, due to geographic differences, it will make economic sense to export crude from the north and import it into the south: crude will continue go to Japan, North Korea, and to Singapore for processing. Second, crude exports will be needed for substantial increases in foreign capital investment and technology acquisition for offshore and some onshore exploration.

Southern and eastern coastal refineries like to handle Indonesian and Persian Gulf crude rather than Daqing's paraffinic crude, which is more

Table 7.4 CNPC's oil exploration and development around the world

Cooperated country	Time	Region or country which oilfields located in	Status	Investment (million US$)	Oil reserves (m bbl)
Turkmenistan	1998	Turkmenistan	Negotiating with government for oil concessions.		
India	January 1998	Kazakhstan	Jointed with India's Oil & Natural Gas Corp. Videsh to set up joint venture to explore for oil in western Kazakhstan where ONGC Videsh has a concession.		
Kazakhstan	August 1997	Kazakhstan	Bought 60% of Uzen field.	1300	1460
Venezuela	June 1997	Venezuela	Bought Caracoles Block.	240	730
Venezuela	June 1997	Venezuela	Bought Intercampo Norte block.	118	
Iraq	June 1997	Iraq	Bought 50% of Al Ahdab field	1200	1022
Kazakhstan	June 1997	Kazakhstan	Bought 60% of Aktyubinskmunai Production Association.	400	1,022
Sudan	May 1997	Sudan	Bought 40% of Heglig field. Signed joint	1000	Unknown
Italy	1997	Central Asia and Africa	venture with Italy's Agip to develop oilfields.		
Iran	1997	Iran	Holds talks with Iran's National Oil Co. on a joint venture to explore for oil in Iranian waters.		
Russia	1997	Russia	Negotiating joint venture with Bashneft state oil company.		
Peru	1993	Peru	Bought Talara block.		Unknown

Source: "Beijing Gusher", *Far East Economic Review*, February 26, 1998, p. 47.

costly to process. Domestic consumption growth in Indonesia and Malaysia, however, may limit their abilities to export crude to China. Thus, China announced that it would buy 7000 b/d of crude oil from Saudi Arabia starting in 1995. China is also negotiating additional long-term crude and product supply contracts with the UAE, reporting that China will import about 10,000 b/d crude per year from UAE, which did not supply a great amount of crude to China before. China has an existing term contract to buy some 20,000 b/d of Abu Dhabi crude under a direct-deal arrangement, and it also buys Abu Dhabi crude on the spot market. China will also seek crude supply from other Middle East countries. As coastal refineries upgrade to run high-sulfur crude, imports of Middle Eastern crude are expected to increase, and their share in total crude imports will also rise.

As will be shown later, we believe Chinese refining capacity will grow faster than demand. As refineries expand, there will be increasing volumes for third-party processing. The inefficiency of China's rail and shipping transport can probably be overcome with variable processing fees, similar to those in Singapore (excluding any profit-sharing element) and favorable freight rates between China and East Asian countries.

Some of China's new refineries would even become export-oriented. While northern and western refineries consume domestic crude, eastern and southern joint-venture refineries will increasingly rely on imported crude. More crude imports for new joint-venture refineries in eastern and southern coastal provinces is likely to mean corresponding refined product exports to meet their foreign exchange balance. Foreign investors and joint-venture partners have an incentive to export refined products to recoup their investments (rather than sell products domestically), since they can minimize currency convertibility and foreign exchange risks if they avoid holding Chinese currency. Therefore, we expect to see more active oil trade in the east and south, led by foreign joint-venture refineries.

Before the mid-1990s, crude oil imports was the only source to fill the domestic gap in crude oil deficit. However, China's oil policy, which has changed since the mid-990s, is targeting foreign oilfields as additional sources of oil supply, i.e. buying oil equity abroad. To secure new energy resource abroad, the China National Petroleum Corporation's (CNPC) oil exploration and development has expanded into South America, Africa, Middle East, the FSU, and Asia (see Table 7.4). This aggressive policy reflects China's concern to fuel its fast-growth economy, and to diversify its energy supplies in the future.

Based on the prediction in Table 7.3, China will have net crude imports as much as 1 million b/d in the late 2010s. Taking some crude exports, which China will continue, into consideration, the actual crude imports will be higher than 1 million b/d. If China can successfully build a crude supply network overseas, those China-owned fields abroad may be able to provide one-third to one-half of crude oil imported to China in the future.

7.2 Petroleum Product Balance

While China has retained its status as a net exporter of crude, it became a net importer of refined products in 1989. In fact, it has been the rate of oil product demand growth that led China to become a net oil importer (combined crude and refined products) in 1993, much earlier than most analysts expected. We will look at the product balance in relation to three factors: product demand, topper (crude distillation unit) capacity and upgrading (secondary unit) capacity.

7.2.1 Petroleum product imports/exports and balance

China has been self-reliant in the supply of nearly all kinds of petroleum products since the middle 1960s. However, as Table 7.5 shows, Chinese imports of all principal products have increased substantially during the past 10 years. In 1985, exports still exceeded imports by more than 75,000 b/d. Since that time, product exports have been halved, while product imports have almost tripled. The first year of net product imports was 1992, but the shortfall in that year was rather small compared with that in 1993, which registered a product deficit of 290,000 b/d. In 1994, product deficit fell by 70,000 b/d over 1993, declining to 225,000 b/d. However, the product deficit jumped to 430,000 b/d in 1997, reaching an all-time high (Fig. 7.3).

The 1993 surge in imports also merits a closer look, since the country recorded a net deficit in virtually every product. This across-the-board deficit differed significantly from previous years, when China was still able to export large amounts of gasoline despite large distillate imports. Light products accounted for about 75 percent of total imports. Not surprisingly, diesel fuel claimed the largest share of the import pie at 54 percent, but it was the strong growth in the lighter and heavier ends of the barrel, which set 1993 apart from other years; gasoline imports increased from about 8000 b/d to over 50,000 b/d, and fuel oil constituted 20 percent of total product imports.

But the rate of product import growth in 1993 was abnormally high, and it is unlikely to serve as a model for the future (as the experience in 1994 confirms). In 1993, falling prices on the international spot markets during the second half of the year encouraged some Chinese regions to import products rather than buy the more expensive product made by China's own refineries. About 355,000 b/d of petroleum products were

Table 7.5 Exports and imports of principal petroleum products, (000 b/d) 1971–1997

	Gasoline			Diesel oil			Fuel oil		
	Exports	Imports	Net exports	Exports	Imports	Net exports	Exports	Imports	Net exports
1971	0.9	0.1	0.8	0.0	0.1	(0.1)	0.1	0.0	0.1
1972	0.0	0.2	(0.2)	0.2	0.2	0.0	0.0	0.2	(0.2)
1973	0.0	0.2	(0.2)	0.6	0.2	0.4	0.0	0.4	(0.4)
1974	0.0	0.2	(0.2)	4.6	0.3	4.3	0.9	0.4	0.5
1975	0.0	0.2	(0.2)	10.0	0.4	9.6	2.6	0.4	2.3
1976	0.0	0.2	(0.2)	9.5	0.3	9.2	3.0	0.3	2.7
1977	0.1	0.2	(0.1)	13.2	0.3	12.9	3.8	0.3	3.6
1978	0.5	0.3	0.1	19.4	0.5	18.9	4.6	0.5	4.1
1979	1.6	0.5	1.2	22.5	0.5	22.0	9.1	0.5	8.7
1980	27.4	0.0	27.4	34.0	0.4	33.6	8.3	7.1	1.2
1981	30.6	0.0	30.6	32.5	0.6	32.0	8.8	9.7	(0.9)
1982	33.1	0.0	33.1	33.3	0.8	32.5	8.2	14.5	(6.3)
1983	32.1	0.0	32.1	26.8	0.9	25.9	11.3	15.3	(3.9)
1984	27.0	0.0	27.0	43.0	0.8	42.2	10.4	13.2	(2.8)
1985	30.3	0.1	30.2	46.1	0.9	45.2	11.8	12.8	(0.9)
1986	25.2	2.3	22.9	43.5	30.4	13.1	13.0	23.6	(10.6)
1987	27.1	3.0	24.1	29.8	34.9	(5.0)	14.3	18.8	(4.6)
1988	25.9	4.2	21.7	29.8	50.1	(20.3)	14.7	19.8	(5.0)
1989	27.7	7.4	20.3	28.9	70.6	(41.6)	10.0	14.7	(4.8)
1990	38.1	3.6	34.5	31.3	40.2	(8.8)	10.5	11.6	(1.1)
1991	47.9	2.5	45.4	23.6	58.5	(34.9)	8.7	21.2	(12.6)
1992	62.4	7.7	54.7	29.4	94.2	(64.8)	8.1	30.9	(22.7)
1993	43.0	50.8	(7.8)	25.6	177.6	(152.0)	3.4	70.7	(67.2)
1994	48.9	24.5	24.4	24.6	123.3	(98.7)	1.2	72.6	(71.5)
1995	43.2	3.7	39.5	26.5	122.2	(95.6)	3.3	109.5	(106.2)
1996	30.5	1.8	28.6	32.1	94.1	(62.0)	5.1	155.8	(150.7)
1997	41.4	2.0	39.4	46.8	150.7	(104.0)	7.0	231.3	(224.2)

Source: Chinese Customs.

imported, and the volume of diesel oil imports soared to 190,000 b/d. This total import volume represented an increase of nearly 50 percent over 1992. From a national perspective, petroleum product imports supplied about 12 percent of total petroleum product need (Table 7.6). The government's reaction to this was negative, swift, and relatively effective.

1994 product imports dropped from 1993's level by 65,000 b/d for crude and 75,000 b/d for petroleum products, to more than 500,000 b/d of total imports (combined crude and products). This was because: (1)

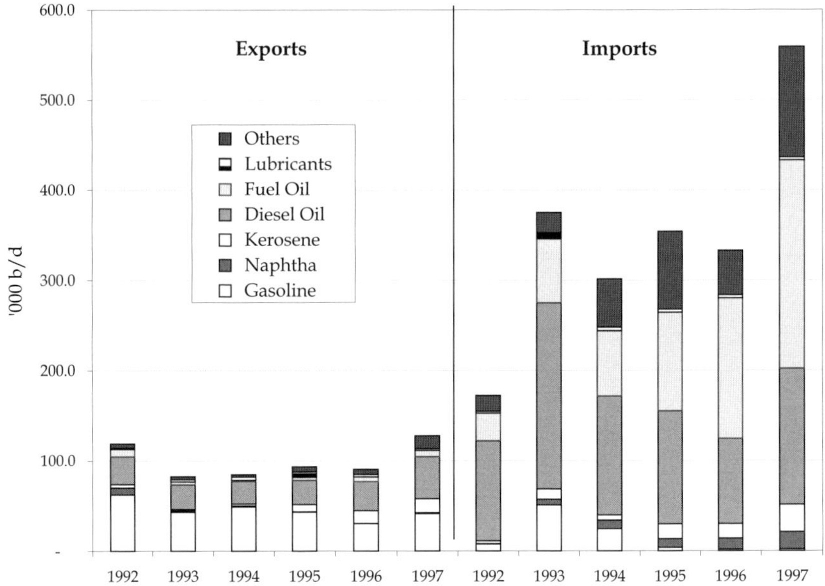

Fig. 7.3 Petroleum product imports and exports
Source: Chinese Customs.

before May 1, 1994, when the central government re-controlled oil prices, refineries and wholesalers stockpiled their inventories waiting for the government to set higher prices; (2) the new regulated prices cut demand; and (3) the central government withdrew some quotas from previously assigned product importers by the end of the year. In 1994, the central government reduced regulated prices several times since they set the rule due to a weak demand.

Among product imports, light diesel amounted to half of total product imports in 1993 and 1994, since the low domestic price of diesel discouraged refineries from producing diesel oil. However, demand for diesel oil is growing, because the low price of diesel also stimulates demand. Without price incentive, imports of diesel could remain high in the foreseeable future.

Product imports in 1997 were quite similar to those in 1993, with a huge jump in total import volumes. However, in 1997, among product imports, light diesel share fell to only one-quarter of total product imports, while fuel oil accounted for more than 40 percent of total imports. After 1994, gasoline changed from a net imported product to a net exported product, evidence that cutting gasoline imports tends to be easier than dampening consumption of industrial and agricultural fuels.

Table 7.6 Petroleum product exports and imports (000 b/d)

	Exports						Imports					
	1992	1993	1994	1995	1996	1997	1992	1993	1994	1995	1996	1997
Total	119.1	82.4	84.5	93.7	90.4	127.8	172.5	375.0	301.5	354.0	332.9	559.3
Gasoline	62.4	43.0	48.9	43.2	30.5	41.4	7.7	50.8	24.5	3.7	1.8	2.0
Naphtha	7.6	1.6	1.0	0.3	0.1	1.1	0.3	6.3	9.5	9.7	12.2	18.9
Kerosene	3.8	1.6	2.3	8.1	14.2	15.4	3.3	11.4	5.7	16.4	16.1	30.3
Diesel oil	31.0	27.1	25.1	26.7	32.0	46.8	110.7	206.7	131.7	125.1	94.5	150.7
Fuel oil	8.1	3.4	1.2	3.3	5.1	7.0	30.9	70.7	72.6	109.5	155.8	231.3
Lubricants	1.4	2.4	3.5	5.9	2.7	1.8	1.9	6.6	4.0	3.4	3.3	3.3
Others	4.7	3.3	2.6	6.2	5.8	14.3	17.6	22.6	53.4	86.1	49.2	122.8

Source: Chinese Customs.

According to the monthly data published by the Chinese Customs, during 1993 to 1997, China's product imports surged in December, nearly doubling the November figure to reach a peak at each year. Imports thus accounted for about 15 to 30 percent of total domestic product demand in December. There are three reasons behind the extreme rise in imports. First, the Chinese usually order imports of petroleum products during the first half of the year, and the physical barrels arrive toward the end of the year. Second, quota signees rushed to spend their quotas by the end of the year in order to secure next year's quota from the central government. Third, Chinese statistical collection is based on a method of monthly cumulation and subtraction, i.e. the current month's year-to-date total is subtracted from last month's year-to-date number to reach the current month's figure. If there are large omissions in previous months, the month when the omission is found would show an unusual rise in the figures. At the end of the year, the Chinese easily find the discrepancy or miscalculation, which took place earlier in the year. In 1994, we heard that Chinese Customs might revise the December 1993 import/export data. However, it failed to do so. We do not yet know the extent of the miscalculation in the data of December 1993. Therefore, the market impact on China's oil imports surge in each December may be not as large as the number shows. In addition, the import surge in December 1993 was also cause by the unified two-tier foreign exchange rate system, which switched the official exchange rate to the foreign exchange swap market rate on January 1, 1994. This encouraged Chinese importers to use their entire foreign exchange quota at the low official rate before the end of 1993. As a result, not only petroleum products, but also crude oil and other commodities poured into China during December 1993. Nevertheless, the end-year import jump is also a real phenomenon, brought about by an inflexible central planning system.

In the course of the transformation to a net product importer, China's trading patterns have grown more complicated, and sometimes the country is a buyer as well as a seller. China both imports and exports products because of its lopsided concentrations of product supply and demand. Refinery output flows from facilities in northeast and east China, while south China's refining capacity cannot keep up with its soaring demand. In addition, because domestic distribution networks remain inadequate, importing products into the south can sometimes be cheaper than sourcing material from domestic refineries. Also, it appears that the long-standing desire to earn foreign currency has provided a strong impetus to product exports from the north.

There are two reasons that the growth of product imports may slow down during the next decade. First, China's total commodity imports exceeded exports in 1993, which led to a severe trade deficit. Customs statistics for 1993 showed that commodity imports surged 28.1 percent to $76.4 billion and exports reached $69.3 billion — a modest 6.4 percent increase compared with 1992. This slide into the red was accelerated by the oil import situation: the cost of net oil product imports was $2.3 billion, almost 30 percent of China's current account deficit in 1993, making the boom in oil product imports the primary cause of China's first annual trade deficit since 1989. The anxiety among Chinese leaders induced by these deficits in the trade and current accounts, and the sheer size of the financial resources necessary to cope with them, should limit the country's ability to further increase petroleum product imports. The continuation of the extremely high ratio of product imports to total product demand would place a heavy burden on the foreign exchange balance and have an adverse effect on the acquisition of other valuable import commodities and technologies.

Second, starting on January 1, 1994, the low official exchange rate (5.8 yuan per US$ in 1993) at which oil import purchases was previously made was abolished in favor of the floating, market-based rate (which averaged 8.7 yuan per US$ in 1993), reflecting a new step for economic reform and China's efforts to join the World Trade Organization. Those state-owned enterprises cannot rely on the import subsidies that the official exchange rate formerly bestowed, they will be forced to cut their imports.

On the other hand, simple economics may encourage strong growth in China's product imports. Low prices in world oil markets may make imports attractive enough on the margin to overcome other concerns, especially considering that some domestic retail prices are much higher than world market prices (f.o.b.). The price incentives for importing (and exporting) are very difficult to project, however, since almost all changes in government policy toward the oil sector have ramifications on these incentives, and government policies have been changing very rapidly.

The government is making efforts to limit imports by utilizing more of the products produced domestically. Under a central plan, which may have important implications for the long term, the government envisages moving products from the north of the country to the industrial heartland in the south instead of engaging in cross-trade. Transportation logistics and lack of storage capacity in the south will make it impossible to supply all of the oil needed. As a result, southern China will be forced to import oil.

In a short-term outlook, China's crude production will continue its pace of low growth, with about 2 percent growth pushing China's crude output above 3.5 million b/d for the first time. Refinery runs will depend on domestic demand and the regulation of imports. Two possible developments could prompt oil demand increases. Loosening import controls could boost supply in coastal areas; on the other hand, relaxing control on domestic product distribution could increase refinery runs over the coming years. If either one or both of these develop, total product demand will rise at an annual rate of 50,000 to 120,000 b/d.

Table 7.7, which summarizes our product balance projections in the long term, is based on the total demand forecast present in Chapter 5, and on the distillation capacity projections presented in Chapter 4. The products yield calculations, upgrading characteristics, and the operating assumptions in our database are spelled out in Chapter 4. We obtain a product yield for each year through 2010 on the basis only of the known projects. These volumes are then subtracted from our forecast level of product demand in Chapter 5. The "balance" numbers in Table 7.7 thus depict the simple difference between demand and supply. A positive number, such as China had from the late 1970s to 1992, means that gross product supply exceeded gross product demand, thereby giving the country an exportable surplus. Conversely, a negative number indicates a deficit of the product and a need to import, drawdown stocks, or raise refinery utilization or configuration.

The petroleum product balance shows that the product deficit, which commenced in 1993, will continue and be more serious than before. This vision of a long period deficit is somewhat misleading, however, since we project that demand will keep growing throughout the time period covered by the book, but we know about fewer and fewer refinery projects. After the year 2000, information about refinery construction is not very reliable, since even the few plans which do exist are subject to change and cancellation based on the assessment of the market over the next few years.

Table 7.7 presents what we call a "first look" at the product balances; it is based on our screening of all the information available on government and industry plans for refinery construction. But China will need

Table 7.7 Projection of principal petroleum product demand, "producibility", and balance

	Gasoline/naphtha			Distillate oil			Fuel oil			Others			Total products		
	Supply	Demand	Balance	Supply	Demand	Balance	Supply	Demand	Balance	Supply	Demand	Balance	Supply	Demand	Balance
1996	1076	1026	50	1055	1111	(56)	425	575	(150)	505	557	(52)	3060	3269	(209)
1997	1156	1074	82	1172	1261	(89)	405	628	(223)	642	701	(59)	3374	3663	(289)
1998	1154	1118	36	1188	1332	(143)	404	631	(227)	651	728	(77)	3398	3809	(412)
1999	1236	1165	71	1277	1407	(130)	422	636	(214)	707	754	(48)	3641	3962	(321)
2000	1266	1214	52	1317	1487	(170)	432	640	(208)	751	780	(29)	3766	4120	(354)
2001	1314	1257	57	1381	1559	(178)	444	645	(200)	786	816	(30)	3925	4277	(351)
2002	1333	1302	31	1410	1635	(225)	444	648	(204)	806	853	(47)	3994	4439	(446)
2003	1379	1349	30	1473	1715	(242)	455	652	(197)	837	891	(54)	4144	4608	(464)
2004	1452	1398	53	1560	1799	(240)	473	656	(184)	884	929	(45)	4368	4783	(415)
2005	1488	1450	38	1612	1888	(276)	477	660	(182)	900	967	(67)	4478	4965	(487)
2006	1520	1503	17	1655	1970	(315)	480	663	(183)	916	1007	(91)	4571	5144	(572)
2007	1555	1559	(4)	1705	2057	(352)	483	666	(183)	924	1047	(124)	4666	5329	(663)
2008	1618	1617	1	1780	2147	(367)	494	669	(175)	949	1088	(139)	4841	5521	(680)
2009	1673	1677	(4)	1852	2241	(389)	501	672	(170)	959	1130	(171)	4985	5719	(735)
2010	1687	1739	(51)	1872	2340	(468)	495	674	(179)	946	1172	(226)	5001	5925	(924)

Notes:
Supply is defined as product "producibility".
Distillate includes diesel oil, kerosene and jet fuel.
Other products include LPG, lubricant, wax, and others.

additional distillation capacity after 2000, and it will almost certainly be built. If the government wants to satisfy internal consumption with domestic refinery production — and it does — then from 1998 to 2005 China will need to have at least 2 million b/d more refinery capacity and achieve a utilization rate of above 80 percent.

Table 7.7 also presents our view of which product imbalances are most likely to occur. The light products appear to be most likely to develop a significant deficit, with an excess of demand over producibility. In contrast, heavy products may become "long" if more distillation capacities were built than we expected. Part of this stems from the fact that more is known about distillation capacity additions than about the upgrading units, which will surely accompany the new distillation columns. Thus, while it is impossible to define the extent of upgrading, the production of fuel oil in the years to come may be lower than our projections, while the supply of light products may be higher than we show.

7.2.2 Gasoline/naphtha imports/exports and balance

Since 1980, China has been a net exporter of gasoline, with volumes ranging from 25,000 b/d to 62,000 b/d. In addition, as noted above, the country remained a net exporter of gasoline until 1993 (see Table 7.8). In 1993, China's gasoline imports soared, exceeding the volume of exports for the first time since 1980. What happened in 1993 did not indicate a turning point for China's gasoline imports and exports. After 1994, gasoline changed from a net imported product to a net exported product again, evidence that cutting gasoline imports tends to be easier than dampening consumption of industrial and agricultural fuels.

China has exported gasoline to the US West Coast, Singapore, and other refining centers for blending, as well as to a variety of end-users throughout the Asia-Pacific region. In recent years, 30–65 percent of gasoline exports has moved to the spot market in Singapore. China started to export gasoline to South Korea in 1992, and exports peaked in 1995 to total 16,000 b/d, about 35 percent of total gasoline exports. Gasoline exports to the US have been relatively small and have dropped rapidly in recent years, falling from 8000 b/d in 1992 to zero from 1994 to 1996. This is due to the imposition of new reformulated gasoline regulations on the US West Coast. At present, China's gasoline exports cannot meet the stringent new requirements.

On the import side, the most striking statistic is the leap in gasoline imports from 8000 b/d in 1992 to 50,000 b/d in 1993 and 25,000 b/d in 1994. The major source of gasoline imports has been Singapore, accounting for 60–85 percent of total imports. The large volumes of Singapore gasoline flowing to China have made it a very important force in the

Table 7.8 Gasoline exports (000 b/d) by destination and imports by source

	1990	1991	1992	1993	1994	1995	1996	1997	1997's shares in total
Gasoline exports by destination:									
Singapore	31.6	41.2	43.8	30.6	32.3	21.0	2.2	12.3	29.7
S. Korea	–	–	6.1	8.6	12.4	16.3	14.9	9.0	21.7
Hong Kong	0.4	0.1	2.3	1.7	3.0	2.3	6.4	10.1	24.4
US	3.5	4.7	7.9	1.2	–	–	–	0.8	1.9
N. Korea	–	–	0.3	0.3	0.3	0.2	0.1	0.4	1.0
Burma	–	0.1	0.3	0.1	0.1	0.1	0.2	0.2	0.6
Japan	2.1	1.8	1.2	–	0.1	2.4	4.0	1.1	2.7
Australia	–	–	–	–	–	0.6	–	0.8	2.0
Russia	–	–	–	–	–	0.2	0.7	–	–
Vietnam	–	–	–	–	–	–	–	4.1	9.8
Saudi Arabia	–	–	–	–	–	–	–	1.4	3.4
Others	0.5	–	0.5	0.4	0.8	0.3	2.0	1.2	2.8
Total exports	38.1	47.9	62.4	43.0	48.9	43.2	30.5	41.4	100.0
Gasoline imports by sources:									
Singapore	2.8	1.6	5.7	31.0	14.2	1.9	1.6	1.7	86.0
Japan	–	–	–	2.8	0.9	–	–	–	–
S. Korea	–	–	–	1.7	1.4	–	–	–	–
Thailand	–	–	–	1.1	–	–	–	–	–
Philippines	–	0.4	0.4	0.8	–	–	–	–	–
Saudi Arabia	–	–	–	–	–	–	–	0.3	13.8
Others	0.1	0.1	0.3	2.8	0.5	0.1	–	–	–
Sub-total in Asia-Pacific	2.9	2.0	6.4	40.2	17.0	2.0	1.6	2.0	99.8
Russia	0.6	0.4	1.0	3.5	5.5	1.5	0.2	0.0	0.2
Romania	–	–	–	3.4	–	–	–	–	–
US	–	–	0.1	1.6	1.8	0.1	–	–	–
Others	–	–	0.1	2.2	0.2	–	–	–	–
Total imports	3.6	2.5	7.7	50.8	24.5	3.7	1.8	2.0	100.0

Source: Chinese Customs.

regional spot market. The slowdown in Chinese imports in 1994 has thus hit the Singapore market especially hard.

Naphtha demand for petrochemical plants boomed in the past decade in reaction to ongoing economic growth and capacity expansion. So tremendous was the increase in petrochemical capacity that — in an investment boom that may foreshadow that of refining — petrochemical margins went into a tailspin. China's naphtha demand skyrocketed, more than doubling during the last decade, from 130,000 b/d in 1987 to an estimated 300,000 b/d in 1997.

China imported naphtha of about 12,000 b/d in 1996 and 19,000 b/d in 1997. During the years from 1993 to 1997, naphtha imports remained a strong up trend. Although total product imports in 1994 fell by 20 percent, naphtha imports went up by about 50 percent, one of the highest growths of product imports in 1994. The imported naphtha was used as ethylene feedstock in some coastal refineries, such as Qilu, Yangzi and Maoming. Spurred by the ethylene production needs, naphtha imports in the near future will continue to rise, until domestic supply can meet the demand.

If we employ a somewhat constrained GDP/naphtha ratio of 0.5 (thus, a 0.5 percent increase in naphtha demand for each 1.0 percent increase in economic growth), our forecast generates an increase in naphtha demand from 300,000 b/d in 1997 to about 500,000 b/d in 2005, and 630,000 b/d in 2010. The projected averaging annual growth rate is 6 percent, compared with about a 8 percent growth rate during the last decade.

Table 7.8 shows the gasoline/naphtha balance over the next decade. It shows a range of surplus in the market, swinging from a maximum surplus of 70,000 b/d in 1999 to a small deficit in 2010. We believe that the absolute numbers in this calculation are less important than the annual change in the balance. The supply calculations assume somewhat constrained operating levels, and they contain simplifications about the crude inputs to distillation units.

The three most important messages in these calculations are that: (1) the surplus of gasoline and naphtha will not increase significantly from its 1997 level; (2) there will be no gasoline and naphtha deficit over the next decade; and (3) there will be a need for sizable additions to gasoline/naphtha production capacity beyond the assumption of our supply predication.

7.2.3 Middle distillate imports/exports and balance

China's demand slate is heavily weighted toward diesel oil. And, for five years (1993–1997), diesel has seen greater demand growth than major principal products. In contrast, diesel supply growth has trailed demand growth for many years due to capacity constraints and inflexible price controls.

China's refining capacity is characterized by large catalytic cracking capacity (equivalent to about 25 percent of distillation capacity). This refining structure favors the production of gasoline rather than diesel oil. Furthermore, such a high catalytic cracking ratio will not change easily in the near future, and this will restrict increases in diesel oil output.

Principal product prices, such as gasoline, diesel oil and fuel oil, are controlled. Ex-plant gasoline is priced at the equivalent of US$ 32 per barrel — $5–6 higher than the ex-plant diesel oil. There is no seasonal adjustment for these prices.

Low diesel prices are designed to stimulate agricultural development, on which China places a high priority in its economic agenda. But the gasoline–diesel price difference has only two consequences: it encourages refiners to produce more gasoline at the expense of diesel and causes consumers to use more diesel. These effects increase the diesel deficit.

Any attempt on the part of government to impose refinery production controls in order to reduce the diesel deficit would be unlikely to meet with success. In 1996, Chinese refiners produced 1.17 times more diesel oil than gasoline vs. a diesel–gasoline ratio of about 2.27 for Singaporean refineries. Whenever the market requires more diesel, refiners respond not by reducing gasoline yields, but by increasing throughput. This constrains diesel output.

In a bid to redress the mismatch, in 1996 China increased regulated diesel prices to align more closely with world prices and promote diesel production. But the price adjustment was too small to encourage refiners to increase diesel output much.

In many ways, 1996 was a nightmare year for China's diesel market, which faced one of its most serious shortages ever. Robust economic growth and a loosening of government restrictions stimulated diesel demand growth of 5 percent. And summer flooding increased demand from the transportation and industrial sectors, further increasing diesel demand. Even with higher domestic diesel prices, the drain in supplies at times made the fuel available only to government-designated end-users.

From the early 1970s to the mid-1980s, China was a net exporter of diesel oil. As Table 7.9 shows, diesel oil was the first refined product in surplus (over 10,000 b/d) in the mid-1970s, and exports ranging from 10,000 b/d to 46,000 b/d lasted almost 20 years. Meanwhile, diesel oil imports rose at a dramatic rate (averaging 20 percent per year after 1990) to reach 180,000 b/d in 1993 and 150,000 b/d in 1997. In addition to healthy demand growth, transportation bottlenecks prevented product from flowing north to south and made importing diesel fuel more attractive in the south.

Due to a surge in world diesel prices, China's 1996 diesel imports were less than expected. Despite strong demand, imports fell short of 1995 levels.

Chinese industry experts estimated that more than 80,000 b/d of diesel import quotas went unused during the first half of 1996, when Singapore diesel prices averaged $25.66/bbl — almost $4/bbl higher than the same period in 1995. Correspondingly, China's cost, insurance, and freight price for imported diesel in the first half of 1996 also jumped $2.50/bbl from 1995 levels.

After tariffs and taxes, the price of imported diesel oil rose to about $33.90/bbl — $3.20/bbl higher than the domestic ex-plant prices. In the Guangdong province the diesel wholesale price surged as high as $40/bbl.

Table 7.9 Light diesel oil exports (000b/d) by destination and imports by source

	1990	1991	1992	1993	1994	1995	1996	1997	1997's shares in total
Light diesel oil exports by destination:									
Singapore	14.0	9.8	10.4	13.6	10.1	1.5	0.2	1.7	3.6
Hong Kong	9.5	9.0	9.9	6.4	4.5	15.6	22.3	20.6	43.9
S. Korea	–	0.2	0.5	2.4	5.2	5.2	4.5	1.2	2.5
Japan	5.3	2.8	2.0	1.5	2.9	0.4	1.0	2.8	5.9
Thailand	–	–	3.0	0.6	1.1	0.5	0.3	0.1	0.2
Philippines	0.4	–	1.3	0.4	0.1	0.6	–	1.0	2.1
Macao	1.2	1.6	0.9	0.3	–	–	0.3	0.9	1.8
N. Korea	–	0.1	0.9	0.3	0.4	0.2	0.3	0.7	1.5
Burma	–	0.1	0.1	–	0.1	0.1	0.1	0.2	0.5
Vietnam	–	–	–	–	–	–	–	6.9	14.6
Taiwan	–	–	–	–	–	–	–	3.9	8.3
Russia	–	–	–	–	–	–	–	1.2	2.6
Others	0.8	–	0.4	–	0.3	2.5	3.2	5.8	12.4
Total exports	31.3	23.6	29.4	25.6	24.6	26.5	32.1	46.8	100.0
Light diesel oil imports by sources:									
Singapore	28.4	36.6	41.1	85.8	77.9	68.1	70.8	85.8	56.9
S. Korea	–	0.2	3.5	21.3	8.9	18.0	6.4	37.2	24.7
Japan	4.2	3.3	6.6	10.4	5.6	12.9	5.1	12.5	8.3
Indonesia	–	–	0.9	1.0	0.1	–	–	–	–
Philippines	–	0.1	–	0.2	0.8	0.5	0.1	–	–
Hong Kong	–	–	–	–	–	–	0.6	–	–
Saudi Arabia	–	–	–	–	–	–	–	1.8	1.2
Others	2.2	1.6	4.0	4.7	1.1	2.1	0.5	4.3	2.8
Sub-total in Asia-Pacific	34.8	41.8	56.1	123.3	94.4	101.5	83.6	141.6	93.9
US	0.4	7.5	23.7	27.2	10.6	3.3	1.0	2.5	1.7
Russia	2.8	8.1	11.1	15.6	12.7	15.0	8.8	5.5	3.7
Mexico	–	–	0.7	4.1	0.3	–	–	–	–
Canada	–	0.1	1.4	2.5	–	0.4	–	0.1	0.1
Romania	–	–	–	1.3	0.6	–	–	–	–
Italy	0.6	–	–	1.1	2.4	–	–	0.5	0.4
Saudi Arabia	–	–	–	–	–	–	0.3	–	–
Others	1.5	0.9	1.3	2.5	2.3	1.9	0.4	0.5	0.3
Total imports	40.2	58.5	94.2	177.6	123.3	122.2	94.1	150.7	100.0

Source: Chinese Customs.

Guangdong imported the most diesel oil. (Most users of imported diesel are Sino–foreign join ventures.)

When Singapore diesel prices were lower than domestic prices, joint ventures imported diesel to cut fuel costs and generate profits by resale. It happened in 1994–1995 and 1998. With high world market prices, there is no incentive for them to do so.

As can be seen, China's diesel imports depend not only on industrial and economic activity, but also on the price differential between the domestic and world markets. If world market prices plus tariffs and taxes are less than controlled prices, demand will encourage imports and smuggling. Otherwise, diesel imports will be sluggish despite strong demand.

Given domestic prices, imports are feasible only when the Singapore diesel price is less than $23/bbl. Like all refiners, those in China will respond to price signals rather than domestic demand. As a result, imported diesel will take more market share in China when Singapore prices are low.

This situation will not change unless much additional refining capacity comes on stream early next century. In the short term, unless there is a shift in oil policy, the diesel market will remain unbalanced.

Several Asian countries favor Chinese low-sulfur diesel oil. While exporting 0.2 percent sulphur diesel, China also imports a lower grade of diesel oil with a maximum of 0.5 percent sulphur. Before 1994, about 50 percent of total diesel oil export flowed to the spot market in Singapore. China started to export diesel oil to South Korea in 1991, but the absolute volume is rather small, averaging only 5000 b/d from 1994 to 1996. Diesel oil exports to Hong Kong averaged about 10,000 b/d during the period 1989–1992 dropped in 1993 and 1994, then rose again to more than 20,000 b/d in 1996 and 1997.

The bulk of diesel oil imports come from Singapore refineries, which account for more than 50 percent of total diesel imports. The preeminence of Singapore among sources of diesel oil imports became very clear when imports doubled in 1993. Other major sources of diesel oil were the US and Russia before 1994, but were substituted by South Korea and Japan after 1995. After 1995, diesel oil imports from South Korea and Japan rose significantly, and started to increase their Chinese market shares. Given the lack of clean product transport and storage infrastructure in China, the diesel trade record suggests that coastal regions are the major diesel oil-consuming areas in China.

As of 1997, China's distillate deficit was about 90,000 b/d, the second largest deficit of any refined product. Our projection of future increases in demand versus increases in "producibility" forecasts a sustainable deficit over 10 years. Thus, since domestic distillate production will not meet the projected distillate demand, robust distillate demand will continue to necessitate substantial imports. China will badly need distillate to fuel its domestic industry and transportation sectors in the next few years. China's distillate balance will be hard to experience any period of surplus during the next decade. After 2003, the deficit in distillate balance will widen. Demand keeps growing, but the pace of supply growth slows down in the absence of refinery construction knowledge. Our refinery yields assume that Chinese refiners will seek to maximize distillate yields, since the distillate deficit will be the largest of any fuel. Even in a more optimistic scenario, where refinery capacities reaches 4.9 million b/d by 2000 and 5.7 million b/d by 2005, it would be difficult to meet domestic demand for distillate without substantial imports, even if refinery distillate yields are increased through additional cracking capacity additions.

7.2.4 Fuel oil imports/exports and balance

China has been a net importer of fuel oil since the early 1980s. Fuel oil imports have continued to rise during recent years, reaching 12,000 b/d in 1990 and 230,000 b/d in 1997 (Table 7.10). On the other hand, fuel oil

Table 7.10 Fuel oil exports (000 b/d) by destination and imports by source

	1990	1991	1992	1993	1994	1995	1996	1997	1997's shares in total
Fuel oil exports by destinations:									
Japan	6.6	4.7	5.5	2.6	0.6	–	1.1	0.1	0.8
Hong Kong	3.9	3.9	2.6	0.8	0.5	1.5	1.5	1.8	25.7
Russia	–	–	–	–	–	–	0.5	–	–
Panama	–	–	–	–	–	–	1.0	1.9	27.1
Singapore	–	–	–	–	–	–	–	0.7	9.5
Liberia	–	–	–	–	–	–	–	0.4	5.3
Others	–	0.1	0.1	–	–	1.8	1.0	2.2	31.6
Total exports	10.5	8.7	8.1	3.4	1.2	3.3	5.1	7.0	100.0
Fuel oil imports by sources:									
Singapore	10.9	18.4	21.5	42.9	46.9	68.0	88.3	70.4	30.4
S. Korea	–	–	4.8	16.8	13.5	23.3	34.6	62.7	27.1
Japan	–	0.6	2.6	3.4	3.1	4.0	15.9	33.1	14.3
India	–	–	0.7	1.1	0.2	–	–	–	–
Thailand	–	–	–	–	–	–	0.7	–	–
Taiwan	–	–	–	–	–	–	1.7	–	–
Philippines	–	–	–	–	–	–	–	4.4	1.9
N. Korea	–	–	–	–	–	–	–	2.1	0.9
Others	0.6	0.7	0.6	1.3	2.1	3.1	2.0	5.7	2.4
Sub-total in Asia-Pacific	11.6	19.7	30.1	65.5	65.8	98.4	143.2	178.2	77.1
Russia	–	0.8	0.6	4.7	5.2	8.3	8.5	52.2	22.6
US	–	0.7	0.1	0.3	0.4	1.8	0.9	0.8	0.3
Others	–	–	0.1	–	1.2	1.1	3.1	–	–
Total imports	11.6	21.2	30.9	70.7	72.6	109.5	155.8	231.3	100.0

Source: Chinese Customs.

exports fell at the same time from 11,000 b/d in 1990 to 7000 b/d in 1997. Overall, net fuel oil imports have risen significantly from 1000 b/d in 1990 to 220,000 b/d in 1997. Before 1996, the destinations of nearly all of China's fuel oil exports were Japan and Hong Kong. Imports too have been concentrated among a very small number of countries; China receives about 80 percent of its fuel oil imports from Singapore, South Korea, and Japan and as much as 20 percent of its fuel oil imports comes from Russia.

In contrast, as new refineries in Japan and South Korea have increased runs, the product surplus (especially fuel oil) in Japan and South Korea has been targeted toward China's market. Japan and South Korea have competed with Singapore to sell refined products in northern Asia. During recent years, China's product deficit has consisted mainly of light diesel and fuel oil. Given the rapidly growing demand in China's southern regions, product imports from Singapore have taken great advantage of easier geographic access and superior transportation systems. Light diesel and fuel oil from Singapore have made up more than half of China's total imports in recent years. In addition, the volumes of light diesel and fuel oil that were shipped from South Korea and Japan to northern China also have increased steadily. For example, 1997 statistics indicate that China received 60,000 b/d of fuel oil from South Korea, compared to 13,000 b/d in 1994, and 40,000 b/d of light diesel thus far, compared to 9000 b/d in 1994.

In our fuel oil balances, residual fuel oil, like middle distillates, will not avoid a net shortfall for most of the next 10 years. This assessment comes as no surprise, given the rate at which China's refining upgrading system will grow. Our experience with markets worldwide shows that increases in "producibility" in excess of increases in demand translate into bearish developments for refiners. If coal supplies cannot consistently meet demand, increased domestic use of fuel oil as a direct fuel could be well justified by low prices. In the meantime, we estimate China's fuel oil deficit to stay at about 200,000 b/d before 2003. Then, the fuel oil deficit will drop to below 200,000 b/d, and reach the level of 180,000 b/d in 2010.

At this point, however, we must take leave of the "first look" based on information about refining and upgrading projects available as of 1997. As mentioned previously, the refinery construction projects known are weighted heavily toward distillation capacity and include only a fraction of the upgrading units they will eventually have. This is unavoidable, since the size and nature of upgrading units is often left out of initial refinery plans. Therefore, when forecasting refinery production, we have obtained only a "first look" at output, containing all known information. As expected, this shows large deficit of distillate oil in China. However, many of these refineries will be built with more upgrading capability than is being publicized. Accordingly, in our study we also take a "second look" at the ramifications for individual product balances of the addition of various amounts of cracking capacity.

How will China deal with distillate oil producibility, if it is unable to overreach growth in demand? How much upgrading capacity will actually be added? The oil industry never has a definitive answer to this question. Economic theory dictates that if demand of a product increases faster than supply in a free market, prices will rise. In China, this would mean that distillate oil prices would climb. In theory, this would continue until the price differential expanded enough to justify the addition and operation of cracking units.

Will this take place in China? This question is extremely difficult to answer for two reasons. The first is that the amount of cracking capacity added around the turn of the century depends on the assessment of future, not present, market conditions. If a glut of refined products causes poor simple refinery margins for a number of years, this might dampen enthusiasm for building new refineries and boost cracking capacity additions. The second reason that the theoretical market mechanism is perhaps inappropriate to apply in estimating the extent of secondary unit construction after the year 2000 is that the Chinese petroleum complex is not truly a market. In China, the government still makes the oil market — if not directly, then through price controls and trade restrictions, then at least through supervision and guidance. And although liberalization programs will affect the oil industry, we do not see the industry achieving substantial deregulation over the next five to 10 years. Indeed, we believe that the oil sector will be one of the last sectors to be liberalized. Of course, the government does have an incentive to encourage upgrading units, if only to limit the cost of imports and to produce higher value fuels. And China's level of complexity is already one of the highest in Asia. But, as we have seen in other Asian countries, matching product supply and demand is only one of the goals of government energy policy.

If significant cracking capacity expansions join the refinery construction queue, however, they might prevent the reappearance of the large deficits in distillate oil and widen the existing deficit of fuel oil. We have no doubt that there will be more upgrading investments than have been announced thus far. Nevertheless, it is useful to look at the effects of various investment programs, given the demand trends that we believe exist in China. On the basis of a cost of $10,000 per daily barrel of installed capacity, this represents an investment requirement of $3.6 billion (current value). Where, when, and how much of this investment will be made? That depends on how the Chinese government defines the rules of the game.

- Singapore is the major exporter of asphalt to China. The countries which receive most of China's asphalt exports, are North Korea, South Korea, Japan, and the US
- Between 1983 and 1993, the growth of asphalt production in China averaged 8.4 percent, compared to refinery output growth of 9.2 percent during the same period.
- The largest asphalt producers in China are located in the east. Three of them: Qilu Petrochemical, Thick Oil Processing Plant in Shengli Oilfield, and Yangzi Petrochemical accounted for about 40 percent of total national production in 1996.

During recent years, China has been rapidly developing its highway system, even though the total mileage is still limited. In 1996, the total distance of paved highways was four times the level of 1992, while other graded roads have had only moderate growth. Up to the late 1990s, China has had very low road density (kilometer per million persons). If we include the second- and third-class roads as part of the total for the paved roads, China's road density was about 250 kilometers in 1996, while Thailand's was 510 kilometers (in 1993) and Brazil's was 700 kilometers (in 1988).

7.3 Can China Afford Import Expenses?

The export of crude oil used to be a major source of hard currency during the 1980s. The amount of exported crude was equivalent to more than one-quarter of the country's export earnings in 1985, before the collapse of oil prices. In 1996, as Table 7.11 shows, oil dollars amounted to less than 2 percent of the annual export revenue, as other Chinese commodities have entered the world market.

In 1996, crude and product imports consumed $6 billion in foreign exchange — the highest price tag ever (see Table 7.11). China imported virtually its entire demand increment. The petroleum "problem" reinforced an already worrisome situation: in 1993, China appeared to be headed for a $12 billion trade deficit, its first since 1989, compared with a surplus of $9 billion in 1991 and $4 billion in 1992. The government's response, which we see as an important indicator of both its control of the oil sector and its fear of losing that control, was to impose an import ban on petroleum products and price changes. These tactics, coupled with a few other measures in the first half of 1994, quickly returned China to a net trade surplus. This basic pattern is likely to recur under different disguises over the years to come.

Another interesting finding concerns the average earnings of crude exports and the cost of crude imports. The price of crude imports has

Table 7.11 Export earnings and import expenses of crude and petroleum products

	1989	1990	1991	1992	1993	1995	1996
Crude exports							
Volumes (000 b/d)	488	480	452	433	391	376	376
Value (m US$)	2750	3402	2957	2774	2409	2239	2777
As a % of total exports	3.0	5.5	4.1	3.3	2.6	1.5	1.8
Crude imports							
Volumes (000 b/d)	66	59	120	229	315	341	341
Value (m US$)	467	424	926	1724	2323	2356	3407
As a % of total imports	0.4	0.8	1.5	2.1	2.2	1.8	2.5
Product exports							
Volumes (000 b/d)	101	112	103	119	80	83	85
Value (m US$)	758	877	823	865	629	792	883
As a % of total exports	0.8	1.4	1.1	1.0	0.7	0.5	0.6
Product imports							
Volumes (000 b/d)	114	68	99	164	354	288	338
Value (m US$)	969	610	902	1394	2985	2065	2612
As a % of total imports	0.9	1.1	1.4	1.7	2.9	1.6	1.9
Total oil exports							
Volumes (000 b/d)	589	592	555	552	471	459	461
Value (m US$)	3508	4279	3780	3640	3038	3031	3661
As a % of total imports	3.8	6.9	5.3	4.3	3.3	2.0	2.4
Total oil imports							
Volumes (000 b/d)	180	126	219	393	669	629	679
Value (m US$)	1435	1034	1828	3118	5309	4422	6018
As a % of total imports	1.4	1.9	2.9	3.9	5.1	3.3	4.3
Total exports (m US$)	91,763	62,090	71,840	84,940	91,740	148,780	151,060
Total imports value (m US$)	103,950	53,350	63,790	80,590	103,960	32,080	138,840
Average crude export price ($/bbl)	15.45	19.43	17.92	17.67	16.98	16.27	18.71
Average crude import price ($/bbl)	19.48	19.75	21.11	20.65	20.21	18.76	20.49
Average product export price ($/bbl)	20.49	21.38	21.93	20.58	21.68	24.48	26.57
Average product export price ($/bbl)	23.26	24.76	25.08	23.27	22.01	18.40	19.81

Sources: Chinese Customs, and *China Statistical Yearbooks, 1990–1997*, China Statistical Publishing House.

been higher than the cost of exported crude during the year from 1993 to 1997.[4] Except for 1990, the price differences between crude exports and imports have been very high, ranging from $2.50/bbl to $4.00/bbl. This suggests that China has not taken the profits that it might have garnered

[4] Crude export price is f.o.b. and crude import price is c.i.f.

by selling its light sweet crudes (e.g. Daqing) and buying lower-grade crudes. Thus, crude exports and imports appear not to be based on considerations of profitability, but rather respond to the shorter-term needs of end-users in different locations.

What is China's crude import policy, and how seriously does the government take the volume of imports? The former president of CNPC, Wang Tao, indicated, "As oil is vital for the national economy, it is only necessary and normal to import an appropriate amount of crude to keep up with the need." But how much is an appropriate amount? Mr. Wang declines to identify an explicit amount.

One of the key assumptions in developing our outlook is the trend in oil prices. Clearly, the future path of crude prices will depend upon a variety of factors including the rate at which crude production capacity in OPEC and non-OPEC expands. The important point here is that lower oil prices means generally low import costs for crude. This will help countries such as China that are somewhat sensitive to foreign exchange limitations. However, as a oil importer, China benefited from the low oil prices. When world oil dropped in 1998, China emerged in a much stronger position, and saved huge oil dollars. The low world oil prices will stimulate China's oil imports in the future.

We envision that China could spend huge sums on crude imports — as much as $8–10 billion (in current dollars) per year by the turn of this century. Can China afford this? We believe so, but only under two conditions: (1) high average economic growth must occur; and (2) exports from other commodities must continue to boom. Otherwise, the only way to curtail imports is to cut demand, which will, in turn, cut into economic growth. In our demand forecast, we did not take a severe supply shortage into consideration, because this has not occurred during the past 20 years (the time span upon which we built our empirical demand forecast models).

Another factor that could disrupt the oil balance is a depreciation of Chinese currencies against the US dollar. Constant, high inflation would result sooner or later in the depreciation of the Chinese yuan, making oil imports extremely expensive. China's leaders are vigilant and have demonstrated a willingness to slow economic activity when inflation alarms sound. The greatest potential for uncontrolled or uncontrollable inflation, therefore, probably lies in the potential confusion that may emerge with the passage of power from the present generation of Communist leaders. Until that time, we expect inflation to be confined to large cities and to be carefully watched.

To help pay for imports and to hedge some of the country's price vulnerability, China has decided to establish a risk fund for crude oil and product imports. This fund is similar to the oil risk fund used in other countries. The risk fund is built up from price differentials between import costs and regulated wholesale prices. When domestic wholesale prices exceed import costs, the margin will replenish the risk fund; when they are lower than import prices, the risk fund will be drawn down to make up the margin. Local pricing administrations have been assigned to handle fund management and supervision. The high wholesale prices set by the central government may help to start such a fund. However, it is not clear how the fund can work if price liberalization is brought back again.

An ongoing question for China's decision-makers will be the precise trade-off between importing crude for domestic processing capacity and importing refined products for direct marketing. It is estimated that one ton of crude creates at least 10 times more value in China if it is used to produce petroleum or petrochemical products. It can also be expected that revenues from growing exports of value-added petrochemicals will exceed outlays for rising crude and feed imports over time, reducing the drain on government hard currency coffers. From an economic point of view, therefore, it makes more sense for China to import crude and produce its own refined products.

Nevertheless, China's economy will suffer from periodic shortages of oil. This will encourage the government to continue its product-allocation system and/or state-imposed pricing mechanisms to balance supply and demand.

7.4 Regional Oil Balance[5]

Oil production and demand data for 1995 reveal the regional oil balance in China, and indicate the possible oil flow among the provinces. The regional oil dynamics in China can be characterized as an increasing crude deficit in the east, and a growing diesel oil and fuel oil deficit in the south, primarily in Guangdong province. China's regional oil imbalances have exacerbated the difficulties of oil transportation and distribution for the country's central planners and provided opportunities for foreign investors.

In the *crude* balance, only two regions within China have a crude surplus: the northeast and the west (Fig. 7.4). The northeast, with Daqing oilfield, is the primary crude supplier in the country, exporting about 700,000 b/d of crude per year over the past 10 years. These surplus barrels are available for exporting to Japan and North Korea, and for feeding the refineries in the north and the east of China.

[5] We follow the regional geographic division given by the Chinese government, which divides 30 provinces and metropolitan areas into six regions: the north, the northeast, the east, the central & south, and the west.

Xinjiang, which covers the Tarim basin, is the only place which significantly increased its crude oil surplus during the years from 1991 to 1995. However, the volume of crude available for other regions is still relatively small (between 40,000 b/d to 100,000 b/d per year). Considering the increasing need for crude oil in neighboring provinces, such as Gansu province, the total availability of crude for the rest of country rose at a slow rate.

The two crude oil deficit regions are the east and the south-central. The east includes large consuming cities and provinces, such as Shanghai, Jiangsu, and Zhejiang, and the second largest oil producer (Shengli) of Shandong province, while the south-central includes the economically booming province of Guangdong. The crude deficit in the east has grown much faster than the south-central over the past few years, due to reduced production in the Shengli oilfield and refinery expansion in the region. In contrast, the crude oil deficit in the south has remained around 200,000 b/d since the early 1990s. Slow refinery expansion and rapid increases in offshore oil exploration in the region helped to ease the crude oil requirement.

The southwest was self-sufficient; there is little crude production and a small refinery. China's internal pattern of crude flow is not expected to change in the near future: crude from offshore E&P may offset the

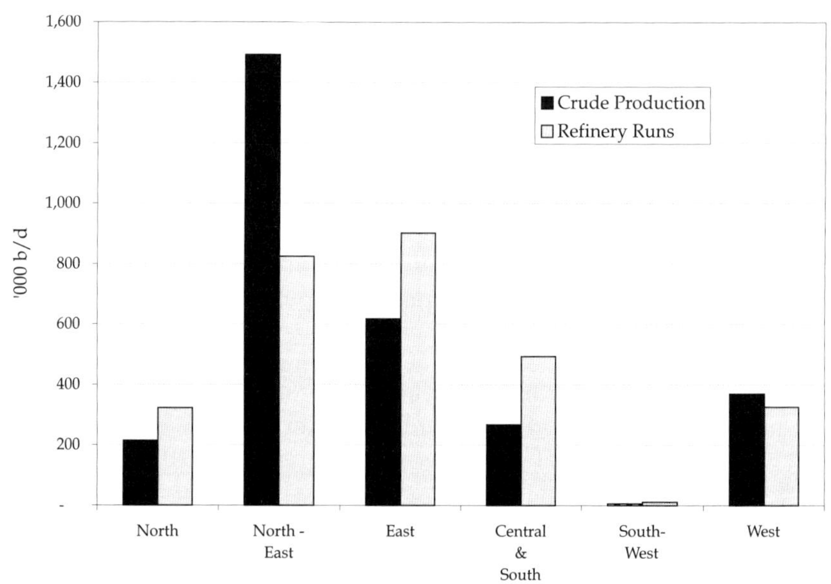

Fig. 7.4 Crude production and refinery runs by region in 1995

decline of crude output in the eastern China, and crude supplies for existing and new refineries in the eastern and southern coastal regions will depend increasingly on the world oil market, especially if domestic supplies fail to increase or are blocked by transportation bottlenecks.

The future of the regional crude oil balances will depend much more on refinery expansion and debottlenecking than changes in crude oil production. This is because the annual ups and downs of crude oil production will appear marginal, but additional refinery capacity in the region would quickly alter the oil balance.

If the refinery expansion plan (including additions of some joint-venture refinery and debottlenecking, as well as revamping existing facilities) in the east and the south can be fulfilled, the crude oil deficit in these areas will rise dramatically. The world crude oil market should keep an eye on these Chinese regions.

With the rapid growth in crude production, Xinjiang in western China is still primarily self-sufficient. If and when small refineries on the periphery of oilfields in Xinjiang achieve their expansion targets, they should absorb additional barrels being developed in the region. This could further limit Xinjiang's ability to ship more oil to central and southwestern China.

On the *refined product* side, the regional pattern of petroleum products as a whole is similar to that for crude oil, since the government has established refineries near large crude production areas. However, the refinery competitive advantage in the north will be reduced when other capacity goes on stream in south and east China (Fig. 7.5).

As the economy in the south (particularly Guangdong province) continues to run on the fast track, the rising shortage of diesel and fuel oil will require more imports. For the other regions, the product market could be generally balanced or move to balanced during the coming years.

The northeast, which holds one-quarter of the country's distillation capacity totaling 1.1 million b/d in the three provinces, has the greatest surplus in principal petroleum products, while the north and the south-central have the largest deficit. During the years from 1991 to 1995, the south-central has had the most dramatic change in its product supply and demand balance, and its deficit in principal products tripled from 50,000 b/d to more than 150,000 b/d. The southwest, which includes the most populated province, Sichuan, has relatively low demand, but is also short in all products due to limited refinery capacity there.

Gasoline (including MTBE (methyl *t*-butyl ether)) and diesel oil are two major products of which China has a surplus in the northeast region. With the limitations of domestic transportation infrastructure and high transportation costs, the destination for these surplus products has not

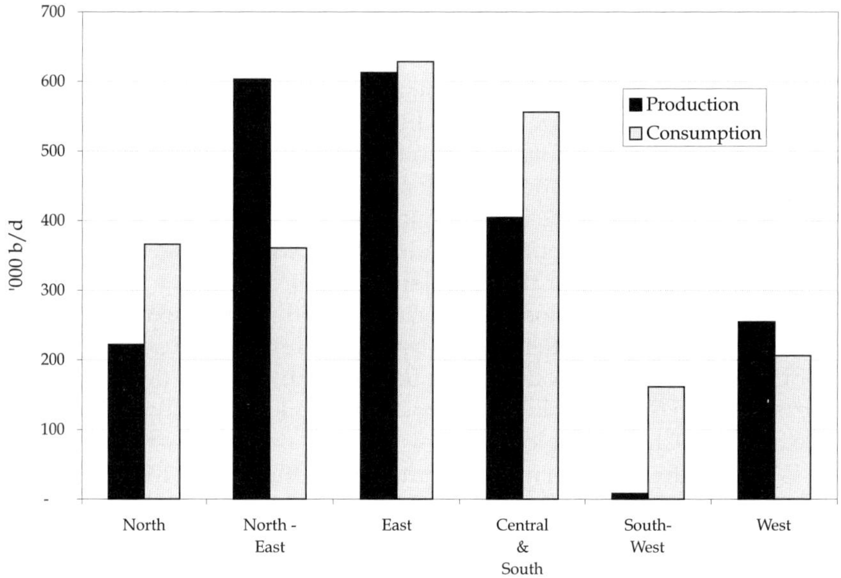

Fig. 7.5 Production and consumption of principal petroleum product by region in 1995

been the product-consuming regions of southern China, but the northern Asian countries: Japan, South Korea, and North Korea. Although the amount of these product exports did increase slightly during the past few years, we do not believe that China will become a major product exporter to north Asia because of the demand growth in its own northern regions. Still small compared to the product export potential of Japan and South Korea, China is unlikely to become their competitor in the near future.

In terms of individual products, the patterns mentioned above apply to gasoline and diesel, but not to kerosene and fuel oil. *Gasoline* supply and demand is more balanced in the major consuming regions, as regional supplies basically meet the regional demand. The largest gasoline-consuming province is Sichuan, which has the largest population. Since there is no sizable refinery in Sichuan, a surge in gasoline demand has caused the demand/supply gap to widen over the last decade. In contrast, gasoline demand in the eastern and southern coastal provinces has kept only slightly ahead of supply, and the market remains balanced to tight.

The *diesel oil* market in China is much more imbalanced than gasoline. Diesel oil deficits are seen in the north and the south-central, where Beijing and Guangdong are both seriously running in the red. The diesel oil deficit overwhelmed the long-term constant surplus in the northeast and the west. The north's requirement of diesel oil could be easily supplied by its neighbor, the northeast. The south-central has to rely on foreign diesel oil imports. Guangdong province is the biggest consumer in the country. In 1995, Guangdong consumed 130,000 b/d of diesel oil, accounting for 15 percent of the national total. The diesel oil deficit jumped to 60,000 b/d in 1995 from only 7000 b/d in 1990, as Guangdong's economy grew at an annual rate of about 15 percent, the highest growth among the regions.

In 1995, the north was self-sufficient in *kerosene*, while the northeast and the south-central had a kerosene surplus to supply the markets in the east, the southwest and the west. Since kerosene demand and production were small relative to other products, inadequate supplies in some regions were not as serious as for other products.

In recent years, *fuel oil* has emerged as a tight product, as fuel oil continued to require imports into China. Except for the west and the north, which could essentially meet their fuel oil demand by themselves, all other regions needed fuel oil imports from the world market. As a result, China's net fuel oil imports were about 110,000 b/d, accounting for 30 percent of total refined product imports in 1995.

Fuel oil as a major shortage appeared in Guangdong. While other regions' fuel oil is balanced to a bit tight, the expansion of oil-fired power plants and industrial boilers in Guangdong will determine how much fuel oil China will import. In 1995, Guangdong itself accounted for 18 percent of national total fuel oil demand compared with only 6 percent of the national total population.

Chinese statistics show that oil imports are highly concentrated in certain regions. Guangdong province has dominated product imports because its product demand is still constrained by the availability of supply from the province's insufficient refining capacity. In 1995, Guangdong, with 6 percent of the national population, imported about 40 percent of crude and more than 50 percent of refined products (40 percent of gasoline, 50 percent of light diesel, and 65 percent of fuel oil) of the national total. The total product imports in Guangdong were about 100,000 b/d in 1995. The remainder flowed to its neighboring province, Guangxi, a relatively poor and undeveloped province (Guangdong takes responsibility to supply refined products under the central allocation system).

The product demand and supply forecasts by region are heavily reliant on the assumptions of sustainable economic growth and economic policy, which determine the local investment environment. Guangdong and Shanghai have enjoyed special treatment given by the central government, with the result that consumer spending and business capital investment are far more robust than in any other regions. The impact of the regional imbalance in oil will be upward price pressure in consuming areas, and a continuing rise in imports. If consumption remains ahead of supply in major consuming regions, it will accelerate the process of building new refineries, and raise profit margins for investors. What does this mean for potential investors? First, since the pattern of product consumption varies among regions according to the industrial structure, new refineries must be careful to configure themselves appropriately. Second, transportation infrastructure is not expected to improve significantly in the near future, so regions with a shortage in refining capacity are unable to get real help from surpluses in the northeast and the west, both of which have some export options. Third, a sound assessment of government policy toward the Special Economic Zones (SEZ) is critical, since the regions that are or will be defined as SEZs have demonstrated great potential for oil demand increases.

7.5 Conclusion

China's crude oil production is unable to keep up with its rising crude demand, and this is having a dramatic impact on its balance of trade in crude oil and petroleum products. Chinese exports of crude oil reached 24 percent of total production in 1985, but in the past 10 years crude oil exports have fallen from 600,000 b/d to only 400,000 b/d, while crude imports have exploded from 5000 b/d to 700,000 b/d. In 1996, China became a net importer of crude oil, and net import could reach 1 million by 2010 — a major stream in world markets.

Energy substitution in the future will also have a great impact on domestic oil balance. China is studying the feasibility of importing natural gas from Kazakhstan and other central Asian states. All such developments, together with emergence of LNG import terminals, would do much to alleviate the pressure of rising domestic demand for imported crude oil.

China has been a net importer of refined products since 1989, and imports of all major products surged in both 1993 and 1997. Growing demand from the domestic market diverted some products available for export. If China is to satisfy internal consumption with domestic refinery production, then by 2005 it must have at least 5.6 million b/d of refinery distillation capacity and a more than 77 percent utilization rate. Much will depend on the types of upgrading equipment included in this new capacity. Through the near future, China will continue to be a net gasoline exporter and a diesel oil and fuel oil importer. This pattern reflects product balances that vary geographically. Over the next 10 years, distillate deficits will be the largest, while gasoline/naphtha will remain constantly small.

Chapter 8
China's Impact on the Asian Market

One of the most remarkable features of the Asia-Pacific region[1] in recent years has been its rapid economic growth, coupled with its strong petroleum product demand growth. From 1970 to 1996, the area's economic activity increased more than fourfold, compared to the doubling of the world economy as a whole. Over the same period, oil demand in Asia-Pacific has also increased fourfold, while global total oil demand has increased by a mere 40 percent. This Asia-Pacific growth has found support mainly in the rising economic power of Japan and accelerating industrialization in the "four dragons" (South Korea, Taiwan, Hong Kong, and Singapore), as well as in the unprecedented opening of China's economy.

As a nation still in the process of developing a basic industrial infrastructure, China will experience a pattern of economic growth and rising living standards similar to what we saw in Japan in the 1960s, and in Taiwan and South Korea in the 1970s. This, in turn, will cause China to emerge as *the* major consumer of refined products in the Asia-Pacific region, with obvious implications for both crude and product trade. Among Asian countries in 1997, China ranked third in crude oil exports (following Indonesia and Malaysia) and fifth in crude oil imports. Compared with Japan, which must import virtually all of its refinery feedstocks, China cannot yet be considered a very strong trading power. The volume of its foreign trade in 1997 was about 1.1 million b/d (total exports plus imports), comprising about 10 percent of the Asian-Pacific total. Within the Asia-Pacific region, China lags far behind both Japan and South Korea in trade volumes of oil.

However, the growth trends of the past portend a future in which China will become increasingly vital to both regional and global oil trades. It is true that exports, in both absolute and net terms, are shrinking. China's share of total crude exports from Asia-Pacific countries has

[1] In this book, Asia-Pacific region does not include countries in the Middle East or the US West Coast.

fallen from 40 percent in 1985 to 15 percent in 1997. But, as detailed in Chapter 7, China's imports of crude oil explode in the coming years, as the country adds large volumes of crude distillation capacity. International crude markets are thus likely to become somewhat dependent on Chinese demand.

This chapter assesses China's impact on Asia's crude and product market. First, we address Asia's overall crude balance and trade flows. The projected Chinese crude balance over the next 10 years will be integrated

Table 8.1 Oil reserves and production in major countries

	Country	Estimated proved oil reserves (million bbl)	R/P ratio (year)		Country	Producing oil wells Dec. 31, 1996	Estimated oil production 1997 (000 b/d)
1	Saudi Arabia	259,000	83.4	1	Saudi Arabia	1400	7900
2	Iraq	112,500	>100	2	US	572,440	6400
3	Kuwait	94,000	>100	3	Russia	104,150	5900
4	Iran	93,000	69.1	4	Iran	1090	3610
5	UAE	92,200	>100	5	China	72,255	3244
6	Venezuela	71,669	57.5	6	Venezuela	14,694	3175
7	Russia	48,573	22.1	7	Norway	539	3175
8	Mexico	40,000	42.7	8	Mexico	3701	3025
9	Libya	29,500	56.4	9	UK	1591	2620
10	US	22,017	9.7	10	Nigeria	2020	2270
11	China	21,300	18.4	11	UAE	1200	2000
12	Nigeria	16,786	19.9	12	Canada	47,093	1880
13	Norway	10,422	9.3	13	Kuwait	792	1825
14	Algeria	9200	19.4	14	Libya	1869	1415
15	Kazakstan	5417	47.7	15	Indonesia	8527	1380
16	Angola	5412	20.9	16	Oman	2100	895
17	Oman	5239	15.8	17	Algeria	1273	850
18	UK	5003	4.6	18	Brazil	7349	845
19	Indonesia	4980	8.6	19	Argentina	12,564	835
20	Canada	4839	9.4	20	Angola	509	715
21	Brazil	4800	16.5	21	Egypt	1200	675
22	India	4340	16.5	22	India	3437	647
23	Yemen	4000	30	23	Malaysia	804	645
24	Malaysia	3900	15.4	24	Colombia	2905	635
25	Egypt	3835	11.4	25	Qatar	280	608
26	Qatar	3700	22.5	26	Australia	1205	570
27	Azerbaijan	1178	>100	27	Syria	964	560
28	Uzbekistan	594	10.9	28	Neutral Zone	468	530
	World	1,019,546	42.2		World	922,984	64,940

Sources: Oil & Gas Journal, February 1998, *World Energy Statistics*, BP, June 1997, and State Statistical Bureau.

into the Asian crude balance, indicating the importance of China's position in the future oil market. This will include a review of the impact of Chinese oil imports on the formation of regional crude prices. Second, we assess the principal product balances, and discuss Chinese supply/demand influences on the corresponding Asian balances. Finally, we focus on the mutual influences between China's refined product imports, and product/crude spreads and refinery margins in the Singapore market.

8.1 Impact of Crude Imports on Regional Trade Flows and Crude Prices

In the global context, the Asia-Pacific region has a low share of proven and discovered oil reserves (Table 8.1). The generally accepted figures for proven crude reserves in the Asia-Pacific region range from 40 to 50 billion barrels (compared to more than 680 billion barrels in the Middle East), representing just 4–5 percent of global total proven reserves.[2] In 1997, China's 21 billion barrels of proven reserves made up 50–60 percent of the Asian total. The numbers suggest that during the coming decade, if China's crude reserves cannot supply China's need for crude (and we believe they cannot), the only reliable source of exports for the volumes needed is the Middle East.

8.1.1 Crude trade flows

In 1997, the Asia-Pacific region imported about 10 million b/d of crude oil, reflecting a 58 percent dependence on the Middle East. Crude oil production in the entire Asia-Pacific region stood at 7 million b/d in 1997, of which 5 million b/d were consumed within the producing countries, while 2 million b/d were exported. As crude oil production from existing oilfields in the region begins its natural decline, the amount produced obviously will depend on the success of new oilfield development. Significant discoveries of new oilfields are expected in places such as the deep sea area of Indonesia, the region from the northern part of Australia to Papua New Guinea, and the area from the East China Sea to the Vietnamese coast.

However, ever-increasing oil imports are inevitable. Indonesia, the largest oil exporter in the region, will probably change from a net oil exporter to a net importer in the 20th century, due to the surge in its

[2] According to *Oil & Gas Journal*, within Asia-Pacific, India has proven reserves of 4.3 billion barrels, Indonesia 5.0 billion barrels, Malaysia 3.9 billion barrels, Australia 1.8 billion barrels, Brunei 1.4 billion barrels, Vietnam 0.6 billion barrels, and others 1.0 billion barrels. See *Oil & Gas Journal*, February, 1998.

Table 8.2 The Asian and China crude oil balance, 1993–2005

	Actual					Forecast							
	1993	1994	1995	1996	1997	1998	1999	2000	2001	2002	2003	2004	2005
China													
Production (000 b/d)	2898	2928	2967	3145	3244	3278	3343	3421	3433	3495	3529	3564	3613
Demand (000 b/d)	2824	2805	2987	3191	3469	3466	3701	3817	3965	4027	4168	4382	4485
Balance (000 b/d)	74	123	(20)	(46)	(225)	(188)	(358)	(396)	(533)	(532)	(639)	(818)	(872)
Dependence on others (%)	0	0	1	1	6	5	10	10	13	13	15	19	19
Contribution in Asian total deficit/surplus (%)	0	0	0	1	2	2	3	3	4	4	4	5	6
Asia													
Production (000 b/d)	6536	6746	6894	6991	6978	7003	6925	6954	6866	6832	6770	6710	6659
Demand (000 b/d)	13,478	14,088	14,936	15,693	17,080	18,220	19,345	19,741	20,625	20,957	21,293	21,839	22,260
Balance (000 b/d)	(6942)	(7342)	(8042)	(8702)	(10,102)	(11,217)	(12,420)	(12,787)	(13,759)	(14,125)	(14,523)	(15,129)	(15,601)
Dependence on others (%)	52	52	54	55	59	62	64	65	67	67	68	69	70

Note: Crude demand includes refinery runs, direct burn, and losses.

domestic demand. Traditional crude exporters (Malaysia and Brunei) and new exporters (Vietnam and Papua New Guinea) in the region simply cannot fill the gap in the regional supply / demand balance, because none of these countries have a large reserve base.

In our forecasts, the crude dependence of the Asia-Pacific region on the Middle East is expected to rise from 58 percent in 1997 to 68 percent in 2005 (see Table 8.2); Chinese crude dependence commenced in 1996 and will increase from about 10 percent in 2000 to about 20 percent in 2005. The country's crude deficit, while quite large in historical terms, remains small in terms of the regional crude shortfall; by 2005, only about 7 percent of the crude supply deficit in Asia-Pacific will be caused by China. However, China will be responsible for about 15 percent of the *increase* in the regional deficit during the period 1997–2005.

Crude oil trade flows in Asia, with its mix of regulated and unregulated economies, are largely determined by economics. Refiners are constantly scouring the market for crudes that meet their product-yield requirements at the lowest cost. Several factors influence the nature of Asia's crude oil balance and the direction of its trade flows. One factor is the facilities-determined demand for sweet crude. The regional supply and demand for sweet crude have created a market whose prices have in recent years been generated by the forward transactions for Malaysia's Tapis crude.

Another factor is a move by sweet crude producers like Malaysia, Indonesia, or China to adapt existing refineries or build new ones to process sour crude instead of the locally produced sweet crude. This strategy — "buy beer and export champagne" — is behind the program to add a sour train to Malaysia's Malacca refinery, provisional plans to modify Indonesia's Balikpapan refinery, and joint-venture refineries in China's coastal provinces. Such plans are obviously based on the expectation of an increase in sweet / sour, light / heavy differentials, and perceptions about the availability of sweet crude in the long-term. Whether such expectations will survive the current trend — in which these differentials have narrowed instead of widened — remains to be seen. But should it happen, it would make more sweet Asian crude available to trade.

Crude oil flows could also move outside of their "normal" channels by means of investment by oil-exporting countries in refineries in oil-importing countries. For example, Saudi partial ownership of refineries in China and other Asian countries will turn them into sour crude refineries (those that were not so already), and close off that market to other sour crude suppliers. With these characterizations of the structural factors affecting crude oil flows in Asia, how will the Asia crude balance change?

For northeast Asia, which is short of oil and gas resources in per capita terms, oil supply security is and will continue to be high on the agenda.

crude market. It is important to note that this will take place in the context of a regional market that is likely to be much more multi-dimensional than today's market, featuring more players buying more cargoes from outside the region.

8.2 Impact of Product Fundamentals on Regional Supply/demand Balances

Rapidly expanding populations, together with economic development, have boosted oil demand tremendously in Asia. In 1990, petroleum product consumption in the Asia-Pacific region exceeded western Europe for the first time. Japan and China currently dominate the picture, accounting together for 70 percent of oil demand. China's economic ups and downs only compound the market's manic-depressive swings. For instance, the nation's "austerity campaign", begun in 1993 to cool its overheated economy, had no real bite on the Asian market, because the "austerity" of this particular plan did not put a dent in China's demand for petroleum products. Given the centrally planned control that still characterizes China's oil industry, the effect of world oil prices on China's exports/imports of its crude oil and refined products is of indeterminate extent and timing. Political decisions may be almost as much of a factor in the future level of imports as the ability of China's oilfields and refineries to produce.

The long-term pattern for light product import dependence of the region is clear, given the thrust of our refining and demand chapters. The regional deficit of light product has widened during the past years. Although economic/financial crises in 1997 and 1998 cut demand, after then the region will resume the trend and return to rising deficits. For fuel oil, the pattern is inverted. China did not and will not follow the rhythm of the region to a large degree. Moreover, Chinese growth rates for the various products and the refinery yields we expect there mean that the country does not mirror the region. In the following sections we review each principal product market in Asia, and focus on the expected changes over the next decade and China's impact on these changes.

8.2.1 Asia's gasoline/naphtha balance

Gasoline/naphtha demand growth in the Asia-Pacific region outpaced Europe and North America since 1984, growing an average of 6 percent, or 170,000 b/d every year. At 4.3 million b/d, sales in 1997 were 2.2 million b/d higher than in 1983. Large increases in production have almost matched the strong demand in the Pacific over the past 10 years.

Since 1984, regional production has grown by 1.5 million b/d, a 4 percent average annual increase, keeping the regional deficit below 500,000 b/d per year. China's share of gasoline/naphtha consumption in the Asia-Pacific total increased from 16 percent in 1984 to 23 percent in 1997, while its share of Asian production also rose from 20 percent to 27 percent over the same period. This makes China a major player in the Asia-Pacific gasoline/naphtha market.

The Pacific fundamentals grew decidedly tight during the latter part of the 1980s. Both production and demand were increasing steadily, but sales growth outstripped supply growth by over 150,000 b/d as the decade drew to a close. A large part of the regional deficit stemmed from the preference of Japanese refiners to import rather than to use spare capacity. But this tightness dissipated after the Gulf Crisis as additional production capacity and a shift in Japanese refining policy resulted in considerable extra supply for the region.

The Asian gasoline/naphtha balance forecasts show a deficit that ranges from a minimum of less than 200,000 b/d in 2000 to a maximum of 900,000 b/d in 2005 (see Table 8.3). The most significant gasoline surpluses are in Singapore, but Thailand and Indonesia could also become a force in the gasoline export market in a few years. On the other side, Japan and South Korea remain gasoline/naphtha importers, but both nations are likely to increase their upgrading capacity to prevent these deficits from widening.

China was basically self-sufficient in gasoline/naphtha. Only in 1993, did imports for gasoline surpass exports. The major imbalances occur in countries with high naphtha consumption. However, naphtha dependence in Asia is limited compared with other products, and exports of finished gasoline are small compared to naphtha imports.

Unlike the case of diesel, healthy gasoline demand growth in China has not been a bullish factor in the Asian gasoline market. Net gasoline exports from China actually remained constant over the past years. The self-sufficient nature of the Chinese gasoline market shows little sign of change. An important factor for regional markets may be Chinese oil policy, since the government may target gasoline/naphtha when it wishes to achieve import reductions.

In 1997, the regional gasoline/naphtha dependence was only 9 percent, and we expect dependence in Asia will fall to 3 percent by 2000. During the early 2000s, such dependence will rise again, climbing to more than 10 percent by 2005.

Over the next 10 years, China will maintain its moderate surplus in its gasoline/naphtha balance, despite the fact that when a small volume of gasoline exports occurs, naphtha imports may continue to rise. A balanced gasoline/naphtha will be due to refining capacity zooms (despite

Table 8.3 The Asian and China gasoline/naphtha balance, 1993–2005

	Actual					Forecast							
	1993	1994	1995	1996	1997	1998	1999	2000	2001	2002	2003	2004	2005
China													
Production (000 b/d)	999	951	1007	1076	1156	1154	1236	1266	1314	1333	1379	1452	1488
Demand (000 b/d)	961	921	967	1026	1074	1118	1165	1214	1257	1302	1349	1398	1450
Balance (000 b/d)	38	31	40	50	82	36	71	52	57	31	30	53	38
Dependence on others (%)	0	0	0	0	0	0	0	0	0	0	0	0	0
Contribution in Asian total deficit/surplus (%)	n.a.	n.a.	n.a.	n.a.	n.a.	n.a.	n.a.	n.a.	n.a.	n.a.	n.a.	n.a.	n.a.
Asia													
Production (000 b/d)	3617	3713	3842	4091	4306	4658	4828	5163	5204	5366	5394	5424	5500
Demand (000 b/d)	3966	4231	4395	4548	4717	4905	5010	5329	5546	5749	5958	6185	6414
Balance (000 b/d)	(349)	(518)	(553)	(457)	(411)	(247)	(182)	(166)	(342)	(383)	(564)	(761)	(914)
Dependence on others (%)	9	12	13	10	9	5	4	3	6	7	9	12	14

a relatively simple refinery configuration). Even with domestic demand increasing, China is unlikely to become a major market player in the Asian gasoline market before the turn of this century. After 2000 China's deficit of gasoline/naphtha could increase gradually. However China's naphtha deficit is unlikely to exceed South Korea and Japan's.

8.2.2 Asia's middle distillate balance

In Asia, diesel oil consumption is so vital to modernization that demand is difficult to hold back with taxes or trade restrictions. Middle distillate production in the region has been increasing steadily for at least the past 10 years, reaching 6.7 million b/d in 1997. Because the Asia-Pacific region has been a major net importer of diesel oil, never completely meeting demand, refineries tended to maximize output year round in order to meet as much domestic demand as possible and decrease trade deficits. In recent years, a few countries have expanded refinery capacity and begun to produce some gasoil for export. Despite this maximization of output, the relative simplicity of the refining sector in Asia means that diesel oil yields, though increasing slowly, are the lowest of the three major consuming regions, averaging only between 20 and 25 percent.

The Pacific region has the highest kerosene yields in the world, eight and a half time greater than western Europe and 17 times that of North America. In the Asia-Pacific region, while kerosene consumption rose 170,000 b/d over the last five years, kerosene production outpaced consumption growth by rising 200,000 b/d. Japan produces nearly half the kerosene in the region and is also the major consumer.

As of 1997, Asia's notional distillate "deficit" (including diesel, kerosene, and jet fuel) was about 100,000 b/d. Our projections of the increase in demand versus the increase in "producibility" are different from the gasoline story — first an increase and then a decrease in the surplus by 2005, after then deficit begins to widen again as the number of upgrading plans diminishes. Japan appears to be most likely to develop a significant distillate surplus. In addition, South Korea will become "long" in middle distillates in 1998 and remain a very significant source of distillate supply to the rest of Asia until the end of the decade. Taking Japan, South Korea, and Taiwan together, northeast Asia becomes, surprisingly, a large net exporter of distillates. This pattern is good news for China, where we believe the refining sector has not made adequate investments to accommodate the distillate-driven oil demand profile. This suggests an interesting strategy option for China: let Japan and Korea provide the incremental distillate imports. However, we doubt that China will follow such a trend.

Table 8.4 The Asian and China middle distillate balance, 1993–2005

	Actual					Forecast							
	1993	1994	1995	1996	1997	1998	1999	2000	2001	2002	2003	2004	2005
China													
Production (000 b/d)	769	756	889	999	1108	1134	1252	1318	1383	1415	1481	1573	1630
Demand (000 b/d)	885	815	995	1063	1210	1277	1347	1422	1489	1559	1633	1711	1792
Balance (000 b/d)	(116)	(59)	(107)	(64)	(102)	(143)	(95)	(104)	(106)	(145)	(152)	(138)	(163)
Dependence on others (%)	13	7	10	5	7	11	9	11	11	14	14	13	15
Contribution in Asian total deficit/surplus (%)	31	13	21	17	78	n.a.	n.a.	n.a.	n.a.	n.a.	n.a.	n.a.	n.a.
Asia													
Production (000 b/d)	5041	5240	5624	6163	6688	6618	6910	7198	7477	7694	7905	8264	8557
Demand (000 b/d)	5420	5709	6126	6493	6803	6516	6689	6913	7271	7561	7869	8188	8515
Balance (000 b/d)	(379)	(468)	(502)	(330)	(114)	102	222	284	207	133	36	75	42
Dependence on others (%)	7	8	8	5	2	0	0	0	0	0	0	0	0

Environmental regulation will not play a very important role in the regional diesel market. Unlike the sulfur content of fuel oil, countries do not seem to be in a hurry to clean up diesel fuel in a dramatic way. While most countries in Asia do have sulfur content restrictions, refiners are generally able to make the cleaner fuels without undue hardship. Some countries in the region do not have plans to tighten standards. Because most of the consumption growth is likely to occur in countries without stringent regulations, there will continue to be strong demand for higher-sulfur fuel. Asia could become a prime market for those refiners in the West not producing low-sulfur distillate. If a glut of high-sulfur distillate develops, the resulting price differential between the low- and high-sulfur distillates would only encourage poorer countries to continue using the fuel and allow them to increase consumption of high-sulfur material without severely distorting their balance of payments deficit.

The re-entry of China into the regional market was led by distillate oil. If China were to begin to soak up supplies in the manner it did in 1993, the distillate market would tighten noticeably. However, this is unlikely for a number of reasons. Chinese oil policy is likely to emphasize stability and central authority, especially in the near term. In the short run, Asian diesel exporters will count on China's vigorous demand growth. Strong Chinese growth is crucial because the Asian diesel supply/demand balance is expected to be the weakest among the middle distillate products.

In the long run, as noted earlier, demand for middle distillates in China is expected to show a sharp increase over the next decade, and substantial gaps between supply and demand will be observed. China is unlikely to be self-sufficient in middle distillates, though it may come close during the years around 2000, if most new joint-venture refineries come on line. However, China's middle distillate deficit will turn into a constant level for several years after 2000, the same time that the regional surplus will shrink and the deficit will decline (see Table 8.4). Erratic Chinese oil policies will probably not cause the kind of import volatility in percentage terms that we expect to see in other fuels. However, since the volumes of distillate imported are much higher, the effect on the regional market of periodic austerity programs could be great.

In 1997, Asian-Pacific middle distillate dependence was only 2 percent, while the same dependence in China was 7 percent. During the remainder of the century, the total Asian distillate market will be balanced as demand growth slows down as a result of Asian economic crises. Meanwhile, China's dependence on other countries for middle distillates will hover around 11 percent. After 2005, we project that the distillate supply deficit in Asia-Pacific will be primarily caused by China; increases in the deficit during the next decade will come from China's shortfall.

Given this situation, where can China obtain the imports it needs? Although there are plans to build new refineries in the Middle East, two developments will restrict the amount of material that is likely to reach Chinese shores: (1) increases in product demand in the Middle East; and (2) growing demand in southern Asia (especially the Indian subcontinent). Furthermore, because of increasing costs due to environmental regulation, a number of refineries on the US West Coast are expected to close. Therefore, it is probably unrealistic for China and other Asia-Pacific nations to expect a large surplus from either the Middle East or the US West Coast. However, this lack of imports is not a cause for concern before 2000, as any potential Chinese deficit should not be much of a problem. Indeed, since the profits of the Chinese refining sector will probably not depend on exports into an oversupplied market, investments may show higher returns than in other Asian countries.

8.2.3 Asia's fuel oil balance

For years after the 1979 oil shock, fuel oil demand was the focus of anti-oil policies in industrialized countries. However, fuel oil demand in Asia increased significantly during the last decade, rising by about 400,000 b/d, while Japanese demand slipped by 157,000 b/d. Korea, Taiwan, Thailand and China have led this growth in Asia. Demand growth in India, Indonesia, and other countries has been relatively restrained. China and India have purposefully refrained from building oil-fired power plants because other domestic energy sources (coal and hydro) held cost and foreign exchange advantages.

Japan is, by far, the largest consumer of fuel oil in the region. Fuel oil demand in Japan has mounted a comeback since the fall of oil prices in 1986. Japan is likely to see continued demand growth, albeit at a slower rate than the region as a whole, because fuel oil is an attractive fuel when nuclear power cannot expand and LNG and coal facilities are difficult to site. In northeast Asia, Korea and Taiwan are also significant users of fuel oil whose demand for quality fuel oil will continue to increase.

Northeast Asia can be described as a fuel oil "market" only in the loosest sense of the term. It is really a fuel oil complex regulated by "governance structures" including the government, the refining industry, and end-users. To the degree that fuel oil imports — especially low-sulfur imports — are needed, either the refiners or the electric utilities will procure the supply. Outsiders should think twice about make investments on the basis of selling high-quality fuel oil into this complex without making contractual arrangements with these governance structures.

Environmental restrictions may provide a significant hurdle in the future for fuel oil demand in Asia. About 40 percent of fuel oil consumed

Table 8.5 The Asian and China fuel oil balance, 1993–2005

	Actual					Forecast							
	1993	1994	1995	1996	1997	1998	1999	2000	2001	2002	2003	2004	2005
China													
Production (000 b/d)	583	556	509	425	405	374	369	343	327	303	286	274	256
Demand (000 b/d)	681	623	615	575	628	630	633	635	637	638	640	641	643
Balance (000 b/d)	(97)	(66)	(106)	(150)	(223)	(256)	(264)	(292)	(310)	(335)	(354)	(367)	(387)
Dependence on others (%)	14	11	17	26	36	41	42	46	49	53	55	57	60
Contribution in Asian total deficit/surplus (%)	n.a.	n.a.	n.a.	n.a.	n.a.	n.a.	n.a.	n.a.	n.a.	n.a.	n.a.	n.a.	n.a.
Asia													
Production (000 b/d)	3504	3575	3696	3600	3709	3539	3607	3666	3810	3916	4006	4149	4271
Demand '000 b/d)	3,216	3,330	3,386	3,345	3,362	3,250	3,289	3,338	3,456	3,515	3,565	3,617	3,668
Balance ('000 b/d)	288	245	310	255	346	290	319	328	354	400	441	532	603
Dependence on others (%)	9	7	9	8	10	9	10	10	10	11	12	15	16

in Asia is low sulfur, but this proportion is expected to rise as high as 60 percent in the next 10 years. Inland fuel consumption in Japan is predominately low sulfur and accounts for more than one-third of total Asian inland sales. South Korea began to restrict sulfur levels in fuel oil for all users in 1981, and over the past few years the sulfur specifications have tightened further. The trend is catching on in other parts of Asia as well, as countries implement, on a limited and gradual basis, environmental measures that they believe are necessary. For example, Thailand restricted the sulfur content of fuel oil burned in Bangkok, and the Philippine government is moving in the same direction. Some Asian nations (e.g. Malaysia, Indonesia, Australia, and later Vietnam) produce and burn considerable amounts of lower-sulfur fuel oil because of the sweetness of their crude feedstock. Asian countries other than those mentioned have so far not encouraged lower-sulfur fuel oil at the national level.

Looking at the Asia-Pacific region as a whole (see Table 8.5), the fuel oil supply–demand balance over the next 10 years actually "improves" i.e. the gains in fuel oil "producibility" exceed the gains in fuel oil demand. In part, that is because many (although not all) of the forthcoming refining projects start life as grass-roots distillation plants, with the upgrading units — which decrease fuel oil supply — decided upon and announced later. Nevertheless, it is interesting and important that the fuel oil balance in the Pacific is leaning toward developing a larger surplus. The upgrading equation is incredibly complex in Asia: southeast Asia (like parts of the Middle East) remains one of the last strongholds of the majors, who will be formidable competitors to outsiders trying to get in. Northeast Asia is even more of a closed shop, and only joint ventures with local refiners make any sense there. In a sense, only China is a true frontier for refiners.

Fuel oil in China is another principal refined product that will have a significant deficit in China. China will remain as a net fuel oil importer, and could contribute significant reductions in the Asian fuel oil surplus over the next decade. Of course, unannounced but expected additional cracking capacity in China may require more fuel oil imports and consume some of this surplus. During the late 1990s and the 20th century, we project China will need 150,000 to 250,000 b/d of net fuel oil imports, which regional suppliers should be able to provide with little problem.

We also believe that China's low-sulfur fuel oil may find consumers in northeast Asia, because: (1) nuclear power's recovery may be blocked by public opinion; (2) LNG is difficult to handle and costly for new electric power generation projects; and (3) environmentalists generally dislike coal, even low-sulfur coal. Therefore, China may engage in fuel oil quality trade with the countries of northeast Asia, sending out low-sulfur fuel oil and importing high-sulfur material.

8.3 Impact of Product/Crude Spreads

International oil has become a mature and efficient market, with the entire panoply of futures, forwards, and derivative products at its disposal. One of the hallmarks of efficient markets is the chronic downward pressure on margins for both commodity producers and commodity upgraders.

In today's Asian market, with its nexus in Singapore, China is an influential factor in product spreads and margins. Some players believe that strong Chinese growth has the potential to buoy the profits of oil companies for years to come. These analysts point to the market dynamics over the past years: when the Chinese bought cargoes of gasoline or light diesel, prices soared; when China re-regulated its market and banned imports, gasoline and light diesel prices in Singapore tumbled. Reports and rumors of Chinese buying and selling, real or imagined, also move oil prices significantly on a routine basis. While China's influence in the next few years will indeed be strong, refinery expansions will eventually erode this influence by making the country more self-sufficient and lowering trade volumes.

As China's oil market has become more closely tied with the world market, price volatility and fluctuations elsewhere have negatively affected the Chinese market. When world oil prices were low, oil smuggling prevailed and supply surpassed demand, upsetting the equilibrium of the domestic price-regulated system. When world oil prices were high, smuggling and imports declined, resulting in a tight domestic market and soaring oil prices.

On the other hand, the Singapore diesel oil market has become significantly more dependent on Chinese buyers. China is likely to purchase a great deal over the next five years if prices do not increase substantially, but South Korea threatens to take market share from Singapore.

8.3.1 Gasoline/crude spread

Reviewing gasoline/crude spreads over the past years will help us to understand the fundamental trends. The Singapore unleaded/Dubai spread reflects a cyclical pattern caused by a mismatch in the growth of supply and demand. Gasoline spreads rose steadily during the 1980s as demand was brisk and supply fairly tight. Spreads then leveled off and showed wide volatility in the late 1980s and early 1990s. Part of this volatility stemmed from the Persian Gulf War (1990–1991), but demand growth has continued to be a factor.

In coming years, however, the market may more closely resemble the mid-1980s. The Singapore unleaded/Dubai margin is expected to decline from its extraordinarily high level (over \$10 per barrel) during the 1990–1991 period toward the \$6 per barrel level of the 1987–1989 period.

Margins for Singapore refiners, which peaked in 1991 and took a particularly bad beating in 1994, are not likely to recover to their peak levels as long as the incremental Asian supply of gasoline exceeds incremental Asian demand. We observe a downward trend in gasoline / crude spreads in 1994, reflecting oversupply, sluggish growth in certain markets (especially China), and relative calm in the Persian Gulf.

We correlated monthly Chinese gasoline imports from Singapore and Singapore gasoline / Dubai spread from January 1994 to February 1997. In general, the correspondence was not strong that year. Much of this was because imports of gasoline accounted for the smallest quantities of any of the principal products. The fate of Singapore gasoline margins has been more closely linked with the state of the Japanese gasoline market. Nevertheless, Chinese gasoline purchasing has had some effect on the Singapore spread, particularly during the summer. China can offer some periodic support for the Singapore gasoline / Dubai spread, which may help to absorb some of the surplus supplies around the Asian market.

8.3.2 Distillate/crude spread

Since the Asia-Pacific region has become extremely dependent on the Middle East for its supplies of gasoil in the past years, crises in the Middle East have had an immediate impact on the Asian spread. The changing spread between Singapore and the Arab Gulf also reflects the changing supply and demand patterns of the two regions. Except for a few days at the end of the war, Singapore prices ran at a constant premium to Arab Gulf prices. As demand in the Pacific grew from the mid-1988 to the beginning of the Gulf crisis, so did the spread. Since then, increasing refining capacity in the Pacific and the subsequent lower demand for Middle Eastern gasoil have brought the spread down. As refining capacity expansions in the Pacific increase and overtake demand growth, the spread will become more volatile and there may be two-way trade. This growing self-sufficiency in the Pacific should help to insulate it from product-related crises in the Middle East.

Fig. 8.1 compares Chinese light diesel oil imports from Singapore with Singapore gasoil / Dubai spread over the first eight months of 1994. From this, we can glean the impact caused by Chinese light diesel imports from Singapore on the Singapore gasoil / Dubai spread, especially when China reduced its imports from May to July in 1994.

Based on China's distillate balance for the next 10 years, the country seems to have the potential to "make the market" for diesel oil in the Asia-Pacific region — at least in the short-term. But actual diesel oil imports into China depend in large part on how the domestic wholesale price relates to import prices. For example, diesel oil demand from the

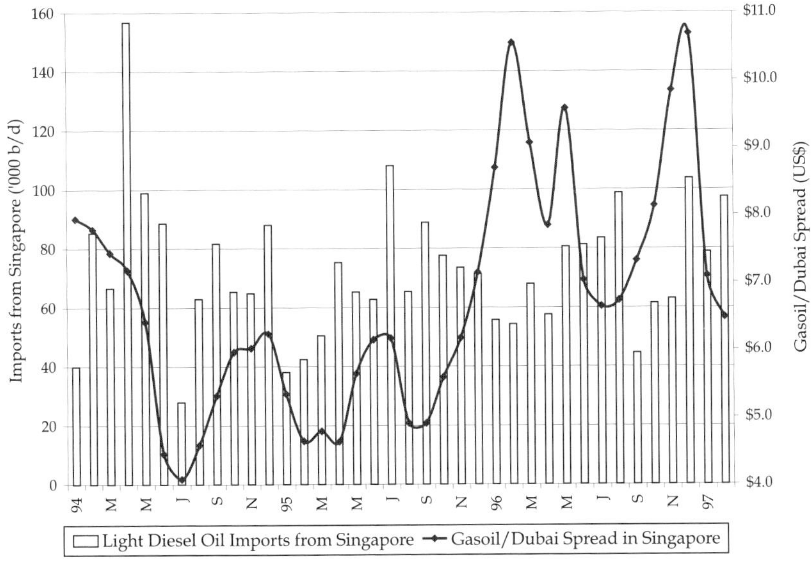

Fig. 8.1 Light diesel imports from Singapore vs. gasoil/Dubai spread in Singapore

domestic market has fared well relative to imports of other fuels, as prices for the product were less affected by 1994's price reforms. The most important part of our analysis on Chinese distillate demand concerns what we do not see happening. We do not see China sucking up all the distillate in the region, as others have suggested. In other words, all of the "export-oriented" refiners hoping to catch the train of runaway Chinese demand may find themselves waiting at the station.

8.3.3 Fuel oil/crude spread

Little more than 10 years ago, Asian refining was in a period of retrenchment. The oil crisis had reduced demand, and fuel oil in particular was on every government's hit list. The problem was most severe in Japan, where the government sought to avoid "destructive" competition by carefully orchestrating the mothballing of a large amount of refining capacity. With reduced capacity and a revival of demand after the collapse of oil prices in 1986, fuel oil spreads in Singapore have performed well in 1990s.

And the good times are not over yet. Our supply-demand outlook indicates that the price of high-sulfur fuel oil in Singapore will strengthen from 1993 levels (when 380 cst averaged $11.47/barrel, about 76 percent of the price of Dubai). However, in later years, when distillation units

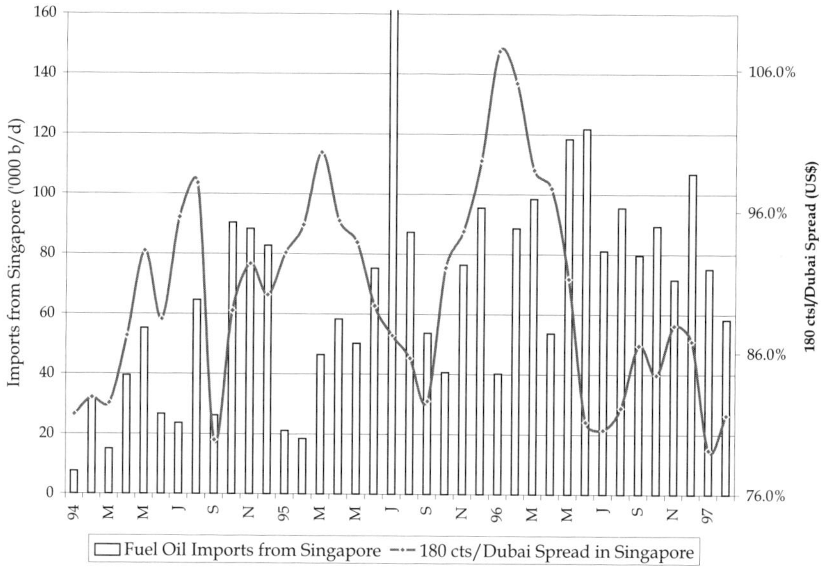

Fig. 8.2 Fuel oil imports from Singapore vs. 180 cts/Dubai spread in Singapore

come on stream with relatively little upgrading capacity, fuel oil spreads are likely to suffer badly. In fact, we see the low-sulfur/high-sulfur fuel oil spread widening in the Pacific, since the marginal barrel of fuel oil produced will come from Middle Eastern crude and will therefore be sour. We project an increase in the spread of low-sulfur versus high-sulfur fuel oil in Singapore of more than $1 per barrel before 2005.

In Fig. 8.2, which compares Chinese fuel oil imports from Singapore with Singapore 180 cst high-sulfur fuel oil/Dubai spread, the impact of China on the Singapore market becomes clear. Note the strong impact caused by Chinese fuel oil purchasing from Singapore on Singapore fuel oil/Dubai spread during the first eight months of 1994 and 1995. This indicates that China's buying interest has increasingly made the Singapore fuel oil market, and provided support for the Asia fuel oil market. This may be a temporary phenomenon due to the lack of demand growth in Japan. In any case, Chinese influence on fuel oil values in Singapore will increase when we move into the next century.

8.3.4 Refinery margins

Singapore provides the best example of an oil-trading area where market forces dictate refinery economics, so we should be able to see a connection between the regional refining balance and Singapore margins. If Asian oil demand grows faster than increases in Asian refining capacity, margins should rise; if new refining "producibility" exceeds demand growth, Singapore margins should fall.

This is a simplification of real life, of course, because outside forces — especially the same refining balance in the Middle East — can also affect Singapore refining margins. But a brief look at history validates our assumption. The increasing Asian refining deficit from 1987 to 1991 does indeed correlate with increases in both simple and complex refining margins, and the falling margins in 1994 can also be traced to the same dynamics. One hesitates to say that the increasing (then decreasing) Asian deficit was the entire cause of the rise (then fall) of Singapore margins, but it stands to reason that the need for products on the part of Asia's importing countries will have been felt first at Asia's primary export refining and export center.

At first glance, the rule of markets would suggest that the pending decline we foresee in Singapore margins would work as a strong disincentive for construction and expansion of refiners all over Asia. We believe it has. The decline since 1991 has already had a dampening effect on the construction of export-oriented refineries, because they live or die by this margin. And it is interesting that in the midst of a refinery planning boom, none of the Singapore refiners have plans to expand their capacity in any significant way.

But a bleak outlook for Singapore margins may have only a modest and indirect effect on construction in China, where regulation is a fact of life and where the government has a refined product supply security/autarky objective. Chinese refiners are only marginally affected by Singapore prices: when products there are cheap, refiners may buy some cargoes on the margin, and when prices are high they may sell some (based on their surplus). But the basic economics of Chinese refiners will continue to be a function of the internal product price structure, which will remain significantly different from that of Singapore until Chinese refineries are truly deregulated.

The outlook for Singapore refinery margins will affect the joint-venture refiners in China who base their product prices on Singapore products. However, state-owned refineries in China are not likely to care about the decline in Singapore margins, because internal prices will continue to be regulated for the indefinite future. This rather low vulnerability to regional market forces will remain one of the principal impediments to the optimization of the Asian refining system.

Fig. 8.3 indicates that the Asian refining margin has remained weak in the first eight months of 1994. Part of the reason may have been a decline of Chinese petroleum product imports from Singapore. From a statistical point of view, the correlation coefficient we get from these eight

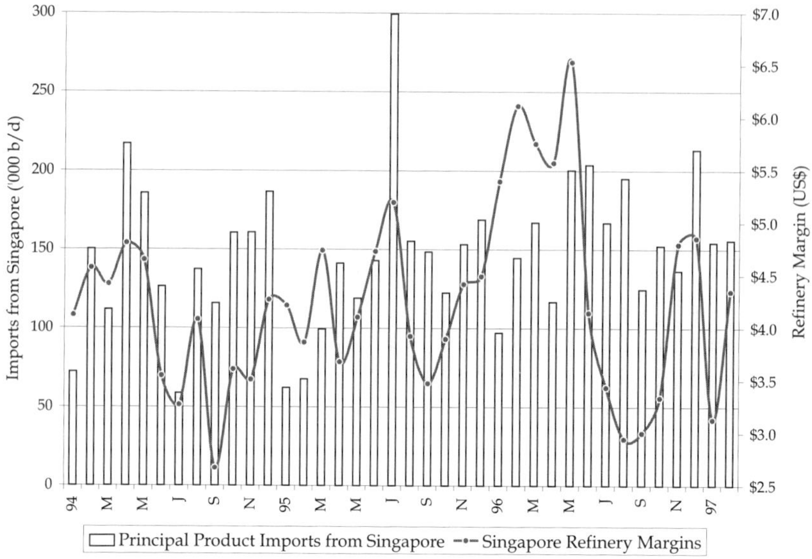

Fig. 8.3 Principal product imports from Singapore vs. Refinery margin in Singapore

months' data is 0.65[3], suggesting a high degree of correlation between Chinese principal product imports (i.e. gasoline, light diesel, heavy diesel, and fuel oil) from Singapore and the refinery margin in Singapore.[4]

Take 1995 as an example. This correlation was clearer in 1994 statistics than it was in 1995. However, the influence of China's purchases in the Singapore market appears to be reflected in the peaks and dips in refinery margins from July to September 1995. In August 1995, notional Singapore refinery margins dropped by more than $1.00 per barrel from July levels, while China's principal product imports from Singapore (gasoline, diesel oil, and fuel oil) were cut in half, from 300,000 b/d in July to 150,000 b/d in August. While the gasoline fundamentals probably reflected refinery problems in Japan, the drop in China's imports from the Singapore market stimulated the decline in the spreads of diesel and fuel oil. China's light diesel imports fell from 108,000 b/d in July to 65,000 b/d in August, while fuel oil imports fell from 188,000 b/d in July to 87,000 b/d in August. Consequently, the diesel spread dropped from

[3] The correlation coefficient should fall between –1 (the most closely negative correlation) and 1 (the most closely positive correlation). And there is no correlation at all when the coefficient is equal to zero.

[4] Singapore refinery margin is calculated by the formulas as (3/10 × gasoline price + 4/10 × gasoil price + 3/10 × 180 cst fuel oil price) – Dubai price. Product prices are Singapore spot prices, and all prices are in US$ per barrel.

$6.17 in July to $4.91 in August, and the fuel oil spread declined from 87.2 percent of the crude price in July to 85.6 percent in August.

Through September 1995's trend of refined product imports in China was very similar to that of 1993. (1994 had an effective import ban in the first half of the year.) In 1995, the amount of product imports increased each quarter. Rising imports were coupled with a significant increase in refinery runs, reflecting the ability of China's market to absorb these additional barrels. Given the higher demand level presented in 1995, the fourth quarter showed a continued rise in product imports, since Chinese refineries were not able to produce enough incremental products to satisfy demand growth during this period. This was good news for Singapore refinery margins, which rose steadily in the fourth quarter as a result of additional demand from China.

However, this correlation does not tell us about causality. It is possible that the China story is only a small part of the larger price and margin determination process. However, it is plausible, based on this basic analysis, to confirm that China does have a direct, and sometimes powerful, influence on the profitability of free-market refining in Asia.

Singapore refining margins slipped in 1994 as a result of refining capacity increases in Malaysia and Indonesia and falling import requirements in China. Refiners are likely to endure more of the same in years to come. In Asia, concern that refining capacity may not keep up with demand may soon give way to concern that demand will not keep up with supply. The falling margins we see in the region reflect the structural change in the Asian oil market, as the region becomes more self-sufficient in products.

One of the keys to the profitability of refining in the future is whether the industry will add more capacity than necessary. Our survey of refinery additions and closures indicates that there is indeed more capacity coming on-line than is strictly necessary in some years, but still too little upgrading capacity in the Asia-Pacific region. We see margins declining for a number of years as completed refinery projects come on-line faster than the market can absorb their production. This bearish view on Singapore refining margins is good news for China as a whole, but bad news for some joint-venture coastal refineries, since the booming domestic market could choose to import product instead of crude.

Caution is advised here: refining investments in China must be influenced by regional economics, and not made on the basis of domestic petroleum demand/supply balances only. Refinery projects for completion after 1997 will continue to be added in the country to keep it self-sufficient. This is especially important in the coastal provinces, which may have a surplus of export-oriented refining capacity after 1997 for a few years.

The slowdown in growth of China's refining industry will fuel the expectation that China will be a net importer of crude and refined products. This is likely to add a few cents to the price of crude and put upward pressure on product prices in the Asian market. Thus, we expect China to have a net bullish effect on regional margins until 1997, then a bearish effect through the end of the century. What happens after that depends mostly on the refinery projects which will be planned over the next few years.

8.4 Conclusion

As China adds refinery capacity in the coming years, its need for crude will outrun its crude production. The resulting explosion of crude imports will make China an increasingly important player in regional and world oil trade. The doubling of Asian crude imports in the next 10 years will probably increase the differential between Dubai (the "Asian" crude) and WTI and Brent. However, the underlying price of crude is less likely to increase dramatically. Prices and spreads for individual products will be strongly influenced by the shape as well as the extent of new refinery capacity in the region.

The relative lack of upgrading capacity in the Asia-Pacific refining sector means that the region is chronically short of light products. The rapid growth of Chinese demand for gasoline, diesel, and fuel oil has influenced the Singapore market in past years, but only additions to refinery capacity will erode this influence by making the country more self-sufficient and thereby lowering trade volumes. However, continuing Chinese imports will provide some support for diesel oil and fuel oil spreads, although their importance for gasoline spreads has largely disappeared.

Chapter 9
Conclusion

In the process of writing this book, the tempered optimism of our view on economic growth, political stability, and petroleum product demand growth in China gave way to a relatively bullish view on the future of Chinese refining and oil market. However, this book is pessimistic in its assessment of future crude oil output to meet demand growth over the next 10 years. The greatest opportunities in the next decade are not in crude production, and the most striking change from the status quo may be either the huge increase in crude imports necessary to feed the mushrooming refining sector or an increasing demand on petroleum products from the world market. Indeed, we believe that China's oil "crisis" will strongly affect the Asian oil market since it will require nearly a million barrels of imports each day some time in the 20th century. There can be no doubt that a workable oil strategy must entail accelerating oil exploration and development (E&D).

To assess the prospects for the success of oil development, foreign investors must address many basic questions about China's policies and regulations. These questions fall into three categories. First, what can we expect regarding the investment incentives for oil exploration and refinery construction in China? How will international oil companies respond to the strong, though not spectacular growth that we see in China's oil demand? Second, how will the restructuring of state-owned oil enterprises and institutions alter their responses to the oil market? Conversely, how do the challenges of restructuring the oil industry affect oil imports/export, transportation, and distribution? Third, how will economic policies evolve, and how will they interact with oil industrial policies?

Based on our observations and on extensive discussions with experts in China, we believe that the oil industry will drag its feet in moving toward free markets until there are substantial and far-ranging reforms in basic economic, legal, and social institutions. Without basic institutional reforms, however, it is doubtful that China will have the capacity to continue the progress made possible by foreign investment. Thus, change will come, but little will take place during the next decade.

In China, policies aimed at increasing efficiency and at increasing the transfer and utilization of foreign technology investment are emphasizing investment in leading industries and high-tech sectors. So far, China has opened its limited oil transportation, distribution and service sectors to foreign investors. Foreign firms could set up joint-venture or cooperative cargo-shipping or sea-shipping agencies. Foreign firms are also allowed to conduct offshore petroleum services in technical consulting, geological survey and marine prospecting with Chinese partners on a cooperative basis. However, the oil industry remains largely closed and tightly controlled, and it would be unwise for foreign businesses to plan or invest in the expectation that the government will relinquish its rigid control of the oil sector, as it has in other industries. A number of industries related to oil (e.g. transportation enterprises such as railways and ports) are also still prohibited from foreign direct investment and banking and foreign trade are also highly regulated. In addition, China still lacks a comprehensive legal framework for settling disputes. Caveat investor. Therefore, there are decided political risks for foreign companies who want to be part of this effort.

China has entered into a difficult period of transition for its oil industry and market, and the government clearly can play a substantial part in meeting the challenges by facilitating investments and setting clear guidelines on policy issues. These policy issues — such as how to encourage exploration and production, what priority to give to restraining consumption, what kind of pricing and distribution system should balance supply / demand, and the optimal amount and configuration of refinery capacity — will continue to require the government's attention for years to come.

Upstream, the government's oil policy must shift its focus from maximizing crude exports to securing a stable supply of crude for domestic refineries. At the same time, the government must create policies that efficiently exploit the real long-term future of China's oil supply — the Tarim Basin. *Crude oil crisis* is a valuable guide to China's ongoing challenges. Our conclusion leads to a pessimistic assessment of the oil future, if the Tarim Basin cannot fill such a big gap in the supply / demand deficit. The best practical possibilities for the next decade are not even for stabilizing the crude production, but merely for reducing the domestic feedstock for the mushrooming refineries. Indeed, China's oil crisis will adversely affect the Asian oil market. There can be no doubt that a workable planetary future will require not only accelerating oil E&D, but also an appreciable moderation of the Chinese oil industry and market.

Downstream, the government must decide how to provide new refining capacity while avoiding over-building or under-building. This will be especially tricky if, as we expect, the region experiences a glut of simple refinery capacity before the turn of the century. A lack of adequate refining capacity might make China heavily dependent on product imports, thereby causing China's import bill to spend more on the value-added commodities. China needs foreign capital investment but is reluctant to open its market to foreign participation. For China, foreign investment is crucial to ensure that the country's oil industry does not breakdown under the strain of relentless demand growth — about 4.0 percent a year — which far outstrips any rise in refinery capacity. However, the interests of foreign capital focus on full and open entry to China's retail markets, not just a limited market in Special Economic Zones. Without such access in the future, the joint-venture refinery will halt. The tightening of oil policy could lose its luster for foreign investments in the refining sector.

In product distribution, an ongoing question will be the precise trade-off between central control/regulation policy and the use of market mechanism: the former could avoid chaos but might dampen growth. Demand surge is inevitable as an economy booms. Yet, there are substantial opportunities for conservation through price reform. As China re-enters the World Trade Organization and moves further toward a market economy, it must gradually liberalize its oil industry and market, just as it has — successfully — in other sectors.

In sum, principal lessons for the oil industry should include, among others: (1) structural adjustment to accelerate its development and attract more foreign investment to fill the gap of financial shortfall domestically; (2) marketize state-owned oil enterprises to improve resource allocation efficiency; (3) reform pricing policies so that oil prices signal both market demand–supply conditions and reflect real social–economic costs of oil production; and (4) integrate into the international trade and market system to remove distortions rooted in oil price differentials and improve efficiency in oil use.

Appendix 1
Institutional Framework — Who's Who in the Oil Industry

There are four major players in the Chinese oil market: China National Petroleum Corporation (CNPC), China Petrochemical Corporation (Sinopec), China National Offshore Oil Corporation (CNOOC), and China National Chemicals Import and Export Corporation (Sinochem). Both CNPC and Sinopec are ministry-level companies and are operated under the supervision of the State Planning Commission (SPC) Fig. A1.1.[1]

CNPC is a successor to the Ministry of Petroleum, established in 1992. It is directly under the State Council and is a ministry-level institution. CNPC assumes responsibility for exclusively exploring and developing onshore oil and natural gas resources. CNPC owns over 20 oilfields and overseas onshore oil and gas reserves. CNPC sells almost one-half of its 2.8 million b/d crude output to the central government at controlled prices. The central government in turn allocates CNPC's crude to Sinopec refineries. CNPC also owns 22 small refineries located near oilfields, with capacities ranging from 1000 to 50,000 b/d. CNPC has been actively involved in downstream business during recent years. It is planning to expand and upgrade its small refineries.

Sinopec, founded in 1983, is a ministry-level state-owned company in charge of most petroleum refining and petrochemical production based on liquid petroleum feedstocks. Sinopec owns 37 refineries in the country. In 1996, Sinopec accounted for 3.3 million b/d of China's 4.1 million b/d in refining capacity, and its output between 1990 and 1996 rose by roughly 3 percent per annum to some 2.3 million b/d. Sinopec has been actively involved in the product retail business since the early 1990s. Sinopec has 80 percent of the petrochemical production and marketing in China, while the Ministry of Chemical Industry owns the remainder.

[1] The name State Planning Commission (SPC) was changed to State Planning and Development Commission (SPDC) in 1998.

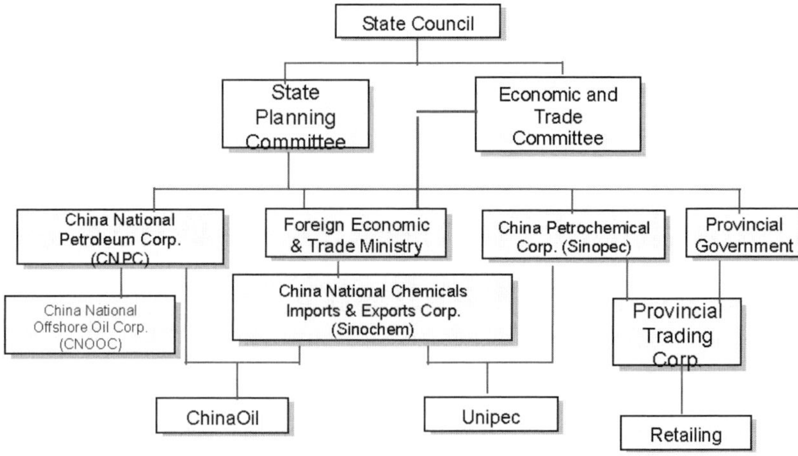

Fig. A1.1 Institutional framework of China's oil industry and the market

Sinopec is also playing a major role in petroleum product distribution through its sales company. It distributes petroleum products to provincial trading companies based on central government plans and quotas. The provincial trading companies sell the products either to retailers or to end-users.

CNOOC was created in 1982, and is in charge of all offshore exploration and development activities. CNOOC has authority to sell its crude at international market prices. Its crude production in 1997 was about 320,000 b/d, and almost half of its offshore production was exported.

Sinochem under the administration of the Foreign Economic and Trade Ministry (FETM) has monopolized China's international trading in crude and petroleum products for more than 40 years, and is China's only state-owned company to get prominent recognition in international business circles.[2] In the past years, as the country's largest foreign trade entity, Sinochem's activity started to diversify into actual refining or processing positions, and it also plans to expand into China's finance, property, transport and retail sectors. The company operates 20 storage facilities with a capacity of 7.5 million barrels and owns a fleet of oil tankers. Sinochem is also buying shares in refineries outside China.

Crude and product sales and purchases transacted between domestic and international markets are profit-making operations under the central planning system. After a long complaint from CNPC and Sinopec, Sinochem had to share its privilege with CNPC and Sinopec in early 1993.

[2] Sinochem ranked 27th in 1991 and 29th in 1992 in *Fortune Magazine's* Global Service 500.

It induced two joint-ventures firms between these three oil giants: China National United Oil Corporation (ChinaOil, also called Sinoil) jointed between CNPC and Sinochem, and China International United Petroleum and Chemicals Corporation (Unipec) joined between Sinopec and Sinochem. Despite some newcomers to break its monopoly in international trading, Sinochem still dominates China's oil trade. For instance, Sinochem's market share includes 40 percent of crude imports and 60 percent of crude exports. In 1993, of the more than 300,000 b/d of crude China imported, Sinochem imported 140,000 b/d.

Approved by the State Council in 1995, Sinochem will convert into a full-fledged trade-investment-financing conglomerate by 2005. Under this 10-year plan, Sinochem will be able to combine trade with industry and financing.

ChinaOil (also called Sinoil) is a 50–50 equity split by Sinochem and CNPC, established in March 1993. ChinaOil is in charge of import and export of crude and refined products, as well as overseas investment in oil and natural gas ventures. ChinaOil has a 40 percent share of the crude export market. In 1993, ChinaOil exported 180,000 b/d of crude. Japan is ChinaOil's biggest customer, accounting for 70 percent of the company's crude exports.

Unipec is a state-owned oil trading company with a 50–50 joint venture of Sinochem and Sinopec. Its business covers crude import, export and import of refined products, petrochemicals, equipment, and technology, as well as project financing, and overseas investments. Unipec is responsible for 60 percent of China's crude imports. In 1993, Unipec imported 100,000 b/d of crude into China, and exported about 20,000 b/d of refined products. The total import of crude and petroleum product is primarily planned and managed by the SPC and the State Economic and Trade Commission. FETM oversees the quotas, which are primarily assigned by the SPC. Sinochem, ChinaOil and Unipec are licensed import agencies to implement transactions under the quota policy.

Appendix 2
Methodology of Demand Forecasts

This appendix presents the methodology used in demand forecasts by principal petroleum products and total petroleum products, followed by a presentation and discussion of the results of model estimates.

I. Data

The data presented in this book are taken from several different data sources. For energy and petroleum data, we supplement Chinese statistics with data from the International Energy Agency, OECD, which provides additional statistical estimates (i.e. fuel oil use as bunker). We also employ GDP data from the World Bank, population data from the UN, and motor vehicles data from the Motor Vehicle Manufacturers Association in the US.

The primary data are taken from various issues of the *Chinese Statistical Yearbook* 1987–1997, and the *Chinese Energy Statistical Yearbook* 1985–1992 published by the China State Statistical Bureau. The data sample consists of 28 provincial regions (excluding Hainan and Tibet) having the required data.

Regional GDP and the gross value of industrial output (GVIO) have been adopted in this book as measures of the overall level of economic activity and industrial development. GVIO excludes gross outputs of construction, transportation, and commercial. Since intermediate inputs measured in GVIO result in double-counting, possible distortions may be introduced into the book.

Petroleum product consumption by region includes use in the industrial, construction, and transportation sectors, and excludes use in agriculture, commercial and residential sectors. We estimate total national consumption of petroleum products by adding imports to production and subtracting exports, while assuming no change in product stock. Using total national consumption of petroleum products and partial consumption by region, we estimate the total regional consumption of petroleum products.

Implied refined product prices used are retrieved as gross output value in the refining industry divided by refinery runs. According to the method used by the Chinese to measure the gross output value, the implied refined product prices obtained here are factory prices, which exclude tax-added and distribution costs before the end-user prices.

Retail price indices are used as a proxy of GDP deflator. According to Chinese statistics, retail price indices are calculated using three kinds of prices, i.e. plan prices, negotiated prices and market prices, and are based on a periodic sample survey.[1] These deflators are used to convert each region's income and industrial output values into constant prices.

Panel data sets are used for two reasons. First, the effectiveness of economic reform and its impact on economic activity and on the oil industry has been seen only during recent years. Longitudinal data allow us to analyze the effects that cannot be addressed using time-series data sets. In particular, pooled data — that is, cross-region and time-series — are generally appropriate to determine demand trends while the regional economies developed unevenly. Second, since there is a lack of time-series data, combining it with cross-regional data sets can give us a large number of data points, and increases the degrees of freedom and improves the efficiency of econometric estimates.

Data on China's oil industry remain quite limited and in some cases unreliable. The inadequacy of the data may lead to some unavoidable bias in the presentation of model estimates and analyses in this book. Consequently, empirical work should be interpreted cautiously; more information will be needed to assess the demand trend in China.

II. Gasoline Demand Forecast

A. Model

There are three different approaches: statistical, engineering, and econometric, which have all been used to predict gasoline demand.

The statistical approach identifies three separate components, i.e. the trend, the cycle, and the seasonal factors. After computing prior trend growth rates, upward or downward cycles, and seasonal indices, demand projections are made.

The engineering approach initiated by Baltagi and Griffin (1983) is based on detailed technical assumptions regarding annual auto registration and

---

[1] Plan prices are determined by government agencies. Negotiated prices in state and nonstate enterprises can be adjusted within upper limits set by the central government. Market prices are determined by supply and demand, although the central government may also set upper limits for certain goods.

their mileage characteristics, miles driven per auto, and an average fleet efficiency of autos. Gasoline demand (*GAS*) is computed by multiplying together the total number of cars in the country (*CAR*), miles driven per auto (*MPA*), and gasoline consumption per mile (*GPM*). It can also be expressed by multiplying the number of cars and miles driven per auto dividing that total by miles driven per gallon (*MPG*).

$$GAS = CAR * MPA * GPM = CAR * MPA * \frac{1}{MPG} \tag{1}$$

The econometric approach was illustrated by Sweeney (1978), who integrated economic factors into the engineering approach. Baltagi and Griffin (1983) developed a forecast performance of the gasoline demand equations in which gasoline demand depends critically on economic stimuli and can be modeled with econometric techniques.

The econometric approach is used to integrate economic factors into the engineering approach. The major factors considered in the econometric model for gasoline demand forecast are: (1) personal income (*INC*); (2) population, specifically those from age 20 to 65 (*POP*); (3) the real gasoline retail price (P^R_{Gas}) adjusted by GDP deflator (P_{GDP}); (4) the number of passenger cars registered (*CAR*); (5) a "saturation effect" — the number of persons per vehicle or the number of vehicles per thousand persons; and (6) fuel efficiency (*EFF*) defined by gasoline consumption per mile (*GPM*). Putting all these factors together, the model is as follows:

$$\ln\left(\frac{GAS}{CAR}\right) = \alpha_0 + \alpha_1 \ln\left(\frac{INC}{POP}\right) + \alpha_2 \ln\left(\frac{P^R_{GAS}}{P_{GDP}}\right) + \alpha_3 \ln\left(\frac{CAR}{POP}\right) + \alpha_4 \ln(EEF) \tag{2}$$

Gasoline retail prices consist of two components, wholesale price (P^W_{Gas}) and tax (T_{GAS}). Theoretically, gasoline wholesale prices follow the movement of crude oil prices and are determined by gasoline supply, demand and inventories in regional markets. Gasoline taxes, in contrast, depend on national energy policies, price regulations and environmental requirements. Price regulations and/or environmental concerns in some countries may cause gasoline retail prices to increase significantly in spite of low gasoline prices in the international market. Government policy also plays a role in setting auto efficiency standards in the gasoline demand model.

Due to lack of data on distances driven by passenger cars, we have been unable to obtain accurate fuel efficiency information. Following McRae (1994), the efficiency variable can be hypothesized as a long-run function of the real price of gasoline and the real income per capita. Then the efficiency variable can be replaced by a lagged-structure of the variables of real price of gasoline and real income per capita.

According to the consideration of gasoline retail price and fuel efficiency, the model can be changed:

$$\ln\left(\frac{GAS}{CAR}\right) = \alpha_0 + \sum_{i=0}^{m} \beta_i \ln\left(\frac{INC}{POP}\right)_{-i} + \sum_{j=0}^{n} \gamma_j \ln\left(\frac{P_{GAS}^{W}}{P_{GDP}}\right)_{-j} + \delta \ln\left(\frac{T_{GAS}}{P_{GDP}}\right) + \eta \ln\left(\frac{CAR}{POP}\right) \quad (3)$$

where m and n demote the degree of the polynomial form.

A great number of studies have been completed by using panel data from several countries for gasoline demand forecasts, due to inadequate time-series data in individual country or statistically unsatisfied for coefficients obtained from individual country model. Houthakker and Kennedy (1978) employed data from 12 OECD countries for a gasoline consumption model. Pindyck (1979) applied the time series data of 11 countries to forecast gasoline demand. Wheaton (1982) used data from 25 countries to obtain price and income elasticity. Baltagi and Griffin (1983 and 1984) based their predictions of gasoline demand in the US on a model pooling the data from 18 OECD countries. Meanwhile, Uri et al. (1985), and Hsing (1990) employed cross-section data in the US to estimate elasticity of gasoline demand. McRae (1994) used individual country data and pooled data from 11 Asian developing countries to estimate price elasticity and GDP elasticity of gasoline demand.

B. Results

Based on geographic differences, regions are divided into six categories, i.e. north, north-east, east, central/south, south-west and west. Each geographic region was estimated separately. Estimates using the polynomial form are not statistically significant for lagged independent variables, probably because there is not adequate data. The results are presented in Table A2.1. The statistical fit of the equations varies by regions. All regions (except west) have a fairly good statistical fit, with an adjusted R^2 ranging from 0.63 to 0.83.

Except for north-east and south-west, the price elasticities have the right sign, but none of them are statistically significant. Income elasticities in these regions are positive and significantly different from zero (except in the north-east), ranging from 0.32 to 0.97. The coefficients for the variable representing the number of vehicles per capita are negative in all the geographic regions, and are also statistically significant in all the geographic regions except the west, ranging from –0.31 to –0.60.

The west region, the lowest income area in the country, has the highest income elasticities (0.97), suggesting that the region may experience a great increase in gasoline demand when its economy takes off. This makes sense, since the undeveloped economy by definition was making the poorest use of resources and had the greatest potential to make large gains quickly. The north-east region has the opposite signs for income and price elasticity, suggesting that this region has less market influence and much more state control in its gasoline consumption.

Table A2.1 Estimated results for gasoline demand model. Dependent variable: gasoline demand

	ln (INC/POP)	ln (P^W)	ln (CAR)	Adj R^2	DW	F
North	0.32	–0.06	–0.45	0.73	1.85	27.70
	(2.22)	(–1.30)	(–3.91)			
North-east	–0.79	0.42	–0.35	0.72	1.33	20.97
	(–3.84)	(6.32)	(–2.88)			
East	0.33	–0.08	–0.31	0.63	2.01	24.09
	(3.82)	(–0.76)	(–4.09)			
Central/south	0.48	–0.19	–0.60	0.84	1.68	49.20
	(3.82)	(–1.45)	(–4.66)			
South-west	0.48	0.02	–0.58	0.77	1.89	19.77
	(2.67)	(0.25)	(–4.06)			
West	0.97	–0.01	–0.18	0.33	2.00	5.85
	(2.16)	(–0.17)	(–0.91)			

Note: Time dummies are also estimated, but not presented in the table. Figures in parentheses are *t*-statistics.

We also use implied product prices as a proxy for gasoline price variables. The test shows that these price elasticities are statistically significant only in the east and central/south regions. It demonstrates that the coastal areas have a dynamic market closer to world oil supply and demand trends.

Due to state-controlled gasoline prices, price is inelastic to gasoline demand. There is no discernible difference between short-run and long-run income effects on gasoline demand. Income elasticities of gasoline demand were found to be lower than unity while income elasticities for high income regions are higher than the low income regions. The implications are that: (1) the stages of economic development affect gasoline consumption; and (2) the income elasticity of demand for gasoline is not constant over income levels, which themselves depend on the stage of development. Regions at different stages of development will have different patterns of gasoline demand.

III. Diesel Oil Demand Forecast

A. Model

We assume diesel oil demand is a function of industrial output, price, the consumption of other competing energy, diesel oil production, and population density:

$$\ln(DIE) = \vartheta_0 + \vartheta_1 \ln\left(\frac{IND}{POP}\right) + \vartheta_2 \ln(P_{DIE}) + \vartheta_3 \ln ENG + \vartheta_4 \ln PRO + \vartheta_5 \ln DEN \quad (4)$$

where *DIE* is diesel demand in barrels per day; *IND* is industrial value in national income; *POP* is population; P_{DIE} is diesel price using implied petroleum product price as a proxy, converted to constant value; *ENG* is other energy consumption excluding oil; *PRO* is diesel production in barrels per day; and *DEN* is population density.

B. Results

Table A2.2 shows the regression results for diesel oil demand in geographic regions. The statistical fit of the equations is striking in all regions, with adjusted R^2 above 0.90. All price elasticities have the right signs, but only the eastern region has a price elasticity statistically significant. Except for the east and the south-west, industrial output elasticities have the right sign and are statistically different from zero, ranging from 0.81

Table A2.2 Estimated results for diesel oil demand model. Dependent variable: diesel oil demand

	ln (INC/POP)	ln (P_{DIE})	ln (DEN)	Ln (ENG)	Adj R^2	DW	F
North	1.21	–0.02	0.05	–1.26	0.95	1.99	76.15
	(2.26)	(–0.79)	(0.27)	(–3.66)			
North-east	2.00	–0.13	–0.96	0.28	0.99	1.90	161.69
	(3.70)	(–0.41)	(–5.74)	(0.79)			
East	0.13	–0.38	0.35	0.48	0.96	2.32	126.71
	(0.67)	(–2.54)	(3.94)	(2.80)			
Central/south	1.30	–0.18	–0.03	–0.51	0.96	1.87	84.49
	(7.91)	(–1.30)	(–0.63)	(–2.51)			
South-west	–0.63	–0.29	–0.89	1.92	0.90	1.71	21.19
	(–0.92)	(–0.36)	(–2.35)	(2.43)			
West	0.81	–0.02	–0.61	0.76	0.96	1.99	77.95
	(4.70)	(–0.64)	(–7.89)	(4.21)			

Note: Time dummies are also estimated, but not presented in the table. Figures in parentheses are *t*-statistics.

in the west to 2.00 in the north-east. Three of these six geographic regions have industrial output elasticities over one, reflecting that diesel oil demand in China will be affected by economic activity much more than other petroleum products. The coefficients for the variable representing the population density vary in the different regions. Except for the north and the east, which have the highest population density, the coefficient reveals that the higher the population density, the lower diesel oil demand per capita. The coefficients for other substituted energy input have the right signs and are statistically significant only in the north and central/south, with –1.26 and –0.51 respectively. It indicates that in most regions of China other energy resources may not be substituted for diesel oil.

IV. Fuel Oil Demand Forecast

A. Fuel oil use in the industrial sector

Most of the demand for fuel oil in the industrial sector is for heat generation in industrial processes and for fueling industrial boilers. It should be a function of industrial output and other input energy in the industrial sector. The model is taken as:

$$\ln RES = \mu_0 + \mu_1 \ln\left(\frac{IND}{POP}\right) + \mu_2 \ln HDC + \mu_3 \ln COK + \mu_4 \ln NGS \qquad (5)$$

where *RES* is fuel oil use in the industrial sector; *IND* is industrial value in national income; *POP* is population; *HDC* is hard coal input in the industrial sector; *COK* is coke input in the industrial sector; and *NGS* is natural gas input in the industrial sector.

The regression results for equation 5 are presented in Table A2.3. Except for the coefficient of hard coal, other selected independent variables are significantly different from zero, while except for the coefficient of natural gas, other selected independent variables have expected signs.

Table A2.3 Estimated results for fuel oil demand model in the industrial sector. Dependent variable: fuel oil consumption in industrial sector

ln (INC/POP)	ln (HDC)	ln (COK)	ln (NGS)	Adj R^2	DW	F
0.56 (3.31)	–0.31 (–0.99)	–1.19 (–4.04)	0.66 (5.75)	0.64	2.31	7.77

Note: Figures in parentheses are *t*-statistics.

B. Fuel oil use in the electric power sector

We assume that fuel oil demand in the electric power sector is a function of electricity demand and the other energy inputs substituted for fuel oil in the power sector. Our estimates show the strong substitution relationship that exists between fuel oil and other inputs to electricity generation.

The availability of energy resources, structural features of economies, and environmental and other policy concerns have a substantial effect on electric power production. Over time, the differences in factor inputs will affect the growth of electricity output. The comparative advantage of energy inputs would determine the energy use in the power sector.

In principle, a production process is a technical concept characterizing the range of possible combinations of input factors to produce a given level of output. A gross output production function can be summarized to four aggregate inputs and can be expressed as:

$$Q = Q(K, L, E, M, t) \tag{6}$$

where Q is gross output; K, L, E, and M are aggregate inputs of capital, labor, energy, and materials, respectively; and t is technology.

To evaluate the effects of fuel oil demand we must first analyze attendant changes in physical inputs and outputs. To study fuel oil use in the electric power sector, and focus on the fuel substitution impact, we will modify the conventional production function to measure the energy input effect on electricity output, and ignore other productive inputs. Since energy is highly complementary with capital, the type of energy used is always associated with a certain amount of capital and technology. So each energy resource used is also a function of capital and technology.

To determine which energy input makes the greatest contribution to electric power production, the model can take the form:

$$ELE = f(RES, HDC, NGS, HYD, NUC) \tag{7}$$

where ELE is electricity output; RES is fuel oil input; HDC is coal input; NGS is natural gas input; HYD is hydroelectric generation; and NUC is nuclear power generation.

To eliminate multicollinearity, we use the growth rate for each variable. Then, the model is taken as:

$$ELE_t^* = \alpha_0 + \alpha_1 RES_t^* + \alpha_2 HDC_t^* + \alpha_3 NGS_t^* + \alpha_4 HYD_t^* \tag{8}$$

where $ELE^*_t = (ELE_t - ELE_{t-1}) / ELE_{t-1}$; $RES^*_t = (RES_t - RES_{t-1}) / RES_{t-1}$; and other variables are in the same transformation form.

Table A2.4 Estimated results for fuel oil demand model in the electric power sector

ln ELE	ln RES	ln HDC	ln NGS	ln HYD	Adj R^2	DW	F
Dependent variable: electricity production							
–	0.09	0.70	0.36	–0.01	0.998	2.22	1389.8
–	(4.74)	(13.50)	(6.31)	(–2.64)			
Dependent variable: fuel oil demand in electricity							
3.25	–	–2.00	–1.46	0.02	0.54	1.68	4.56
(1.97)	–	(–1.73)	(–2.00)	(1.10)			

Note: Figures in parentheses are *t*-statistics.

The regression results for equation 8 are shown in Table A2.4, presenting model estimates. Except for natural gas, all the variables have expected signs and are statistically significant. The coefficient of each independent variable indicates its input elasticity of electricity production, and a high coefficient indicates a high contribution from this type of energy to the electricity generation. The sum of all coefficients of independent variables is greater than one, reflecting that few other input factors are omitted in this model.

From the model estimates, fuel used in the power sector is primarily constrained by the availability of resources in China. Larger contributions to electricity generation come from coal (an elasticity of 0.70) and hydro power (an elasticity of 0.36) which are abundant. Fuel oil input has a low share in total energy inputs in the electric power sector, and makes a small contribution to power generation (0.09). Fuel oil is given its unique role as the quintessential substitutable fuel input in the power sector.

Berndt and Christensen (1973) indicated that relaxing the strong separability assumption among inputs is equivalent to imposing a priori restrictions on the substitution possibilities between the production inputs. If we change the model specification, exchanging the position of electricity output and fuel oil input, we can further estimate the possible substitution of fuel oil with other forms of energy. The assumption for this model is that fuel oil use is determined by electricity generation and availability of other energy input in the electric power sector. The model presents as:

$$RES_t^* = \beta_0 + \beta_1 ELE_t^* + \beta_2 HDC_t^* + \beta_3 NGS_t^* + \beta_4 HYD_t^* + \beta_5 NUC_t^* \qquad (9)$$

where all variables have the same definition used above.

To capture the interfuel substitution effects, we expect a positive coefficient for dependent variables: electricity production and negative coefficients for coal and natural gas input, hydroelectricity generation, and nuclear power generation. We have eliminated the variable of nuclear power generation due to lack of information. Table A2.4 demonstrates the results. In most cases, the coefficients have the right sign.

C. Fuel oil for bunker use

In the absence of major changes in the composition of the tanker fleet and its engine technology, we would expect bunker demand to rise as China's trade continues to increase in the 1990s. The model is simply taken as:

$$\ln BUK_t = \eta_0 + \eta_1 \ln BUK_{t-1} + \eta_2 \ln NEX_t + \varepsilon_t \qquad (10)$$

where *BUK* is fuel oil use in bunker, and *NEX* is value of exports and imports (Table A2.5). Due to lack of information on physical volumes of total exports and imports, we take the sum of the values of imports and exports.

Table A2.5 Estimated results for fuel oil demand model in bunker use. Dependent variable: fuel oil use in bunker

ln NEX	Adj R^2	DW	F
0.50 (10.85)	0.91	1.87	117.68

Note: Figures in parentheses are *t*-statistics.

D. Fuel oil for refinery use

Residual fuel oil is also used to fuel refineries themselves, and a significant amount of fuel oil disappears into this sector. A forecast of the precise amount would obviously be based on the anticipated level of refinery runs. Therefore, if we envision an increase in overall global demand for petroleum products of about 500,000 b/d to 800,000 b/d per year over the next decade, and a corresponding increase in refinery throughput levels, we also envision an increase in the amount of fuel oil used by refineries. We are aware of no technological breakthroughs that might significantly lower the amount of fuel consumed, and only in the Americas, the Middle East, and parts of Europe is something other than fuel oil (i.e. natural gas) likely to be used to fuel refineries. Hence, we take the model as:

$$\ln ERE_t = \upsilon + \sigma \ln ERE_{t-1} + \sum_{i=0}^{m} \delta_{t-i} \ln THR_{t-i} + \varepsilon_t \qquad (11)$$

where *ERE* is fuel oil use in refinery, and *THR* is refinery throughput.

Table A2.6 Estimated results for fuel oil demand model in refinery. Dependent variable: fuel oil use in refinery

	ln NEX	ln THR	Adj R^2	DW	F
Time-series					
T	0.14 (4.20)	–	0.94	1.56	86.80
T_{-1}	0.63 (4.62)	0.19 (4.20)	–	–	–
T_{-2}	0.17 (4.16)	–	–	–	–

Note: Figures in parentheses are *t*-statistics.

Regression results are presented in Table A2.6. The elasticity of refinery throughput to fuel use in refinery is positive and significantly different from zero. The best results for the polynomial form were obtained using a one-year lag for refinery throughput.

V. Jet Fuel Oil Demand Forecast

The econometric demand model built for jet fuel oil demand is based on the following technical identity equation:

$$\text{Jet Fuel Demand} = \frac{\text{Aircraft Revenue Ton - kilometers / Aircraft Load Factor}}{\text{Average Aircraft Efficiency}} \quad (12)$$

Aircraft revenue ton-kilometers is modeled as a function of real GDP per capita and world market jet fuel prices in constant US dollars. Aircraft efficiency equals available ton-kilometers divided by total jet fuel demand. The model is a logarithmic function of these variables, taking the following form:

$$\ln JET_t = \mu_0 + \mu_1 \ln JET_{t-1} + \mu_2 \ln GDP_t + \mu_3 \ln P_{JET,t} + \mu_4 \ln LF_t + \mu_5 \ln EF_t \quad (13)$$

where *JET* is jet fuel oil demand; $P_{JET,t}$ is jet fuel oil prices in constant US dollars; *LF* is average aircraft load factor; and *EF* is average aircraft efficiency. Data in this model (Table A2.7) range from 1978 to 1992.

Table A2.7 Estimated results for jet fuel demand model. Dependent variable: jet fuel oil consumption

ln JET_{t-1}	ln GDP	ln P_{JET}	ln LF	ln EF	Adj R^2	DW	F
0.41	0.39	–0.05	0.26	–0.11	0.89	1.82	178.5
(3.13)	(2.04)	(–1.89)	(2.10)	(–2.18)			

Note: Figures in parentheses are *t*-statistics.

VI. Total Petroleum Product Demand Forecast

A. Model

Since petroleum products are an input of production, the growth rate of petroleum product consumption cannot be seen separately from the rate at which the economy grows. This does not mean that economic growth is the sole factor that determines petroleum product consumption growth. Other factors are energy intensity in consumption and production, structural changes (changes from an agricultural to an industrialized economy) and price elasticity changes. Moreover, exogenous variables such as geography, resource endowment, technology, demographic factors, and economic policies are also necessary to explain changes in energy consumption. It should be realized that not all factors can be fully taken into account in economic analysis, including factors such as politics, environmental concerns, noncommercial energy use, etc. In addition, in terms of total petroleum product, demand forecasts are also driven by consumption of principal petroleum products:

$$TOT_t^* = \pi_0 + \pi_1 GDP_t^* + \pi_2 TEN_t^* + \pi_3 IND_t^* + \pi_4 GAS_t^* + \pi_5 DIE_t^* + \pi_6 RES_t^* \quad (14)$$

where *TOT* is total petroleum product demand; *GDP* is GDP per capita; *TEN* is energy intensity; *IND* is industrial share in total national income; *GAS* is gasoline consumption; *DIE* is diesel oil consumption; and RES is fuel oil consumption. The star (*) means to take transformation, i.e. $TOT^*_t = (TOT_t - TOT_{t-1})/TOT_{t-1}$; $GDP^*_t = (GDP_t - GDP_{t-1})/GDP_{t-1}$; and other variables are in the same transformation form.

B. Results

Regression results are presented in Table A2.8. With unadjusted R^2 of 0.88, it shows an essential relationship between total petroleum product demand and other independent variables. All the coefficients have the right sign, and are significantly different from zero except for gasoline. Among the individual petroleum products, diesel oil has the highest coefficient, indicating its greater contribution to total product demand growth. The level or stage of economic development is one of the forces affecting petroleum product demand. The coefficients of GDP per capita show positive signs, presenting an elasticity of 0.47. Table A2.8 also shows the impact of population density on product consumption, indicating a strong influence. Unlike other independent variables, industrial share in total national income has a negative impact on product demand, suggesting that the change in the nature of industrial share

Table A2.8 Estimated results for total product demand model

ln GDP	ln DEN	ln IND	ln GAS	ln DIE	ln RES	Adj R^2	DW	F
0.47	0.59	–0.34	0.12	0.48	0.30	0.88	1.58	21.97
(2.31)	(2.60)	(–2.53)	(1.07)	(5.17)	(3.82)			

Note: Figures in parentheses are *t*-statistics.

as a part of the economic structure (i.e. change from heavy industry to a greater service sector) could reduce the consumption of petroleum products. Such a result suggests that oil conservation in China could yield significant results.

VII. Assumptions

The general assumptions we take in our demand for gasoline, diesel oil, fuel oil, and jet fuel are as follows:

- Economic growth rates are based on our analysis presented in Chapter 1, but it is varying in provinces;
- The growth of industrial output will continue to surpass the growth rate of GDP;
- The number of motor vehicle registrations is related to the growth rate of GDP per capita;
- Total national population forecasts are based on UN forecasts taken from *The Sex and Age Distribution of the World Populations*, the 1992 revision;
- Aircraft efficiency will increase in the long term, and aircraft load factor will be essentially constant over the next 10 years; and
- Jet fuel prices are assumed to be a function of the West Texas Intermediate (WTI) price.

References

Baltagi, B.H. and James M. Griffin (1984) US gasoline demand: what next? *The Energy Journal*, Vol. 5, No. 1:129–140.

Baltagi, B.H. and James M. Griffin (1983) Gasoline demand in the OECD, *European Economic Review*, 22, pp. 117–137.

Berndt, E.R. and L.R. Christensen (1973) The internal structure of functional relationships: separability, substitution and aggregation, *Review of Economic Studies*, 40, July, pp. 403–410.

Energy Information Administration (1990) *Review of the EIA Short-term Energy Outlook Process and Models* (Draft), December 21.

Energy Security Analysis, Inc. (1993) *The Global Gasoline Market to the Year 2003*, Washington DC.

Energy Security Analysis, Inc. (1993) *The Global Fuel Oil Market to the Year 2003*, Washington DC.

Energy Security Analysis, Inc. (1993) *The Global Middle Distillate Market to the Year 2000*, Washington DC.

Houthakker, H.S., P.K. Verleger, and D.P. Sheehan (1974) Dynamic demand analysis for gasoline and residential electricity, *American Journal of Agricultural Economics*, 56(May): 412–418.

Houthakker, H.S. and Michael Kennedy (1978) Long-range energy prospects, *Journal of Energy and Development*, 4(1) (Autumn): 1–28.

Hsing, Yu (1990) On the variable elasticity of the demand for gasoline, *Energy Economics* (April): 132–136.

Krapels, Edward and Haijiang Wang (1994) Fuel oil demand, policy impact, and energy substitution," Paper presented at IGT Conference, Energy Modeling: Methodology, Models, and Applications, Atlanta.

McRae, Robert (1994) Gasoline demand in developing Asian countries, *The Energy Journal*, Vol. 15, No. 1, pp. 143–155.

Pindyck, R.S. (1979) *The Structure of World Energy Demand*, MIT Press, Cambridge Massachusetts.

Sweeney, John C. (1978) Energy modeling and forecasting: implications for strategic planning, *Business Economics*, 13(4), Sept, pp.21–27.

Uri, Noel D. and Saad A. Hassanein (1985) Testing for stability: motor gasoline demand and distillate fuel oil demand, *Energy economics* (April): 87–92.

Wheaton, W.C. (1982) The long-run structure of transportation and gasoline demand. *Bell Journal of Economics*, Vol. 13, No. 2 (Autumn): 439–445.

Appendix 3
Statistics

Table A3.1 Crude Oil Consumption (000 b/d) by Sector

	1985	1986	1987	1988	1989	1990	1992	1995
Total consumption	1901.9	2042.1	2137.1	2608.7	2385.0	2416.9	2645.2	2977.3
Farming, Forestry, Animal Husbandry, Fishery and	0.2	0.1	0.1	0.2	0.2	0.0	-	2.0
Water Conservancy Industry	1878.0	2018.6	2106.4	2217.4	2301.6	2330.8	2623.1	2943.3
A. By Light and Heavy Industry								
Light Industry	46.4	49.8	52.4	54.2	59.6	73.0	86.6	n.a.
Heavy Industry	1831.6	1968.8	2054.0	2163.2	2242.0	2257.8	2536.5	n.a.
B. By Industrial Sector								
Mining Total	150.5	137.6	158.3	181.4	199.1	213.6	268.9	337.2
Manufacturing Total	1727.5	1881.0	1948.1	2036.0	2102.6	2117.2	2354.1	2592.7
in which								
Coal Mining and Dressing	–	–	–	–	–	–	–	–
Petroleum and Natural Gas Extraction	150.5	137.6	158.3	181.3	199.1	213.6	268.5	337.2
Food, Beverage and Tobacco	0.4	0.3	0.4	0.2	0.5	0.4	0.2	0.4
Textiles	1.2	0.4	0.2	0.4	0.4	0.5	0.3	0.3
Paper-Making and Paper Products	0.5	0.6	0.3	0.2	0.3	0.2	0.2	0.1
Power Generation, Steam and Hot Walter Production and Supply	68.2	62.2	56.6	49.0	37.2	28.8	21.9	13.3
Petroleum Processing and Coking	1403.9	1548.3	1599.2	1691.8	1752.3	1720.8	1945.5	2267.7
Chemicals and Allied Products	197.7	211.5	231.6	231.8	248.4	289.1	298.7	215.8
Medical and Pharmaceutical Products	0.3	0.1	0.2	0.0	0.1	0.1	0.0	0.0
Chemical Fibers	41.5	46.1	48.8	49.3	53.8	69.0	81.5	95.6
Building Materials and Other Non-Metal Mineral Products	56.3	56.3	56.3	56.3	56.3	56.3	56.3	56.3
Smelting and Pressing of Ferrous Metals	4.0	2.8	2.2	5.1	3.4	3.3	0.4	0.6
Smelting and Pressing of Non-Ferrous Metals	3.9	4.5	4.4	0.0	0.3	0.0	0.4	0.1
Machine Building, Electric and Electronic Equipment	1.7	0.8	0.9	2.0	0.9	1.0	0.9	0.4
Construction	14.8	11.3	19.8	12.9	10.2	11.0	10.9	0.5
Transportation, Postal and Telecommunication Services	8.9	12.1	10.0	9.9	9.1	10.4	11.2	31.4
Commerce, Food Services, Material Supply and **Marketing and Storage**	0.0	0.0	0.1	0.0	0.1	0.1	0.0	0.1
Others	0.1	0.0	0.7	0.0	0.2	0.1	–	–
Residential Consumption	–	–	–	–	–	–	–	–

Note: Total consumption used here may differ from the notional consumption presented in the previous chapters which does not include inventory buildup or withdraw.
Sources: *China Statistic Yearbook, 1994, 1997*, and *China Energy Statistic Yearbook, 1986, 1991*.

Table A3.2 Gasoline Consumption (000 b/d) by Sector

	1985	1986	1987	1988	1989	1990	1992	1995
Total consumption	325.2	349.4	378.3	416.8	431.7	442.3	584.5	677.6
Farming, Forestry, Animal Husbandry, Fishery and Water Conservancy	28.5	33.2	34.1	35.8	31.3	34.0	41.6	41.8
Industry	105.1	106.3	117.7	130.5	128.9	137.2	159.4	189.2
A. By Light and Heavy Industry								
Light Industry	32.4	31.5	36.9	41.4	40.0	44.2	51.6	n.a.
Heavy Industry	72.7	74.8	80.8	89.1	88.8	93.0	107.7	n.a.
B. By Industrial Sector								
Mining Total	22.9	22.9	25.4	27.1	27.9	30.1	30.2	31.6
Manufacturing Total	82.2	83.4	92.3	103.4	101.0	107.1	129.2	157.5
in which								
Coal Mining and Dressing	6.0	7.3	8.4	8.5	8.5	9.2	9.8	8.8
Petroleum and Natural Gas Extraction	7.3	6.2	6.4	8.2	8.4	9.2	8.5	13.7
Food, Beverage and Tobacco	6.7	7.9	8.8	9.5	9.9	11.0	11.8	16.8
Textiles	5.0	4.8	5.6	6.7	6.6	7.0	9.3	9.9
Paper-Making and Paper Products	1.7	2.0	2.3	2.6	2.5	2.7	3.3	3.4
Power Generation, Steam and Hot Walter Production and Supply	2.1	2.0	2.5	2.6	2.9	3.2	4.6	9.2
Petroleum Processing and Coking	2.5	1.4	1.6	1.9	2.0	2.0	3.2	6.8
Chemicals and Allied Products	11.0	10.4	11.8	13.7	12.9	13.4	16.4	14.6
Medical and Pharmaceutical Products	1.0	1.0	1.2	1.3	1.4	1.5	1.9	2.1
Chemical Fibers	0.3	0.3	0.5	0.8	0.9	0.8	0.9	1.1
Building Materials and Other Non-Metal Mineral Products	56.3	56.3	56.3	56.3	56.3	56.3	56.3	56.3
Smelting and Pressing of Ferrous Metals	4.3	4.4	5.1	5.4	6.1	6.6	7.5	9.9
Smelting and Pressing of Non-Ferrous Metals	1.4	2.0	1.7	2.0	2.1	2.2	3.1	3.0
Machine Building, Electric and Electronic Equipment	21.5	22.3	23.5	26.7	24.4	25.2	30.5	37.4
Construction	17.0	18.5	20.5	20.3	22.2	20.8	25.2	24.1
Transportation, Postal and Telecommunication Services	111.2	118.4	124.0	131.7	137.8	144.4	188.1	228.8
Commerce, Food Services, Material Supply and Marketing and Storage	5.4	7.2	8.0	10.7	11.2	10.7	11.3	45.9
Others	55.5	63.2	70.6	84.1	97.6	91.0	152.5	132.9
Residential Consumption	2.5	2.6	3.4	3.8	2.7	4.2	6.5	14.8

Note: Total consumption used here may differ from the notional consumption presented in the previous chapters which does not include inventory buildup or withdraw.
Sources: *China Statistic Yearbook, 1994, 1997*, and *China Energy Statistic Yearbook, 1986, 1991*.

Table A3.3 Diesel Oil Consumption (000 b/d) by Sector

	1985	1986	1987	1988	1989	1990	1992	1995
Total consumption	396.4	432.6	472.9	526.3	561.6	550.1	689.4	883.2
Farming, Forestry, Animal Husbandry, Fishery and Water Conservancy	128.6	137.2	149.1	156.8	168.8	180.2	181.0	204.7
Industry	131.6	129.8	138.3	155.2	162.7	148.8	213.5	243.2
A. By Light and Heavy Industry								
Light Industry	23.9	26.3	26.1	29.8	31.8	32.4	39.3	n.a.
Heavy Industry	107.8	103.5	112.2	125.4	130.9	116.4	174.2	n.a.
B. By Industrial Sector								
Mining Total	17.1	20.8	22.6	24.3	26.4	30.3	32.0	46.9
Manufacturing Total	114.6	109.0	115.7	131.0	136.3	118.5	181.5	196.3
in which								
Coal Mining and Dressing	3.2	3.5	3.8	4.8	5.4	5.7	7.0	6.5
Petroleum and Natural Gas Extraction	5.7	8.3	9.4	10.1	11.6	13.0	14.2	30.2
Food, Beverage and Tobacco	6.6	5.9	5.8	6.2	8.0	7.6	8.0	12.5
Textiles	4.9	5.3	5.0	7.2	7.6	6.6	7.6	7.4
Paper-Making and Paper Products	1.1	1.3	1.2	1.4	1.6	1.7	1.9	5.6
Power Generation, Steam and Hot Walter Production and Supply	26.5	15.7	18.6	42.2	48.0	27.3	72.2	48.6
Petroleum Processing and Coking	1.7	1.2	1.5	4.2	3.4	3.4	4.2	10.0
Chemicals and Allied Products	30.7	29.7	31.3	13.5	10.2	14.0	16.2	19.3
Medical and Pharmaceutical Products	0.2	0.3	0.4	0.4	0.6	0.7	0.8	0.8
Chemical Fibers	0.2	0.2	0.4	0.5	0.5	0.5	0.6	1.1
Building Materials and Other Non-Metal Mineral Products	12.1	15.7	15.9	16.7	16.2	17.3	19.1	29.1
Smelting and Pressing of Ferrous Metals	3.5	4.3	5.3	5.0	5.8	6.4	8.3	15.0
Smelting and Pressing of Non-Ferrous Metals	1.4	1.8	1.9	1.8	2.1	2.2	3.6	4.4
Machine Building, Electric and Electronic Equipment	16.3	16.8	17.1	21.1	18.8	17.3	22.8	22.3
Construction	25.5	30.7	31.7	29.1	30.0	27.2	34.7	24.2
Transportation, Postal and Telecommunication Services	92.9	118.7	132.8	141.0	147.4	145.0	169.2	254.8
Commerce, Food Services, Material Supply and Marketing and Storage	2.2	3.3	4.4	6.5	7.8	4.6	17.1	21.2
Others	15.1	12.5	16.0	36.4	44.8	44.4	73.5	132.0
Residential Consumption	0.4	0.4	0.6	1.2	0.2	0.0	0.4	3.3

Note: Total consumption used here may differ from the notional consumption presented in the previous chapters which does not include inventory buildup or withdraw.
Sources: *China Statistic Yearbook, 1994, 1997*, and *China Energy Statistic Yearbook, 1986, 1991*.

Table A3.4 Kerosene Consumption (000 b/d) by Sector

	1985	1986	1987	1988	1989	1990	1992	1995
Total consumption	83.2	83.0	82.2	77.3	77.9	75.8	90.2	110.6
Farming, Forestry, Animal Husbandry, Fishery and Water Conservancy	0.7	0.5	0.6	0.6	0.8	0.7	0.8	0.8
Industry	4.3	4.4	5.0	4.9	4.7	4.4	5.4	9.7
A. By Light and Heavy Industry								
Light Industry	1.1	1.2	1.5	1.6	1.3	1.2	1.8	n.a.
Heavy Industry	3.2	3.2	3.5	3.3	3.4	3.2	3.6	n.a.
B. By Industrial Sector								
Mining Total	0.6	0.6	0.8	0.6	0.6	0.8	0.6	0.6
Manufacturing Total	3.8	3.9	4.2	4.3	4.1	3.7	4.8	9.1
in which								
Coal Mining and Dressing	0.4	0.4	0.5	0.4	0.4	0.6	0.4	0.3
Petroleum and Natural Gas Extraction	0.1	0.1	0.2	0.1	0.0	0.1	0.1	0.1
Food, Beverage and Tobacco	0.1	0.1	0.1	0.2	0.1	0.1	0.1	0.6
Textiles	0.2	–	–	0.3	0.2	0.2	0.2	0.6
Paper-Making and Paper Products	0.0	0.1	0.1	0.1	0.0	0.0	0.1	0.4
Power Generation, Steam and Hot Walter Production and Supply	0.0	–	–	0.0	0.0	0.0	0.0	0.3
Petroleum Processing and Coking	0.2	0.0	0.0	0.0	0.1	0.0	0.0	0.2
Chemicals and Allied Products	0.3	0.5	0.5	0.5	0.6	0.4	1.1	1.7
Medical and Pharmaceutical Products	0.0	0.0	0.0	0.0	0.0	0.0	0.1	0.0
Chemical Fibers	0.0	0.0	0.0	0.0	0.0	0.1	0.0	0.0
Building Materials and Other Non-Metal Mineral Products	0.4	0.6	0.6	0.5	0.5	0.5	0.6	0.6
Smelting and Pressing of Ferrous Metals	0.1	0.1	0.1	0.1	0.1	0.1	0.1	0.1
Smelting and Pressing of Non-Ferrous Metals	0.0	0.1	0.1	0.1	0.1	0.1	0.1	0.1
Machine Building, Electric and Electronic Equipment	1.9	1.7	1.7	1.7	1.6	1.6	1.7	2.1
Construction	0.3	0.4	0.4	0.3	0.3	0.3	0.3	0.8
Transportation, Postal and Telecommunication Services	12.1	11.4	14.6	14.8	15.6	20.2	35.6	54.0
Commerce, Food Services, Material Supply and Marketing and Storage	0.0	–	–	0.4	0.2	0.1	0.3	1.8
Others	39.5	37.3	34.0	30.4	28.5	27.5	29.1	29.6
Residential Consumption	26.3	29.0	27.6	26.0	27.7	22.6	18.8	13.9

Note: Total consumption used here may differ from the notional consumption presented in the previous chapters which does not include inventory buildup or withdraw.
Sources: *China Statistic Yearbook, 1994, 1997*, and *China Energy Statistic Yearbook, 1986, 1991*.

Table A3.5 Fuel Oil Consumption (000 b/d) by Sector

	1985	1986	1987	1988	1989	1990	1992	1995
Total consumption	517.7	548.9	569.4	590.8	618.7	614.5	624.4	674.0
Farming, Forestry, Animal Husbandry, Fishery and Water Conservancy	0.6	0.2	0.7	0.8	0.8	0.5	0.8	1.5
Industry	485.8	513.1	526.6	550.5	572.8	564.1	567.8	621.5
A. By Light and Heavy Industry								
Light Industry	54.0	62.5	72.1	64.7	60.6	60.6	49.4	n.a.
Heavy Industry	431.8	450.5	454.5	485.8	512.2	503.5	518.4	n.a.
B. By Industrial Sector								
Mining Total	11.3	11.1	15.7	16.6	14.9	26.1	32.8	45.0
Manufacturing Total	474.5	501.9	510.9	533.9	557.8	538.0	535.0	576.5
in which								
Coal Mining and Dressing	–	–	–	0.0	0.0	0.0	–	0.2
Petroleum and Natural Gas Extraction	8.8	9.3	14.8	15.1	13.3	24.5	30.9	41.4
Food, Beverage and Tobacco	2.8	2.4	2.4	2.1	2.0	2.2	4.5	6.3
Textiles	5.5	5.2	4.7	3.5	3.7	3.9	3.9	6.4
Paper-Making and Paper Products	5.8	5.2	4.3	3.6	3.7	3.9	3.3	3.0
Power Generation, Steam and Hot Walter Production and Supply	174.0	169.1	169.3	179.6	169.8	148.9	126.5	177.5
Petroleum Processing and Coking	56.9	63.9	68.1	64.3	89.0	85.3	113.3	111.7
Chemicals and Allied Products	73.5	79.4	84.9	91.4	99.4	99.9	87.8	70.9
Medical and Pharmaceutical Products	2.6	2.4	2.8	3.9	2.3	2.2	2.7	7.1
Chemical Fibers	19.1	17.4	15.1	17.7	18.5	21.9	21.1	16.5
Building Materials and Other Non-Metal Mineral Products	45.7	52.6	53.2	55.4	57.8	56.8	58.3	59.3
Smelting and Pressing of Ferrous Metals	56.2	71.6	68.9	74.1	94.9	78.9	78.6	84.8
Smelting and Pressing of Non-Ferrous Metals	5.7	7.5	8.0	8.2	11.3	11.2	12.3	11.3
Machine Building, Electric and Electronic Equipment	17.5	17.2	17.6	17.6	15.9	15.0	15.2	12.4
Construction	3.4	2.8	8.3	4.9	5.8	8.6	11.1	2.6
Transportation, Postal and Telecommunication Services	26.3	31.6	32.1	33.5	36.5	38.0	41.6	41.5
Commerce, Food Services, Material Supply and Marketing and Storage	0.6	0.3	0.4	0.5	0.5	0.3	0.4	1.2
Others	1.1	0.9	1.3	0.6	2.3	2.9	2.7	5.6
Residential Consumption	–	–	–	–	–	–	–	–

Note: Total consumption used here may differ from the notional consumption presented in the previous chapters which does not include inventory buildup or withdraw.
Sources: *China Statistic Yearbook, 1994, 1997*, and *China Energy Statistic Yearbook, 1986, 1991.*

Table A3.6 Oil Balance Sheet (000 b/d)

	1980	1981	1982	1983	1984	1985	1986	1987	1988	1989	1990	1991	1992	1994	1995
Total Oil Available for Consumption	1758.9	1665.5	1653.5	1687.9	1734.4	1838.7	1947.4	2069.9	2220.3	2317.2	2287.0	2473.0	2710.6	2955.8	3208.5
Primary Energy Output	2118.9	2024.4	2042.5	2121.4	2292.3	2497.9	2613.8	2682.8	2740.9	2752.8	2766.1	2819.8	2841.9	2921.6	2995.0
Import (+)	16.5	14.2	31.4	27.0	22.5	18.0	70.0	64.7	101.7	213.0	151.1	249.9	424.9	580.7	734.6
Exports (–)	361.2	376.8	417.9	418.5	573.7	726.1	692.4	658.8	628.5	621.3	622.1	586.1	571.9	476.0	490.9
Change in Inventory (million bbl)	(5.6)	1.4	(0.9)	(15.3)	(2.4)	17.9	(16.1)	(6.9)	2.3	(10.0)	(3.0)	(3.9)	5.7	(18.4)	(11.0)
Total Oil Consumption	1751.5	1661.2	1642.1	1676.5	1731.0	1833.8	1945.6	2062.4	2218.5	2316.7	2297.1	2476.7	2670.7	2991.2	3213.0
Consumption By Sector															
1. Material Production Sectors	1626.8	1541.3	1526.9	1547.5	1599.0	1687.4	1789.8	1896.9	2025.3	2099.5	2088.7	2203.6	2369.5	2618.0	2798.6
1.1. Farming, Forestry, Animal Husbandry, Fishery and Water Conservancy	163.0	155.2	145.2	148.4	159.9	151.7	163.6	176.7	185.8	193.8	206.7	213.6	214.5	217.8	240.6
1.2. Industry	1240.6	1173.7	1152.7	1148.5	1171.1	1234.3	1280.4	1334.9	1446.9	1494.8	1464.3	1526.1	1631.0	1836.3	1869.9
Light Industry	n.a.	n.a.	n.a.	n.a.	n.a.	121.7	134.8	149.1	149.2	146.0	149.4	n.a.	n.a.	n.a.	n.a.
Heavy Industry	n.a.	n.a.	n.a.	n.a.	n.a.	1,112.6	1145.7	1185.8	1297.7	1348.8	1314.9	n.a.	n.a.	n.a.	n.a.
1.3. Construction	35.0	34.0	36.0	42.1	47.5	58.4	60.6	78.0	64.5	65.5	65.5	72.3	79.1	48.3	48.6
1.4. Transportation, Postal and Telecom.	182.3	171.5	185.6	200.5	212.7	235.3	275.1	295.3	311.4	326.9	336.6	374.2	417.0	470.1	572.7
1.5. Commerce, Food Service, Material Supply, Marketing, and Storage	5.8	6.9	7.4	7.9	7.8	7.6	10.0	12.0	16.7	18.4	15.5	17.3	27.9	45.4	66.8
2. Non-Material Production Sectors	96.3	87.7	85.3	92.1	90.1	101.2	103.0	111.1	137.9	158.1	151.6	192.7	233.6	268.3	278.1
3. Residential Consumption	28.4	32.2	29.9	37.0	41.8	45.2	52.9	54.5	55.3	59.1	56.9	64.6	72.2	104.9	136.4
By Use															
1. Processing and Transformation of 2nd Energy	420.4	409.7	382.3	368.7	360.0	349.1	343.5	339.2	374.7	365.4	326.1	330.6	336.6	361.5	446.0
Power Generation	413.1	402.4	374.7	361.2	351.9	285.1	271.7	275.0	302.6	289.1	246.9	254.1	259.9	235.7	271.7
Heating	–	–	–	–	–	57.1	64.3	56.8	64.7	67.9	71.3	67.5	67.0	112.0	80.0
Gas Generation	7.3	7.3	7.7	7.5	8.0	6.9	7.6	7.4	7.5	8.4	7.9	9.1	9.7	13.8	10.3
2. Final Uses	1262.2	1187.5	1198.1	1240.1	1308.9	1412.7	1525.4	1643.3	1754.9	1864.8	1860.9	2036.7	2179.5	2493.2	2735.3
Industry	756.1	703.6	711.7	716.7	751.4	892.4	865.9	921.2	988.9	1048.9	1036.1	1109.3	1144.2	1338.4	1419.1
3. Loses in Oil Refining	16.3	13.9	12.0	14.5	13.4	22.6	29.5	34.4	36.2	36.4	59.2	58.1	102.9	89.1	84.0
4. Other Losses	52.6	50.2	49.6	53.2	48.8	49.4	47.2	45.6	52.7	50.2	50.9	51.3	51.7	47.3	31.7
Balance	7.4	4.4	11.4	11.4	3.4	5.0	1.8	7.5	1.8	0.4	(10.1)	(3.7)	39.9	(15.4)	(4.4)

Note: Total consumption used here may differ from the notional consumption presented in the previous chapters which does not consider inventory buildup or withdraw.
Sources: *China Statistic Yearbook, 1994, 1997*, and *China Energy Statistic Yearbook, 1986, 1991*.

Table A3.7 Crude Oil Balance Sheet (000 b/d)

	1980	1981	1982	1983	1984	1985	1986	1987	1988	1989	1990
Total Crude Oil Available for Consumption	1844.6	1750.2	1747.2	1822.0	1842.9	1903.3	2043.5	2139.2	2240.4	2321.5	2354.1
Production	2118.9	2024.4	2042.5	2121.4	2292.3	2497.9	2613.8	2682.8	2740.9	2752.8	2766.1
Import (+)	7.3	1.4	12.9	7.4	4.9	–	9.1	–	17.1	65.3	58.5
Exports (–)	266.2	275.1	304.1	303.9	445.9	600.6	570.0	544.5	520.9	487.8	479.8
Change in Inventory (million bbl)	(5.7)	(0.2)	(1.5)	(1.1)	(3.1)	2.2	(3.4)	0.3	1.2	(3.2)	(3.0)
Total Crude Oil Consumption	1841.0	1745.1	1737.7	1816.0	1842.0	1901.9	2042.1	2137.1	2240.4	2321.4	2352.4
By Sector											
1. Material Production Sectors	1839.8	1744.1	1737.5	1815.9	1842.0	1901.8	2042.1	2136.4	2240.4	2321.2	2352.3
1.1. Farming, Forestry, Animal Husbandry, Fishery and Water Conservancy	1.6	1.4	0.9	0.0	0.0	0.2	0.1	0.1	0.2	0.2	0.0
1.2. Industry	1822.4	1727.0	1721.8	1799.7	1823.7	1878.0	2018.6	2106.4	2217.4	2301.6	2330.8
Light Industry	n.a.	n.a.	n.a.	n.a.	n.a.	46.4	49.8	52.4	54.2	59.6	73.0
Heavy Industry	n.a.	n.a.	n.a.	n.a.	n.a.	1,831.6	1,968.8	2054.0	2163.2	2242.0	2257.8
1.3. Construction	5.8	6.5	6.6	8.3	10.1	14.8	11.3	19.8	12.9	10.2	11.0
1.4. Transportation, Postal and Telecom.	10.0	9.2	8.1	7.8	8.1	8.9	12.1	10.0	9.9	9.1	10.4
1.5. Commerce, Food Service, Material Supply, Marketing, and Storage	–	0.0	0.0	0.0	0.0	0.0	0.0	0.1	0.0	0.1	0.1
2. Non-Material Production Sectors	1.2	1.1	0.2	0.1	0.0	0.1	0.0	0.7	0.0	0.2	0.1
3. Residential Consumption	28.4	32.2	29.9	37.0	41.8	45.2	52.9	54.5	55.3	59.1	56.9
By Use											
1. Processing and Transformation of 2nd Energy	1688.6	1603.4	1602.6	1687.3	1717.2	1785.9	1914.2	2011.6	2097.3	2169.6	2221.4
Power Generation	114.8	108.2	96.0	79.1	67.0	55.9	49.2	51.8	46.8	32.6	24.9
Heating	–	–	–	–	–	12.3	13.1	4.9	2.5	4.6	4.2
Refining	1573.8	1495.2	1506.6	1608.2	1650.2	1717.8	1852.0	1955.0	2047.9	2132.5	2192.2
2. Final Uses	99.9	91.9	85.7	75.9	76.7	70.1	81.5	80.3	90.6	102.2	80.4
Industry	85.9	77.1	72.5	63.8	60.1	51.0	63.6	55.0	73.2	87.9	66.7
3. Loses in Oilfield	52.5	49.9	49.3	52.7	48.0	45.9	46.4	45.2	52.5	49.6	50.6
Balance	3.6	5.1	9.6	6.0	0.9	1.4	1.4	2.1	0.0	0.1	1.7

Note: Total consumption used here may differ from the notional consumption presented in the previous chapters which does not include inventory buildup or withdraw.
Sources: *China Energy Statistic Yearbook, 1986, 1991.*

Table A3.8 Gasoline Balance Sheet (000 b/d)

	1980	1981	1982	1983	1984	1985	1986	1987	1988	1989	1990
Total Gasoline Available for Consumption	232.7	220.0	231.0	257.8	283.9	325.9	349.4	378.3	417.0	431.7	438.8
Primary Energy Output	251.3	258.9	259.4	294.4	314.4	342.8	392.4	404.6	440.7	481.5	06.1
Import (+)	–	–	–	–	–	0.1	2.3	3.0	4.2	11.1	3.9
Exports (–)	(27.4)	(30.6)	(33.1)	(32.1)	(27.0)	(30.3)	(25.2)	(27.1)	(25.9)	(46.3)	(54.4)
Change in Inventory (million bbl)	8.9	(8.3)	4.7	(4.5)	(3.4)	13.3	(20.1)	(2.1)	(2.1)	(14.7)	(16.9)
Total Gasoline Consumption	232.6	218.9	231.3	255.0	279.4	325.2	349.4	378.3	416.8	431.7	442.3
1. Material Production Sectors	187.4	174.2	186.0	206.2	225.9	267.2	283.5	304.3	328.9	331.4	347.2
1.1. Farming, Forestry, Animal Husbandry, Fishery and Water Conservancy	12.4	13.4	14.1	16.8	20.8	28.5	33.2	34.1	35.8	31.3	34.0
1.2. Industry	63.6	60.7	64.0	71.3	81.9	105.1	106.3	128.2	130.5	128.9	137.2
Light Industry	n.a.	n.a.	n.a.	n.a.	n.a.	32.4	31.5	36.9	41.4	40.0	44.2
Heavy Industry	n.a.	n.a.	n.a.	n.a.	n.a.	72.7	74.8	80.8	89.1	88.8	93.0
1.3. Construction	12.6	11.5	12.8	14.3	15.0	17.0	18.5	20.5	20.3	22.2	20.8
1.4. Transportation, Postal and Telecom.	94.3	84.0	89.7	98.0	102.5	111.2	118.4	124.0	131.7	137.8	144.4
1.5. Commerce, Food Service, Material Supply, Marketing, and Storage	4.5	4.6	5.4	5.7	5.8	5.4	7.2	8.0	10.7	11.2	10.7
2. Non-Material Production Sectors	45.1	44.7	45.3	48.8	53.5	55.5	63.2	70.6	84.1	97.6	91.0
3. Residential Consumption	–	–	–	–	–	2.5	2.6	3.4	3.8	2.7	4.2
Balance	0.2	1.0	(0.2)	2.8	4.6	0.8	–	–	0.2	–	(3.6)

Note: Total consumption used here may differ from the notional consumption presented in the previous chapters which does not include inventory buildup or withdraw.
Sources: *China Energy Statistic Yearbook, 1986, 1991.*

Table A3.9 Diesel Oil Balance Sheet (000 b/d)

	1980	1981	1982	1983	1984	1985	1986	1987	1988	1989	1990
Total Diesel Oil Available for Consumption	339.9	336.1	324.4	336.6	358.2	397.3	431.5	478.5	528.0	566.9	549.7
Primary Energy Output	373.6	363.4	356.9	389.0	397.8	413.5	455.9	483.5	502.8	527.6	533.2
Import (+)	0.4	0.6	0.8	0.9	0.8	0.9	30.4	34.9	50.1	82.1	47.8
Exports (–)	(34.0)	(32.5)	(33.3)	(26.8)	(43.0)	(46.1)	(43.5)	(29.8)	(29.8)	(33.2)	(34.7)
Change in Inventory (million bbl)	(0.0)	1.7	(0.0)	(9.7)	1.0	10.6	(4.1)	(3.7)	1.8	(3.5)	1.2
Total Diesel Oil Consumption	339.9	336.1	324.4	336.6	358.2	396.4	432.6	472.9	526.3	561.6	550.1
By Sector											
1. Material Production Sectors	328.1	324.9	315.6	324.3	347.9	380.9	419.7	456.4	488.7	516.6	505.7
1.1. Farming, Forestry, Animal Husbandry, Fishery and Water Conservancy	153.1	144.0	133.7	136.0	144.1	128.6	137.2	149.1	156.8	168.8	180.2
1.2. Industry	93.5	101.7	95.9	91.9	100.0	131.6	129.8	138.3	155.2	162.7	148.8
Light Industry	n.a.	n.a.	n.a.	n.a.	n.a.	23.9	26.3	26.1	29.8	31.8	32.4
Heavy Industry	n.a.	n.a.	n.a.	n.a.	n.a.	107.8	103.5	112.2	125.4	130.9	116.4
1.3. Construction	15.6	14.8	16.2	19.4	21.9	25.5	10.2	31.7	29.1	30.0	27.2
1.4. Transportation, Postal and Telecom.	64.6	62.0	67.6	74.7	79.9	92.9	118.7	132.8	141.0	147.4	145.0
1.5. Commerce, Food Service, Material Supply, Marketing, and Storage	1.3	2.4	2.1	2.3	2.1	2.2	3.3	4.4	6.5	7.8	4.6
2. Non-Material Production Sectors	11.8	11.2	8.7	12.3	10.3	15.1	12.5	16.0	36.4	44.8	44.4
3. Residential Consumption	–	–	–	–	–	0.4	0.4	0.6	1.2	0.2	0.0
By Use											
1. Processing and Transformation of 2nd Energy	14.8	24.6	14.8	11.2	14.3	22.2	12.3	18.9	40.1	45.8	25.9
Power Generation	14.8	24.6	14.8	11.2	14.3	21.2	12.0	18.9	40.0	45.7	25.4
Heating	–	–	–	–	–	58.4	65.7	58.0	66.1	69.4	72.8
2. Final Uses	325.2	311.5	309.5	325.5	343.9	373.5	420.3	454.0	486.2	515.8	524.2
Industry	78.7	77.1	81.1	80.7	85.7	108.8	117.5	119.4	115.1	116.9	122.9
3. Loses	–	–	–	–	–	0.7	–	–	–	–	–
Balance	–	–	–	–	–	1.0	(1.1)	5.5	1.7	5.2	(0.5)

Note: Total consumption used here may differ from the notional consumption presented in the previous chapters which does not include inventory buildup or withdraw.
Sources: *China Energy Statistic Yearbook, 1986, 1991.*

Table A3.10 Kerosene Balance Sheet (000 b/d)

	1980	1981	1982	1983	1984	1985	1986	1987	1988	1989	1990
Total Kerosene Available for Consumption	77.5	72.9	73.1	75.3	80.5	82.7	83.2	82.4	77.3	77.9	75.8
Primary Energy Output	86.0	79.2	82.9	88.6	88.1	87.5	89.9	90.3	83.2	85.3	84.7
Import (+)	1.1	1.7	1.9	2.1	2.6	3.3	3.5	4.8	4.8	4.2	5.6
Exports (–)	(10.1)	(9.7)	(11.9)	(12.0)	(10.7)	(9.9)	(10.0)	(11.6)	(12.2)	(11.7)	(12.0)
Change in Inventory (million bbl)	0.5	1.6	0.2	(3.3)	0.5	1.9	(0.1)	(1.0)	1.5	0.2	(2.6)
Total Kerosene Consumption	79.0	74.9	73.7	79.5	83.0	83.2	83.2	82.4	77.3	77.9	75.8
1. Material Production Sectors	10.9	11.4	12.5	13.1	15.7	17.5	17.0	20.8	21.0	21.7	25.7
1.1. Farming, Forestry, Animal Husbandry, Fishery and Water Conservancy	0.5	0.6	0.6	0.6	0.6	0.7	0.5	0.6	0.6	0.8	0.7
1.2. Industry	3.4	3.0	3.3	3.5	3.7	4.3	4.4	5.0	4.9	4.7	4.4
Light Industry	n.a.	n.a.	n.a.	n.a.	n.a.	1.1	1.2	1.5	1.6	1.3	1.2
Heavy Industry	n.a.	n.a.	n.a.	n.a.	n.a.	3.2	3.2	3.5	3.3	3.4	3.2
1.3. Construction	0.2	0.1	0.2	0.2	0.2	0.3	0.4	0.4	0.3	0.3	0.3
1.4. Transportation, Postal and Telecom.	6.8	7.7	8.4	8.9	11.2	12.1	11.4	14.6	14.8	15.6	20.2
1.5. Commerce, Food Service, Material Supply, Marketing, and Storage	0.0	0.0	0.0	0.0	0.0	0.0	0.2	0.3	0.4	0.2	0.1
2. Non-Material Production Sectors	46.8	38.6	39.5	39.4	35.3	39.5	37.3	34.0	30.4	28.5	27.5
3. Residential Consumption	21.3	24.9	21.7	26.9	32.0	26.3	29.0	27.6	26.0	27.7	22.6
Balance	(1.5)	(2.0)	(0.5)	(4.1)	(2.4)	(0.5)	–	–	–	–	–

Note: Total consumption used here may differ from the notional consumption presented in the previous chapters which does not include inventory buildup or withdraw.
Sources: *China Energy Statistic Yearbook, 1986, 1991.*

Table A3.11 Fuel Oil Balance Sheet (000 b/d)

	1980	1981	1982	1983	1984	1985	1986	1987	1988	1989	1990
Total Fuel Oil Available for Consumption	564.9	519.9	514.7	526.8	525.6	519.7	549.8	569.5	590.8	614.2	605.9
Production	573.3	514.2	510.7	528.5	521.3	517.4	544.9	571.2	587.3	590.7	596.3
Import (+)	7.1	9.7	14.5	15.3	13.2	12.8	23.6	18.8	19.8	36.3	30.5
Exports (–)	(8.3)	(8.8)	(8.2)	(11.3)	(10.4)	(11.8)	(13.0)	(14.3)	(14.7)	(16.3)	(17.7)
Change in Inventory (million bbl)	(7.2)	4.9	(2.3)	(5.6)	1.5	1.3	(5.7)	(6.4)	(1.5)	3.5	(3.2)
Total Fuel Oil Consumption	560.8	519.9	512.9	521.2	525.3	517.7	548.9	569.5	590.8	618.7	614.5
By Sector											
1. Material Production Sectors	559.5	518.7	512.1	520.3	524.7	516.6	548.0	568.2	590.2	616.4	611.6
1.1. Farming, Forestry, Animal Husbandry, Fishery and Water Conservancy	0.4	0.7	0.6	0.3	0.4	0.6	0.2	0.7	0.8	0.8	0.5
1.2. Industry	536.0	494.4	484.6	492.4	495.3	485.8	513.1	526.6	550.5	572.8	564.1
Light Industry	n.a.	n.a.	n.a.	n.a.	n.a.	54.0	62.5	72.1	64.7	60.6	60.6
Heavy Industry	n.a.	n.a.	n.a.	n.a.	n.a.	431.8	450.5	454.5	485.8	512.2	503.5
1.3. Construction	2.7	2.7	2.2	2.2	2.7	3.4	2.8	8.3	4.9	5.8	8.6
1.4. Transportation, Postal and Telecom.	19.9	20.4	24.1	24.8	25.6	26.3	31.6	32.1	33.5	36.5	38.0
1.5. Commerce, Food Service, Material Supply, Marketing, and Storage	0.5	0.5	0.6	0.6	0.7	0.6	0.3	0.4	0.5	0.5	0.3
2. Non–Material Production Sectors	1.3	1.2	0.8	0.9	0.6	1.1	0.9	1.3	0.6	2.3	2.9
3. Residential Consumption	–	–	–	–	–	–	–	–	–	–	–
By Use											
1. Processing and Transformation of 2nd Energy	265.6	253.1	248.0	254.2	254.5	236.5	243.7	239.3	253.0	252.0	241.8
Power Generation	258.9	246.4	241.0	247.3	247.1	190.2	190.8	185.9	195.7	191.1	178.3
Heating	–	–	–	–	–	40.0	46.0	46.7	50.4	53.3	56.3
Gas Generation	6.7	6.7	7.0	6.9	7.3	6.3	6.9	6.8	6.8	7.6	7.2
2. Final Uses	295.2	266.8	264.9	267.0	270.9	280.8	305.2	330.2	337.9	366.7	372.7
Industry	270.3	241.3	236.6	238.2	240.8	248.8	269.3	287.3	297.5	320.8	322.3
3. Loses	–	–	–	–	–	0.5	–	–	–	–	–
Balance	4.1	0.0	1.9	5.6	0.3	1.9	0.9	–	–	(4.5)	(8.6)

Note: Total consumption used here may differ from the notional consumption presented in the previous chapters which does not include inventory buildup or withdraw.
Sources: *China Energy Statistic Yearbook, 1986, 1991.*

Table A3.12 Population by Region (million)

	1985	1986	1987	1988	1989	1990	1991	1992	1993	1996
National Total	1041.1	1067.9	1073.9	1088.9	1110.4	1131.9	1153.9	1165.7	1184.8	1223.9
North	119.5	122.6	123.9	125.7	127.5	129.3	133.6	134.8	137.4	141.1
Beijing	9.6	9.8	10.7	10.8	10.8	10.9	11.0	11.0	11.1	12.6
Tianjin	8.1	8.2	8.3	8.4	8.6	8.8	9.1	9.2	9.4	9.5
Hebei	55.5	57.1	57.1	58.0	59.1	60.3	62.2	62.8	64.0	64.8
Shanxi	26.3	26.9	27.2	27.6	27.9	28.2	29.4	29.8	30.5	31.1
Inner Mongolia	20.1	20.6	20.7	20.9	21.0	21.1	21.8	22.1	22.4	23.1
North-east	93.0	95.1	95.5	96.6	98.2	99.9	100.7	101.6	102.7	104.5
Liaoning	36.9	37.7	37.8	38.2	39.1	40.0	39.9	40.2	40.5	41.2
Jilin	23.0	23.5	23.5	23.7	24.4	25.2	25.1	25.3	25.6	26.1
Heilongjiang	33.1	33.9	34.2	34.7	34.7	34.8	35.7	36.1	36.5	37.3
East	304.8	311.7	313.4	317.6	324.4	331.1	336.7	339.7	344.9	350.5
Shanghai	12.2	12.4	12.5	12.6	13.1	13.5	13.4	13.5	13.6	14.2
Jiangsu	62.1	63.3	63.5	64.4	66.3	68.2	68.5	69.1	70.1	71.1
Zhejiang	40.3	41.0	41.2	41.7	41.3	40.8	42.0	42.4	42.7	43.4
Anhui	51.6	52.7	53.0	53.8	55.0	56.3	57.6	58.3	59.5	60.7
Fujian	27.1	27.8	28.1	28.5	29.5	30.6	30.8	31.2	31.7	32.6
Jiangxi	34.6	35.5	35.6	36.1	37.2	38.3	38.6	39.1	39.8	41.1
Shandong	77.0	79.1	79.6	80.6	82.0	83.4	85.7	86.1	87.5	87.4
South-central	283.9	291.9	293.5	298.1	305.9	313.7	319.3	323.0	328.9	337.1
Henan	77.1	79.2	79.7	80.9	83.5	86.1	87.6	88.6	90.4	91.7
Hubei	49.3	50.7	51.2	51.9	53.3	54.8	55.1	55.8	56.7	58.3
Hunan	56.2	57.6	57.9	58.9	59.8	60.6	62.1	62.7	63.7	64.3
Guangdong	62.5	64.3	58.2	59.3	61.2	63.2	64.4	65.3	66.6	69.6
Guangxi	38.7	40.0	40.2	40.9	41.7	42.5	43.2	43.8	44.5	45.9
Hainan	–	–	6.3	6.3	6.4	6.4	6.8	6.9	7.0	7.3
South-west	167.6	172.1	172.7	175.1	176.6	178.1	182.2	184.2	186.8	192.7
Sichuan	101.9	104.4	104.5	105.8	106.1	106.4	109.0	110.0	111.3	114.3
Guizhou	29.7	30.5	30.7	31.3	32.0	32.7	33.1	33.6	34.2	35.6
Yunnan	34.1	35.1	35.3	35.9	36.3	36.8	37.8	38.3	39.0	40.4
Tibet	2.0	2.1	2.1	2.1	2.2	2.2	2.2	2.3	2.3	2.4
West	72.3	74.5	74.9	75.8	77.8	79.9	81.4	82.5	84.1	87.1
Shaanxi	30.0	30.9	30.9	31.4	31.9	32.5	33.6	34.1	34.8	35.4
Gansu	20.4	21.0	21.2	21.4	22.1	22.9	22.8	23.1	23.5	24.7
Qinghai	4.1	4.2	4.3	4.3	4.4	4.4	4.5	4.6	4.7	4.9
Ningxia	4.2	4.3	4.4	4.5	4.6	4.7	4.8	4.9	5.0	5.2
Xinjiang	13.6	14.1	14.2	14.3	14.8	15.4	15.6	15.8	16.2	16.9

Sources: *China Statistical Yearbook, 1989, 1994, 1997.*

Table A3.13 GDP by Region (billion yuan)

	1985	1986	1987	1988	1989	1990	1991	1992	1993	1996
National Total	1124.5	1395.2	1559.5	1721.5	1973.7	2376.3	3422.7	4538.4	5799.1	6858.4
North	149.5	185.5	209.5	233.5	265.4	317.4	433.5	553.2	709.0	846.1
Beijing	32.1	41.0	45.6	50.1	59.9	70.9	86.4	108.4	139.5	161.6
Tianjin	22.6	26.0	28.3	31.1	34.3	41.1	53.6	72.5	92.0	110.2
Hebei	53.3	65.4	74.9	83.6	96.0	115.5	169.8	218.7	285.0	345.3
Shanxi	23.9	29.6	35.0	40.0	43.2	52.0	70.5	85.4	109.2	130.6
Inner Mongolia	17.6	23.5	25.7	28.7	32.1	37.9	53.3	68.2	83.3	98.5
North-east	140.6	170.3	186.6	200.9	223.2	266.8	393.2	501.7	593.7	689.7
Liaoning	68.4	84.0	92.2	95.6	107.3	129.8	201.1	246.2	279.3	315.8
Jilin	28.1	34.8	36.1	39.4	42.4	51.5	71.8	93.7	112.9	133.7
Heilongjiang	44.0	51.5	58.3	65.9	73.5	85.6	120.3	161.9	201.5	240.3
East	384.9	476.3	525.5	571.6	672.4	827.3	1212.5	1669.0	2155.4	2547.6
Shanghai	56.5	67.1	69.7	74.5	85.8	111.4	151.2	197.2	246.3	290.2
Jiangsu	91.9	113.2	123.8	131.6	147.1	197.2	299.8	405.7	515.5	600.4
Zhejiang	57.9	72.2	79.2	83.7	98.4	122.1	190.9	266.7	352.5	414.6
Anhui	41.6	51.3	57.3	60.7	80.0	73.0	107.0	148.8	200.4	233.9
Fujian	26.3	35.4	41.7	45.9	54.7	69.5	113.3	168.5	216.1	260.7
Jiangxi	25.9	32.1	36.3	42.0	46.5	56.0	72.3	94.8	124.5	151.7
Shandong	84.7	105.1	117.5	133.3	159.9	198.2	277.9	387.2	500.2	596.0
South-central	276.3	342.0	391.6	434.9	495.7	594.2	874.2	1161.1	1529.4	1807.9
Henan	61.0	73.8	82.6	89.6	99.3	121.3	166.3	222.4	300.3	368.3
Hubei	54.2	62.5	70.1	79.1	85.7	100.1	142.4	187.9	239.1	297.0
Hunan	47.5	58.4	64.1	70.3	78.6	92.0	127.8	169.4	219.6	264.7
Guangdong	83.6	109.9	131.2	147.2	176.1	229.4	322.5	424.1	573.4	651.9
Guangxi	24.4	30.0	34.9	39.3	45.3	37.2	89.4	124.2	160.6	187.0
Hainan	5.5	7.5	8.7	9.5	10.8	14.2	25.8	33.1	36.4	39.0
South-west	112.1	142.5	157.5	182.0	203.4	236.6	332.9	431.9	540.7	649.1
Sichuan	72.9	92.5	99.8	114.5	128.1	149.1	209.6	277.8	353.4	421.5
Guizhou	16.8	21.1	23.6	25.5	29.0	33.2	41.6	52.1	61.1	72.0
Yunnan	20.7	26.8	31.9	39.6	43.3	51.0	77.9	97.4	120.7	149.2
Tibet	1.6	2.0	2.2	2.4	3.1	3.3	3.7	4.6	5.6	6.5
West	61.2	78.5	88.8	98.6	113.6	133.9	176.3	221.4	270.8	317.9
Shaanxi	23.3	30.2	33.7	37.4	42.8	49.3	67.1	81.7	99.5	117.5
Gansu	15.3	19.2	21.7	23.4	25.5	30.2	37.2	45.2	55.3	71.4
Qinghai	4.2	5.5	6.0	6.5	7.3	8.4	11.0	13.8	16.5	18.4
Ningxia	3.6	4.8	5.6	6.1	6.8	7.9	10.4	13.4	17.0	19.4
Xinjiang	14.7	18.9	21.7	25.2	31.2	38.2	50.6	67.4	82.5	91.2

Sources: *China Statistical Yearbook, 1989, 1994, 1997.*

Table A3.14 Gross Domestic Product Indices by Region (100 for previous year)

	1985	1986	1987	1988	1989	1990	1991	1992	1993	1994	1995	1996
National Total	100.0	105.5	108.4	120.5	118.4	104.2	109.1	114.1	113.1	112.6	109.0	109.7
North												
Beijing	100.0	106.7	108.7	121.9	118.5	105.4	107.5	111.6	112.1	113.5	112.4	109.2
Tianjin	100.0	107.2	106.9	117.7	115.1	102.5	104.4	111.7	112.1	114.3	114.9	114.3
Hebei	100.0	105.2	108.3	118.1	118.4	104.7	109.0	113.9	117.7	114.9	113.9	113.5
Shanxi	100.0	105.3	107.5	121.0	119.1	105.1	103.3	112.8	112.2	109.4	111.1	111.0
Inner Mongolia	100.0	105.0	108.1	116.3	115.9	108.2	107.5	111.1	110.6	110.1	109.1	112.7
North-east												
Liaoning	100.0	106.0	109.0	119.3	118.4	100.3	105.5	111.7	114.9	111.2	107.1	108.6
Jilin	100.0	105.4	107.5	119.9	116.9	103.5	104.8	113.0	112.8	114.3	109.7	113.7
Heilongjiang	100.0	105.9	109.6	117.8	114.0	104.5	103.9	106.5	107.6	108.8	109.6	110.5
East												
Shanghai	100.0	106.7	108.8	121.3	116.7	103.5	107.0	114.9	114.9	114.3	114.1	113.0
Jiangsu	100.0	106.5	109.3	121.7	118.0	104.4	108.9	126.2	120.7	116.5	115.4	112.2
Zhejiang	100.0	106.0	109.5	122.1	117.8	104.0	115.4	118.6	122.0	120.0	116.7	112.7
Anhui	100.0	105.2	109.7	121.8	117.1	103.2	96.3	117.4	121.0	120.7	114.3	114.4
Fujian	100.0	105.9	109.4	126.5	118.8	107.0	114.7	120.8	125.2	121.7	115.2	115.4
Jiangxi	100.0	105.8	106.9	121.8	118.6	104.9	108.2	114.8	113.7	117.0	114.5	113.4
Shandong	100.0	104.2	108.0	118.3	117.1	105.3	113.9	119.5	118.5	116.3	114.2	112.2
South-central												
Henan	100.0	105.0	108.1	120.2	118.3	104.5	106.9	113.7	115.8	113.8	114.8	113.9
Hubei	100.0	104.2	107.6	119.5	117.0	102.5	104.5	112.1	114.3	115.2	114.6	113.2
Hunan	100.0	104.8	110.6	125.9	118.1	104.0	107.8	112.3	113.1	111.0	110.9	112.6
Guangdong	100.0	104.8	111.7	130.2	121.0	111.5	117.3	122.0	122.3	119.1	114.9	110.7
Guangxi	100.0	105.1	108.0	121.0	121.3	107.0	112.7	118.3	121.2	116.0	115.3	110.3
Hainan	100.0	104.8	111.7	127.8	126.8	109.5	113.4	133.3	120.9	111.9	104.3	104.8
South-west												
Sichuan	100.0	103.9	107.5	120.0	118.3	103.7	107.7	112.4	113.9	111.1	110.0	109.8
Guizhou	100.0	105.3	107.3	120.2	117.4	104.2	109.9	109.1	109.9	108.5	107.5	108.9
Yunnan	100.0	105.0	106.6	119.6	119.3	108.7	106.6	110.9	110.6	111.6	111.2	110.4
Tibet	100.0	106.1	107.3	118.3	117.7	108.9	101.6	107.1	108.2	115.6	117.9	113.2
West												
Shaanxi	100.0	105.2	108.6	119.0	118.8	104.3	110.9	108.7	113.3	108.5	109.0	110.2
Gansu	100.0	106.0	107.4	118.6	116.4	105.5	105.5	109.7	111.6	110.4	109.9	111.5
Qinghai	100.0	106.1	107.3	118.3	117.7	103.7	104.7	107.4	109.6	108.2	108.0	108.6
Ningxia	100.0	104.9	108.0	117.5	117.8	103.8	104.2	107.5	110.1	108.2	109.0	110.5
Xinjiang	100.0	106.7	107.1	114.6	116.7	109.2	112.9	112.1	110.3	110.9	109.0	106.4

Sources: *China Statistical Yearbook, 1989, 1994, 1997.*

Table A3.15 Crude Oil Production by Region (000 b/d)

	1985	1986	1987	1988	1989	1990	1991	1992	1993	1994	1995
National Total	2498	2614	2683	2741	2753	2766	2820	2858	2898	2928	2967
North	281	283	250	219	210	208	223	222	224	217	214
Beijing	–	–	–	–	–	–	–	–	–	–	–
Tianjin	75	83	91	93	94	94	95	98	103	116	110
Hebei	206	200	159	126	116	114	108	104	100	101	104
Shanxi	–	–	–	–	–	–	–	–	–	–	–
Inner Mongolia	–	–	–	–	–	–	20	20	20	–	–
North-east	1333	1363	1398	1431	1449	1458	1456	1460	1472	1497	1492
Liaoning	184	204	229	253	269	274	275	277	286	298	309
Jilin	43	47	57	63	68	71	68	69	68	68	68
Heilongjiang	1106	1111	1111	1114	1111	1112	1112	1113	1118	1130	1115
East	552	603	646	682	684	688	690	686	671	632	618
Shanghai	–	–	–	–	–	–	–	–	–	–	–
Jiangsu	11	12	13	15	17	18	19	17	17	18	20
Zhejiang	–	–	–	–	–	–	–	–	–	–	–
Anhui	0	0	0	1	0	0	1	–	–	–	–
Fujian	–	–	–	–	–	–	–	–	–	–	–
Jiangxi	–	–	–	–	–	–	–	–	–	–	–
Shandong	541	590	632	666	667	670	671	669	654	614	597
South–central	182	203	217	227	218	205	218	233	240	250	267
Henan	159	176	186	196	191	176	170	158	153	136	118
Hubei	20	21	20	20	19	17	15	15	16	17	17
Hunan	–	–	–	–	–	–	–	–	–	–	–
Guangdong	2	5	9	9	7	10	31	58	70	96	130
Guangxi	1	1	1	2	2	2	3	1	1	1	3
Hainan	–	–	–	–	–	–	–	–	–	–	–
South-west	2	2	2	2	3	3	3	6	6	6	6
Sichuan	2	2	2	2	3	3	3	3	3	3	4
Guizhou	–	–	–	–	–	–	–	–	–	–	–
Yunnan	–	–	–	–	–	–	–	3	3	2	2
Tibet	–	–	–	–	–	–	–	–	–	–	–
West	148	161	170	181	189	204	231	251	286	326	370
Shaanxi	4	5	7	9	11	14	17	19	23	26	34
Gansu	31	31	31	30	29	29	32	43	30	48	52
Qinghai	4	7	12	13	15	16	20	21	22	22	24
Ningxia	8	7	6	6	6	5	5	4	4	6	8
Xinjiang	100	110	115	123	129	139	157	164	207	223	251

Sources: *China Statistical Yearbook, 1994, China Energy Statistical Yearbook, 1991,* and State Statistical Bureau.

Table A3.16 Refinery Throughput by Region (000 b/d)

	1985	1986	1987	1988	1989	1990	1991	1992	1993	1994	1995
National Total	1718	1852	1955	2173	2269	2321	2433	2529	2651	2639	2881
North	212	226	241	273	280	277	276	291	320	312	323
Beijing	111	119	123	135	137	136	134	133	133	128	130
Tianjin	72	66	71	79	82	79	78	76	87	83	89
Hebei	30	41	48	58	59	60	64	66	84	84	88
Shanxi	–	–	–	1	2	2	0	–	–	–	0
Inner Mongolia	0	0	0	0	0	0	0	16	16	17	16
North-east	624	656	676	732	750	767	778	761	786	766	825
Liaoning	391	407	420	460	471	470	474	488	504	490	503
Jilin	68	76	85	86	87	92	90	85	83	82	93
Heilongjiang	165	172	171	187	192	205	213	189	199	194	229
East	461	503	556	616	665	679	730	795	829	829	902
Shanghai	132	137	143	153	157	160	172	180	185	184	193
Jiangsu	93	102	127	155	157	164	174	179	183	184	199
Zhejiang	37	44	45	49	56	59	69	82	93	105	116
Anhui	40	43	49	52	51	51	51	53	53	48	56
Fujian	–	–	–	–	–	–	–	33	33	32	44
Jiangxi	15	22	25	29	30	31	36	42	46	40	47
Shandong	144	155	166	179	213	213	228	227	236	236	248
South-central	285	321	324	373	381	390	416	435	445	437	493
Henan	20	25	25	40	38	42	48	43	52	57	73
Hubei	83	90	85	99	102	101	97	99	97	97	100
Hunan	53	56	57	63	63	64	67	73	69	68	70
Guangdong	129	149	155	169	177	182	204	218	224	213	242
Guangxi	1	1	1	1	1	1	1	1	2	2	9
Hainan	–	–	–	0	0	0	0	–	–	–	–
South-west	2	2	2	2	2	3	3	4	4	4	11
Sichuan	2	2	2	2	2	3	3	4	4	4	5
Guizhou	–	–	–	–	–	–	–	–	–	–	–
Yunnan	–	–	–	–	–	–	–	–	–	–	6
Tibet	–	–	–	–	–	–	–	–	–	–	–
West	133	144	155	177	190	205	230	243	267	290	326
Shaanxi	3	3	4	6	7	11	17	15	17	18	24
Gansu	72	77	80	87	91	95	101	104	107	116	134
Qinghai	3	3	3	4	5	5	4	3	3	17	17
Ningxia	–	–	–	–	–	–	3	8	11	11	13
Xinjiang	55	61	68	79	87	94	105	113	129	129	139

Sources: *China Statistical Yearbook, 1994, China Energy Statistical Yearbook, 1991*, and State Statistical Bureau.

Table A3.17 Principal Product Consumption by Region (000 b/d)

	1985	1986	1987	1988	1989	1990	1991	1992	1993	1994	1995
National Total	1349	1472	1535	1601	1661	1700	1794	1912	2088	1977	2107
North	199	192	205	195	200	205	204	224	252	230	222
Beijing	108	105	112	91	92	97	94	96	96	80	65
Tianjin	58	54	56	59	62	62	61	64	73	65	72
Hebei	33	33	37	46	46	47	49	54	73	70	72
Shanxi	–	–	–	–	0	0	0	0	0	–	0
Inner Mongolia	–	–	–	–	0	0	0	10	11	15	13
North-east	529	523	528	551	566	577	583	593	633	589	603
Liaoning	364	347	348	361	371	374	374	379	411	371	358
Jilin	46	50	57	60	61	65	65	66	65	66	71
Heilongjiang	118	126	124	130	134	138	145	148	157	152	175
East	327	388	419	445	462	465	512	566	612	573	613
Shanghai	92	90	89	95	98	105	109	114	116	103	112
Jiangsu	84	86	102	113	111	112	118	125	132	120	124
Zhejiang	35	39	43	43	51	43	64	74	86	86	90
Anhui	38	37	42	43	35	41	42	42	46	40	46
Fujian	0	0	0	0	0	0	0	25	26	25	36
Jiangxi	10	22	25	27	28	29	34	38	44	37	41
Shandong	68	114	117	124	139	135	146	149	163	162	165
South-central	189	263	270	286	298	304	327	351	383	358	405
Henan	18	15	23	34	32	36	38	38	46	49	64
Hubei	51	79	72	81	82	82	79	81	87	80	81
Hunan	38	45	46	48	49	49	51	56	57	55	54
Guangdong	66	124	127	122	134	136	157	174	188	168	199
Guangxi	16	1	1	1	0	1	1	2	4	7	7
Hainan	–	–	–	–	–	–	–	–	–	–	–
South-west	1	1	1	1	2	2	2	3	3	3	9
Sichuan	1	1	1	1	2	2	2	3	3	3	3
Guizhou	–	–	–	–	–	–	–	–	–	–	–
Yunnan	–	–	–	–	–	–	–	–	–	–	5
Tibet	–										
West	104	104	111	123	133	147	166	177	204	223	255
Shaanxi	3	4	5	6	7	11	13	13	15	17	21
Gansu	51	48	52	51	53	55	63	63	72	78	98
Qinghai	4	3	3	3	2	1	3	2	2	11	14
Ningxia	1	1	2	2	2	2	4	6	8	8	12
Xinjiang	44	48	50	62	69	78	84	93	108	109	110

Note: Principal products includes gasoline, kerosene, diesel oil, and fuel oil.
Sources: *China Statistical Yearbook, 1994, China Energy Statistical Yearbook, 1991*, and State Statistical Bureau.

Table A3.18 Estimated Principal Product Consumption by Region (000 b/d)

	1985	1986	1987	1988	1989	1990	1991	1992	1993	1994	1995
National Total	1312	1402	1493	1605	1673	1650	1784	1929	2259	2063	2280
North	257	268	288	318	301	304	316	334	408	320	366
Beijing	127	127	129	145	124	126	130	139	182	134	187
Tianjin	59	62	68	73	69	69	65	68	78	52	47
Hebei	33	35	41	45	49	49	54	57	70	65	69
Shanxi	19	21	25	26	29	30	33	33	36	36	36
Inner Mongolia	19	22	25	29	29	30	34	36	42	33	27
North-east	273	283	298	327	335	337	341	358	398	415	361
Liaoning	158	162	161	169	177	177	177	184	206	189	183
Jilin	30	30	32	41	42	39	38	41	55	50	49
Heilongjiang	86	91	105	117	116	121	127	133	137	176	128
East	349	385	408	430	453	446	483	531	609	569	629
Shanghai	119	135	132	153	141	143	145	161	156	154	168
Jiangsu	62	67	74	71	87	85	97	108	133	131	127
Zhejiang	32	37	42	37	44	45	51	53	68	65	67
Anhui	24	28	28	29	31	29	31	31	37	28	34
Fujian	16	16	20	21	20	19	21	25	29	27	35
Jiangxi	17	20	20	23	22	22	23	26	35	34	42
Shandong	78	83	92	96	109	102	115	126	150	131	157
South-central	256	285	301	312	352	326	375	409	511	464	556
Henan	36	40	46	53	56	53	58	60	71	67	76
Hubei	62	72	69	71	74	68	72	78	94	87	91
Hunan	35	35	39	41	42	38	40	43	53	45	45
Guangdong	107	121	126	119	150	138	167	192	248	226	290
Guangxi	15	17	21	24	25	24	30	30	36	34	36
Hainan	–	–	–	3	5	5	7	7	9	6	19
South-west	87	90	98	111	109	106	117	124	152	128	161
Sichuan	49	58	56	61	63	62	69	75	93	83	90
Guizhou	11	11	16	17	16	15	16	15	22	15	26
Yunnan	22	21	27	32	30	29	33	35	37	30	45
Tibet	6	n.a.	n.a.	n.a.	n.a.	n.a.	n.a.	n.a.	n.a.	n.a.	n.a.
West	90	92	101	108	123	132	152	173	181	167	207
Shaanxi	19	22	24	26	24	24	26	30	33	28	44
Gansu	24	24	25	26	27	30	32	35	38	46	39
Qinghai	8	7	8	8	8	9	10	10	7	6	9
Ningxia	3	4	4	5	7	8	9	16	13	12	14
Xinjiang	35	35	40	43	57	61	74	82	90	76	100

Note: Consumption is estimated based on national total production, imports/exports, and partial regional consumption data, and is notional consumption excluding inventory buildup or withdraw.
Sources: *China Statistical Yearbook, 1994, China Energy Statistical Yearbook, 1991*, and State Statistical Bureau.

Table A3.19 Gasoline Production by Region (000 b/d)

	1985	1986	1987	1988	1989	1990	1991	1992	1993	1994	1995
National Total	342.8	392.4	404.5	442.1	459.3	506.1	559.8	627.8	736.0	664.7	708.6
North	42.7	54.5	59.0	62.7	63.2	70.8	69.5	78.8	94.6	84.5	80.9
Beijing	20.4	26.5	29.5	28.4	28.9	33.0	31.4	32.1	32.9	27.8	23.1
Tianjin	12.7	13.8	14.2	14.6	14.1	16.3	16.5	16.9	21.6	20.0	20.9
Hebei	9.6	14.3	15.3	19.8	20.2	21.4	21.6	23.8	33.8	30.6	31.5
Shanxi	–	–	–	–	–	–	–	–	–	–	–
Inner Mongolia	–	–	–	–	–	–	–	6.1	6.3	6.1	5.4
North-east	104.1	121.8	117.3	134.2	138.5	154.5	168.7	179.1	207.7	187.8	189.3
Liaoning	64.4	73.4	69.6	81.9	85.2	92.7	100.8	108.3	125.7	106.0	93.8
Jilin	13.9	18.6	20.0	22.5	23.1	26.9	30.7	31.9	32.1	31.4	34.6
Heilongjiang	25.9	29.8	27.7	29.7	30.2	34.9	37.2	38.9	50.0	50.4	61.0
East	80.4	97.9	101.8	111.8	116.9	121.3	141.8	173.5	201.1	167.8	180.0
Shanghai	14.3	19.3	18.0	20.2	21.6	25.1	26.3	26.8	29.6	20.0	26.0
Jiangsu	15.8	18.9	20.2	23.8	20.5	21.9	23.9	28.3	36.5	27.7	25.7
Zhejiang	9.6	11.6	11.2	11.7	16.3	13.9	26.2	32.0	36.5	30.3	29.0
Anhui	8.9	10.5	10.9	12.6	12.9	13.6	13.6	15.5	17.0	14.5	17.3
Fujian	–	–	–	–	–	–	–	13.8	14.4	14.0	18.8
Jiangxi	5.7	8.9	9.5	10.1	!0.0	11.2	13.3	16.3	18.6	14.3	15.1
Shandong	26.0	28.6	32.1	33.3	35.6	35.6	38.6	40.7	48.4	47.0	48.1
South-central	74.8	78.5	81.9	87.7	94.3	106.6	118.1	127.2	149.6	135.0	153.3
Henan	6.7	3.6	11.2	16.8	17.4	19.0	20.0	20.6	25.2	25.2	32.0
Hubei	26.4	26.8	23.1	27.0	26.8	28.1	27.8	28.4	34.5	29.2	29.5
Hunan	15.1	14.7	17.1	17.3	16.8	18.9	20.2	22.0	24.4	23.1	21.1
Guangdong	26.5	33.2	30.3	26.5	33.3	40.4	49.9	55.5	63.8	54.3	67.1
Guangxi	0.1	0.1	0.1	0.1	0.1	0.3	0.3	0.6	1.7	3.1	3.5
Hainan	–	–	–	–	–	–	–	–	–		
South-west	0.4	0.3	0.4	0.4	0.6	1.0	1.1	1.1	1.2	1.3	4.1
Sichuan	0.4	0.3	0.4	0.4	0.6	1.0	1.1	1.1	1.2	1.3	1.2
Guizhou	–	–	–	–	–	–	–	–	–	–	–
Yunnan	–	–	–	–	–	–	–	–	–	–	3.0
Tibet	–	–	–	–	–	–	–	–	–	–	–
West	40.4	39.3	44.1	45.2	45.8	52.0	60.6	68.1	81.7	88.2	100.9
Shaanxi	1.0	1.2	1.8	2.3	2.9	4.1	4.4	6.1	7.1	8.4	9.0
Gansu	18.3	15.2	18.2	17.8	16.0	15.6	21.2	22.3	26.8	29.2	38.2
Qinghai	0.5	0.5	0.4	0.4	0.3	0.7	0.8	0.7	0.7	6.1	6.8
Ningxia	0.3	0.4	0.6	0.6	0.8	0.8	1.6	3.0	3.8	3.7	5.5
Xinjiang	20.2	22.0	23.1	24.2	25.8	30.7	32.8	35.9	43.4	40.8	41.4

Sources: *China Statistical Yearbook, 1994, China Energy Statistical Yearbook, 1991*, and State Statistical Bureau.

Table A3.20 Estimated Gasoline Consumption by Region (000 b/d)

	1985	1986	1987	1988	1989	1990	1991	1992	1993	1994	1995
National Total	325.9	349.4	378.2	418.4	432.4	438.8	514.5	569.5	693.3	625.4	669.0
North	49.8	54.6	58.8	64.4	66.1	68.4	78.9	85.0	98.3	88.5	80.0
Beijing	8.9	9.8	11.3	14.1	10.6	12.3	13.4	14.7	17.6	13.9	12.3
Tianjin	7.2	7.8	8.2	8.5	9.4	9.1	10.5	11.0	13.1	13.6	7.8
Hebei	13.1	14.5	15.0	15.9	19.0	19.3	22.9	24.3	28.4	27.7	28.0
Shanxi	10.9	11.2	12.2	12.4	14.0	14.0	15.6	16.7	17.1	17.7	17.2
Inner Mongolia	9.8	11.3	12.1	13.6	13.0	13.7	16.5	18.3	22.1	15.6	14.7
North-east	49.6	55.1	58.5	66.2	67.9	68.0	82.5	89.1	120.1	109.8	105.9
Liaoning	18.8	20.9	22.0	23.2	26.5	25.8	30.9	34.4	43.5	37.1	39.3
Jilin	9.9	10.9	12.2	12.8	13.6	13.3	15.8	17.4	28.3	27.5	21.3
Heilongjiang	21.0	23.4	24.2	30.2	27.8	28.8	35.8	37.2	48.3	45.1	45.3
East	75.4	82.2	90.1	95.1	102.9	105.8	124.4	143.2	177.3	164.6	171.8
Shanghai	10.1	11.3	12.8	15.5	13.4	13.8	16.2	19.7	22.4	24.6	27.0
Jiangsu	15.3	17.4	19.6	18.7	22.9	23.1	28.0	31.9	41.7	39.7	37.7
Zhejiang	8.8	9.8	10.8	9.8	11.8	13.9	15.0	16.0	21.6	18.6	20.5
Anhui	9.3	9.9	9.9	11.0	11.4	11.7	14.1	15.4	18.8	12.9	14.5
Fujian	7.1	6.5	7.4	8.9	7.8	7.7	9.1	9.6	!0.0	9.9	7.0
Jiangxi	8.6	9.4	9.3	10.9	9.9	!0.0	10.9	11.7	15.6	15.2	14.8
Shandong	16.1	18.0	20.3	20.4	25.7	25.6	31.1	39.0	47.4	43.7	50.3
South-central	63.8	69.7	75.5	86.8	88.2	85.4	100.5	109.6	135.3	126.9	142.5
Henan	13.6	14.7	15.9	19.0	19.4	19.0	22.3	25.5	28.9	28.6	32.7
Hubei	15.4	17.1	18.8	21.5	22.0	21.4	24.7	27.1	30.2	30.4	29.5
Hunan	11.5	11.6	13.8	15.2	15.3	14.6	17.1	18.8	24.3	20.3	18.5
Guangdong	17.1	19.1	18.9	19.1	20.1	18.8	22.8	24.3	36.6	34.8	42.1
Guangxi	6.3	7.3	8.0	9.3	8.9	9.2	!0.0	10.9	11.7	!0.0	11.2
Hainan	–	–	–	2.6	2.6	2.3	3.6	2.9	3.5	2.7	8.5
South-west	49.5	48.9	53.6	62.9	59.6	61.4	70.8	75.6	94.5	71.0	86.7
Sichuan	24.6	26.4	29.3	33.1	33.0	34.1	39.3	42.9	55.5	43.2	46.9
Guizhou	7.0	7.3	7.4	8.6	7.7	7.7	8.7	8.5	15.2	8.8	15.9
Yunnan	14.7	15.2	17.0	21.2	19.0	19.6	22.8	24.3	23.8	19.0	23.9
Tibet	3.2	n.a.	n.a.	n.a.	n.a.	n.a.	n.a.	n.a.	n.a.	n.a.	n.a.
West	37.7	38.8	41.8	43.0	47.7	49.9	57.5	67.1	67.8	64.7	82.1
Shaanxi	9.2	9.9	10.7	11.9	11.3	11.6	13.5	15.0	17.4	13.1	16.6
Gansu	8.6	9.5	9.8	11.8	10.9	11.1	12.5	13.7	13.8	20.0	24.1
Qinghai	3.8	3.4	3.8	4.1	3.9	4.2	4.9	4.9	3.9	3.4	6.0
Ningxia	2.2	2.2	2.4	2.8	2.4	2.9	2.9	6.7	4.3	3.8	4.5
Xinjiang	14.0	13.8	15.1	12.3	19.1	20.1	23.8	26.9	28.3	24.4	30.9

Note: Consumption is estimated based on national total production, imports/exports, and partial regional consumption data, and is notional consumption excluding inventory buildup or withdraw.
Sources: *China Statistical Yearbook, 1994, China Energy Statistical Yearbook, 1991,* and State Statistical Bureau.

Table A3.21 Diesel Oil Production by Region (000 b/d)

	1985	1986	1987	1988	1989	1990	1991	1992	1993	1994	1995
National Total	413.5	455.9	483.5	502.8	527.6	533.2	583.2	621.7	724.1	711.2	794.4
North	56.7	61.1	66.9	50.3	54.2	52.3	52.3	57.1	73.2	70.2	74.8
Beijing	35.3	36.5	38.7	17.9	19.3	19.3	18.1	16.6	20.0	21.7	16.1
Tianjin	13.6	13.8	15.4	15.9	17.5	17.1	17.0	18.0	22.9	16.5	22.8
Hebei	7.8	10.9	12.8	16.4	17.4	16.0	17.1	18.7	26.1	27.6	31.9
Shanxi	–	–	–	–	0.0	0.0	0.0	0.0	0.0	–	0.0
Inner Mongolia	–	–	–	–	0.0	0.0	0.0	3.9	4.2	4.4	4.0
North-east	149.0	161.1	171.1	181.9	185.5	190.2	202.5	204.2	230.3	219.6	233.8
Liaoning	103.6	108.7	113.7	117.8	119.9	119.4	127.2	127.7	148.4	136.2	137.1
Jilin	8.2	9.9	12.7	14.3	14.5	16.3	16.9	18.3	18.8	18.7	20.4
Heilongjiang	37.2	42.5	44.7	49.7	51.1	54.5	58.3	58.2	63.1	64.7	76.4
East	107.1	118.6	126.2	135.3	146.0	139.4	158.8	175.6	204.6	197.6	225.1
Shanghai	19.1	20.6	20.6	22.5	23.3	21.1	23.0	22.4	24.7	21.0	26.0
Jiangsu	26.9	30.1	32.4	34.3	34.3	32.2	34.5	35.5	39.2	37.0	42.9
Zhejiang	9.3	11.0	12.0	12.7	16.1	14.6	23.5	27.5	36.3	38.0	40.9
Anhui	11.1	11.9	13.4	15.8	16.6	16.5	17.2	17.5	19.0	16.2	20.2
Fujian	0.0	0.0	0.0	0.0	0.0	0.0	0.0	10.7	11.6	11.3	16.9
Jiangxi	4.0	5.9	7.3	8.2	9.5	9.4	12.2	13.8	16.3	13.1	17.1
Shandong	36.7	39.0	40.5	41.8	46.3	45.5	48.5	48.1	57.5	61.1	61.0
South-central	69.1	78.9	80.8	92.8	93.6	97.6	109.9	119.9	136.9	133.6	156.2
Henan	5.4	8.4	8.8	12.8	11.8	14.1	15.1	14.6	16.7	18.6	24.1
Hubei	20.7	22.1	20.9	25.4	24.8	25.3	25.0	24.5	29.0	28.1	30.9
Hunan	14.3	15.1	16.7	18.9	18.9	18.1	18.6	21.5	21.9	20.6	21.3
Guangdong	28.6	33.1	34.2	35.5	38.0	39.8	51.0	58.5	67.7	63.5	76.9
Guangxi	0.2	0.2	0.2	0.2	0.2	0.3	0.3	0.8	1.7	2.8	3.0
Hainan	–	–	–	–	–	–	–	–	–	–	–
South-west	0.2	0.2	0.2	0.2	0.3	0.5	0.6	0.7	0.8	0.7	3.3
Sichuan	0.2	0.2	0.2	0.2	0.3	0.5	0.6	0.7	0.8	0.7	1.0
Guizhou	–	–	–	–	–	–	–	–	–	–	–
Yunnan	–	–	–	–	–	–	–	–	–	–	2.3
Tibet	–	–	–	–	–	–	–	–	–	–	–
West	31.5	36.0	38.4	42.2	47.9	53.3	59.0	64.2	78.3	89.5	101.1
Shaanxi	0.8	1.0	1.5	2.2	3.0	4.0	4.5	5.4	6.4	6.9	8.0
Gansu	12.5	14.4	14.5	15.1	17.1	18.2	20.2	20.6	25.5	30.2	38.0
Qinghai	0.8	0.7	0.8	0.7	0.6	0.7	1.0	0.9	1.0	4.6	4.2
Ningxia	0.5	0.5	0.5	0.6	0.7	0.7	1.3	2.4	3.9	3.8	5.0
Xinjiang	16.8	19.4	21.0	23.6	26.6	29.6	31.9	34.8	41.5	43.9	45.9

Sources: *China Statistical Yearbook, 1994, China Energy Statistical Yearbook, 1991*, and State Statistical Bureau.

Table A3.22 Estimated Diesel Oil Consumption by Region (000 b/d)

	1985	1986	1987	1988	1989	1990	1991	1992	1993	1994	1995
National Total	397.3	431.5	478.5	528.0	566.9	549.7	612.3	655.1	827.8	762.4	892.8
North	101.2	96.3	110.3	133.7	122.1	123.7	129.2	136.8	175.5	152.5	195.2
Beijing	53.8	50.9	54.1	65.7	55.7	58.2	60.3	64.8	83.0	68.5	117.8
Tianjin	25.8	21.2	20.2	26.6	20.7	19.9	19.9	20.3	29.8	25.3	21.0
Hebei	10.2	10.7	17.1	19.6	20.9	20.3	21.9	23.6	32.0	30.5	31.8
Shanxi	5.7	7.5	10.1	10.8	12.8	13.5	14.4	14.0	15.4	15.0	15.5
Inner Mongolia	5.7	6.0	8.7	11.1	12.1	11.8	12.7	14.0	15.3	13.2	9.1
North-east	51.8	56.0	63.8	74.7	81.2	86.9	94.0	100.3	111.8	113.6	116.5
Liaoning	22.6	24.7	28.8	32.9	34.7	33.5	39.4	42.4	54.4	49.7	50.1
Jilin	4.4	4.9	5.5	5.9	6.6	6.4	7.5	7.5	13.2	9.0	13.6
Heilongjiang	24.8	26.3	29.5	35.9	39.9	47.0	47.1	50.4	44.1	43.5	52.8
East	102.7	117.9	127.1	127.1	145.9	139.0	148.1	160.3	203.7	208.1	218.3
Shanghai	32.4	38.6	37.4	38.7	39.8	39.8	40.2	45.5	47.1	46.1	52.6
Jiangsu	22.0	22.6	26.9	25.9	33.7	29.1	30.0	32.3	46.0	44.8	43.7
Zhejiang	9.3	10.7	10.7	8.3	12.5	12.0	13.4	14.5	19.9	18.4	22.1
Anhui	6.4	8.5	8.7	8.2	10.0	11.1	11.6	10.4	12.3	10.0	11.5
Fujian	6.0	6.7	8.3	8.0	7.9	7.4	7.9	10.5	13.6	9.8	15.1
Jiangxi	5.0	5.4	6.5	7.3	7.7	7.7	7.7	9.5	13.0	12.5	20.9
Shandong	21.6	25.4	28.5	30.6	34.3	31.9	37.4	37.5	51.9	45.8	52.4
South-central	87.5	102.6	109.4	116.5	138.9	123.2	153.0	161.7	229.6	245.3	251.0
Henan	16.1	18.2	22.1	26.3	28.7	26.3	27.7	26.4	32.9	27.9	31.8
Hubei	20.8	28.2	23.8	22.7	25.1	21.9	23.2	24.4	38.4	35.1	41.0
Hunan	12.2	12.1	14.3	14.6	15.3	13.3	13.4	13.5	17.6	14.4	15.8
Guangdong	31.1	36.2	39.5	40.8	53.9	46.7	68.0	76.6	112.8	119.4	134.3
Guangxi	7.4	8.0	9.8	11.5	13.3	12.2	16.9	17.3	22.3	21.0	22.2
Hainan	–	–	–	0.7	2.6	2.8	3.6	3.5	5.6	3.0	5.9
South-west	27.6	31.5	36.6	39.2	41.4	37.4	39.2	40.0	47.8	50.2	51.9
Sichuan	17.4	25.2	21.4	22.5	25.1	22.8	25.2	26.6	32.3	33.1	35.2
Guizhou	2.0	2.2	6.9	7.2	6.5	5.9	5.3	4.3	4.7	4.5	9.7
Yunnan	6.0	4.1	8.2	9.5	9.7	8.7	8.7	9.1	10.8	7.6	7.0
Tibet	2.2	n.a.	n.a.	n.a.	n.a.	n.a.	n.a.	n.a.	n.a.	n.a.	n.a.
West	26.7	27.3	31.2	36.9	37.4	39.5	48.9	55.9	59.4	60.3	59.9
Shaanxi	4.5	4.9	5.9	5.6	6.4	6.3	6.7	8.2	8.3	7.3	7.4
Gansu	4.4	4.7	5.2	6.1	6.2	6.7	8.7	9.4	12.3	15.1	12.4
Qinghai	2.2	1.9	2.5	2.6	2.5	2.7	2.8	2.9	1.9	1.9	2.8
Ningxia	0.9	1.2	1.0	1.2	1.3	1.4	1.4	4.1	2.6	2.9	5.2
Xinjiang	14.6	14.7	16.6	21.3	21.0	22.5	29.2	31.3	34.3	27.1	32.1

Note: Consumption is estimated based on national total production, imports/exports, and partial regional consumption data, and is notional consumption excluding inventory buildup or withdraw.
Sources: *China Statistical Yearbook, 1994*, *China Energy Statistical Yearbook, 1991*, and State Statistical Bureau.

Table A3.23 Kerosene Production by Region (000 b/d)

	1985	1986	1987	1988	1989	1990	1991	1992	1993	1994	1995
National Total	75.4	78.5	75.7	68.4	69.8	64.6	59.2	58.2	44.6	44.7	94.3
North	2.3	2.7	3.2	2.9	2.9	2.8	2.6	2.4	2.0	4.6	2.9
Beijing	0.2	0.2	0.5	0.5	0.5	0.5	0.4	0.4	0.2	0.2	0.1
Tianjin	1.6	1.8	1.7	1.6	1.7	1.7	1.5	1.3	1.1	3.8	2.0
Hebei	0.5	0.6	1.0	0.9	0.7	0.7	0.7	0.8	0.7	0.6	0.8
Shanxi	–	–	–	–	–	–	–	–	–	–	–
Inner Mongolia	–	–	–	–	–	–	–	–	–	–	–
North-east	31.7	30.9	28.4	22.4	24.5	23.0	19.8	20.1	15.0	12.9	23.5
Liaoning	26.8	26.4	24.5	18.2	20.3	19.0	16.4	16.4	12.4	10.6	18.2
Jilin	0.1	0.1	0.1	0.1	0.1	0.0	0.0	0.1	0.0	0.0	0.4
Heilongjiang	4.8	4.5	3.8	4.2	4.2	3.9	3.4	3.6	2.7	2.4	4.8
East	16.6	18.0	19.5	18.9	18.2	16.6	15.1	14.4	10.5	12.4	30.4
Shanghai	6.5	6.7	6.5	6.5	6.4	5.8	5.3	5.7	4.0	3.6	7.8
Jiangsu	4.1	5.6	8.0	7.9	7.6	7.2	6.2	5.2	3.5	4.3	10.2
Zhejiang	0.7	0.8	0.5	0.5	0.6	0.5	0.5	0.4	0.6	2.5	7.0
Anhui	–	–	–	–	–	–	–	–	–	–	–
Fujian	–	–	–	–	–	–	–	–	–	–	–
Jiangxi	0.4	0.3	0.1	0.1	0.1	0.2	0.7	0.8	0.5	0.3	0.5
Shandong	4.9	4.7	4.3	3.8	3.6	2.9	2.3	2.2	1.9	1.9	5.0
South-central	16.1	17.6	15.9	16.2	16.9	15.5	15.2	14.7	11.7	9.6	24.8
Henan	–	0.1	0.1	0.1	–	0.7	0.8	0.8	1.6	1.4	4.6
Hubei	3.6	3.8	3.4	3.8	3.3	3.2	2.6	2.4	1.4	1.3	3.0
Hunan	3.5	3.5	1.7	1.5	1.7	1.2	1.6	1.5	0.9	0.9	1.6
Guangdong	9.0	10.2	10.8	10.8	11.9	10.5	10.1	!0.0	7.8	6.1	15.7
Guangxi	–	–	–	–	–	–	–	–	–	0.0	–
Hainan	–	–	–	–	–	–	–	–	–	–	–
South-west	0.3	0.2	0.2	0.2	0.2	0.2	0.1	0.1	0.1	0.1	0.6
Sichuan	0.3	0.2	0.2	0.2	0.2	0.2	0.1	0.1	0.1	0.1	0.4
Guizhou	–	–	–	–	–	–	–	–	–	–	–
Yunnan	–	–	–	–	–	–	–	–	–	–	0.1
Tibet	–	–	–	–	–	–	–	–	–	–	–
West	8.4	9.0	8.5	7.8	7.0	6.5	6.3	6.5	5.2	5.0	12.1
Shaanxi	0.1	0.1	0.1	0.1	0.1	0.1	0.0	0.0	0.0	0.0	0.4
Gansu	6.8	7.0	6.6	5.9	5.1	4.6	4.7	4.7	3.5	3.4	8.4
Qinghai	–	–	–	–	–	–	–	–	–	–	–
Ningxia	–	–	–	–	–	–	–	–	–	–	–
Xinjiang	1.5	2.0	1.8	1.8	1.8	1.8	1.6	1.8	1.7	1.6	3.4

Sources: *China Statistical Yearbook, 1994, China Energy Statistical Yearbook, 1991*, and State Statistical Bureau.

Table A3.24 Estimated Kerosene Consumption by Region (000 b/d)

	1985	1986	1987	1988	1989	1990	1991	1992	1993	1994	1995
National Total	68.7	71.8	67.2	64.8	64.4	55.6	54.2	54.1	57.0	52.6	102.6
North	23.7	25.9	22.0	25.5	19.3	15.0	14.0	12.7	18.0	14.2	23.0
Beijing	21.8	23.8	20.5	23.8	17.8	14.0	13.2	12.1	17.5	13.1	21.4
Tianjin	0.3	0.3	0.2	0.2	0.1	0.2	0.1	0.1	0.0	0.6	0.2
Hebei	0.8	1.1	0.7	0.8	0.8	0.4	0.3	0.3	0.3	0.2	0.6
Shanxi	0.7	0.6	0.5	0.7	0.5	0.3	0.3	0.2	0.2	0.2	0.7
Inner Mongolia	0.1	0.2	0.1	0.1	0.1	0.1	0.0	0.0	0.1	0.1	0.1
North-east	4.6	4.7	5.1	5.1	4.9	4.0	4.0	3.9	5.5	3.7	9.8
Liaoning	2.7	2.6	3.1	2.7	2.9	2.5	2.6	2.5	3.8	2.2	5.9
Jilin	0.2	0.5	0.6	0.7	0.6	0.5	0.4	0.4	0.5	0.7	1.4
Heilongjiang	1.7	1.7	1.5	1.6	1.3	1.1	1.0	1.0	1.2	0.7	2.5
East	14.1	13.5	13.6	14.2	14.7	16.0	16.1	16.0	14.7	16.9	33.4
Shanghai	9.2	7.5	7.9	9.2	9.9	12.5	12.4	12.3	9.3	9.6	19.4
Jiangsu	0.9	1.0	0.8	0.6	0.6	0.4	0.3	0.3	0.3	1.5	1.9
Zhejiang	1.1	1.2	1.2	1.4	1.2	0.7	1.2	0.9	1.7	0.9	1.9
Anhui	0.3	0.7	0.6	0.5	0.5	0.3	0.3	0.4	0.6	0.1	0.2
Fujian	0.5	0.7	0.8	1.0	0.9	1.0	0.7	1.1	1.9	3.5	7.5
Jiangxi	0.8	0.9	0.6	0.4	0.4	0.2	0.3	0.3	0.3	0.6	1.0
Shandong	1.3	1.6	1.7	1.1	1.2	0.9	0.8	0.7	0.6	0.8	1.5
South-central	12.6	14.0	14.6	6.6	14.0	10.8	10.7	10.6	5.1	4.0	12.1
Henan	0.6	0.7	0.4	0.5	0.5	0.3	0.2	0.2	0.2	0.5	1.3
Hubei	0.9	1.7	1.2	1.3	1.1	0.8	0.6	0.8	0.7	0.9	2.9
Hunan	0.6	0.8	0.6	0.8	0.7	0.4	0.4	0.3	0.3	0.4	0.7
Guangdong	10.3	10.7	10.7	2.0	10.6	8.3	8.4	9.1	3.6	2.1	3.9
Guangxi	0.3	0.1	1.6	2.0	1.1	1.0	1.0	0.2	0.2	0.1	0.4
Hainan	–	–	–	0.0	0.0	0.0	0.0	0.0	0.0	0.0	2.9
South-west	7.7	7.2	5.6	6.1	5.9	5.0	4.4	5.5	5.5	6.2	10.3
Sichuan	5.3	5.0	3.8	4.4	4.1	3.7	2.9	4.0	3.8	3.8	6.4
Guizhou	1.1	0.8	0.7	0.6	0.9	0.6	0.5	0.5	0.2	0.1	0.2
Yunnan	1.2	1.4	1.2	1.1	0.9	0.8	1.0	1.0	1.6	2.3	3.7
Tibet	0.2	n.a.	n.a.	n.a.	n.a.	n.a.	n.a.	n.a.	n.a.	n.a.	n.a.
West	6.0	6.3	6.2	7.3	5.6	4.7	4.9	5.4	8.1	7.5	13.9
Shaanxi	1.2	3.1	3.0	3.7	2.6	2.5	2.3	2.4	3.2	3.3	4.7
Gansu	2.9	1.0	1.1	0.8	0.7	0.5	0.4	0.5	0.8	0.7	1.0
Qinghai	0.1	0.0	0.0	0.0	0.0	0.0	0.0	0.0	0.0	0.0	0.0
Ningxia	0.0	0.1	0.1	0.1	0.0	0.0	0.0	0.0	0.0	0.0	–
Xinjiang	1.8	2.1	2.1	2.6	2.1	1.7	2.1	2.5	4.0	3.5	8.1

Note: Consumption is estimated based on national total production, imports/exports, and partial regional consumption data, and is notional consumption excluding inventory buildup or withdraw.
Sources: *China Statistical Yearbook, 1994, China Energy Statistical Yearbook, 1991*, and State Statistical Bureau.

Table A3.25 Fuel Oil Production by Region (000 b/d)

	1985	1986	1987	1988	1989	1990	1991	1992	1993	1994	1995
National Total	517.4	544.9	571.2	587.3	603.8	596.3	591.9	604.5	583.1	556.3	509.2
North	97.7	73.8	76.0	79.3	79.6	79.0	80.0	85.1	82.5	70.9	63.3
Beijing	52.4	42.2	42.8	43.9	43.7	43.8	44.3	46.7	42.7	30.3	26.2
Tianjin	29.8	24.4	24.9	26.6	28.5	26.8	26.2	27.4	27.4	24.6	25.8
Hebei	15.5	7.1	8.3	8.7	7.4	8.4	9.5	11.0	12.3	11.2	7.8
Shanxi	–	–	–	–	–	–	–	–	–	–	–
Inner Mongolia	–	–	–	–	–	–	–	–	–	4.8	3.4
North-east	244.5	208.8	211.7	212.7	217.0	209.5	192.0	189.2	180.2	168.7	156.4
Liaoning	169.6	138.3	139.8	142.7	145.3	142.6	129.3	126.4	124.3	118.1	108.5
Jilin	24.2	21.3	23.9	23.4	22.9	22.0	17.1	15.8	14.5	16.3	15.5
Heilongjiang	50.7	49.2	48.0	46.6	48.8	44.9	45.6	47.0	41.4	34.3	32.4
East	122.5	153.7	171.4	178.6	180.9	188.1	196.0	202.9	196.0	195.6	177.6
Shanghai	52.0	43.3	43.8	45.5	47.0	53.1	54.7	59.2	57.6	58.5	52.6
Jiangsu	36.8	31.7	41.7	46.8	48.2	50.7	53.1	55.8	52.6	51.1	44.7
Zhejiang	15.3	15.5	19.5	18.3	18.1	14.4	14.0	13.7	12.4	15.1	12.9
Anhui	18.2	14.8	17.7	14.3	5.9	11.1	10.7	9.2	10.2	9.3	8.8
Fujian	–	–	–	–	–	–	–	–	–	–	–
Jiangxi	–	6.9	8.1	8.2	8.1	7.8	7.4	7.4	8.5	9.4	7.9
Shandong	0.1	41.4	40.6	45.5	53.6	51.0	56.2	57.6	54.6	52.1	50.6
South-central	29.0	88.1	91.3	88.8	93.3	84.1	83.5	88.8	85.2	79.9	70.3
Henan	5.5	2.8	3.0	3.8	2.9	2.6	2.5	2.4	3.0	3.6	3.2
Hubei	0.1	26.6	25.0	25.1	27.6	25.0	23.9	25.3	22.4	21.4	17.3
Hunan	5.5	11.2	11.0	10.6	11.3	11.1	11.0	10.7	10.1	10.5	10.1
Guangdong	2.1	47.2	51.9	49.0	51.2	45.2	45.9	50.1	49.2	43.7	39.1
Guangxi	15.7	0.4	0.4	0.4	0.2	0.3	0.2	0.3	0.5	0.6	0.6
Hainan	–	–	–	–	–	–	–	–	–	–	–
South-west	0.4	0.4	0.3	0.3	0.4	0.4	0.4	0.6	0.6	0.6	0.8
Sichuan	0.4	0.4	0.3	0.3	0.4	0.4	0.4	0.6	0.6	0.6	0.8
Guizhou	–	–	–	–	–	–	–	–	–	–	–
Yunnan	–	–	–	–	–	–	–	–	–	–	–
Tibet	–	–	–	–	–	–	–	–	–	–	–
West	23.3	20.1	20.5	27.5	32.7	35.1	40.0	37.8	38.6	40.6	40.9
Shaanxi	1.5	1.5	1.2	1.6	1.5	2.7	3.6	1.3	1.1	1.5	3.4
Gansu	13.1	11.6	12.2	11.9	14.8	16.0	16.6	15.4	16.2	15.7	13.9
Qinghai	2.5	1.9	2.1	1.8	1.5	–	1.4	–	–	–	2.5
Ningxia	0.6	0.4	0.4	0.4	0.4	0.3	0.9	0.3	0.2	0.3	1.8
Xinjiang	5.6	4.7	4.5	11.9	14.5	16.1	17.5	20.8	21.2	23.1	19.3

Sources: *China Statistical Yearbook, 1994*, *China Energy Statistical Yearbook, 1991*, and State Statistical Bureau.

Table A3.26 Estimated Fuel Oil Consumption by Region (000 b/d)

	1985	1986	1987	1988	1989	1990	1991	1992	1993	1994	1995
National Total	519.7	549.8	569.5	593.8	608.9	605.9	602.8	650.6	680.6	622.7	615.2
North	82.1	91.5	96.9	94.0	93.1	96.7	93.6	99.3	116.5	65.1	67.8
Beijing	42.8	43.0	42.8	41.7	40.2	41.4	43.4	47.5	63.8	38.5	35.9
Tianjin	25.4	33.0	39.4	37.5	38.3	39.6	34.3	36.6	35.4	12.9	17.7
Hebei	8.5	9.1	8.6	9.0	8.6	9.2	9.1	9.2	9.8	6.6	8.7
Shanxi	1.8	1.9	1.9	2.0	2.0	2.0	2.3	2.6	2.9	2.6	2.2
Inner Mongolia	3.6	4.7	4.3	3.8	3.9	4.6	4.4	3.5	4.7	4.4	3.3
North-east	167.4	167.0	170.4	180.8	180.8	178.0	160.8	164.9	161.0	199.1	128.6
Liaoning	113.5	113.8	107.2	109.9	113.1	115.6	103.8	104.6	104.1	!00.0	88.2
Jilin	15.0	14.1	13.9	21.3	21.2	18.5	14.1	16.2	13.2	12.8	12.7
Heilongjiang	38.8	39.1	49.3	49.6	46.5	43.9	42.9	44.1	43.7	86.3	27.7
East	156.7	171.7	176.8	193.4	189.8	184.8	194.2	211.3	212.8	200.0	205.2
Shanghai	67.8	77.5	73.5	89.3	78.1	77.4	76.4	83.3	77.4	73.2	69.0
Jiangsu	23.7	25.8	26.3	25.5	29.7	32.5	38.3	43.6	45.5	44.9	43.6
Zhejiang	12.9	15.6	19.4	17.6	18.0	18.4	21.1	21.9	24.9	26.7	22.0
Anhui	8.3	8.7	9.1	9.7	9.0	6.1	4.9	5.3	5.8	5.5	7.5
Fujian	2.1	2.5	3.0	2.8	3.0	2.8	3.4	3.3	3.6	3.5	5.0
Jiangxi	3.1	3.9	3.8	4.5	4.5	4.3	4.0	4.9	5.9	5.8	5.4
Shandong	38.8	37.7	41.7	44.0	47.6	43.4	46.2	48.9	49.8	40.5	52.5
South-central	91.8	98.1	101.9	102.4	110.5	106.4	111.4	127.2	140.8	112.2	150.4
Henan	5.7	6.5	7.1	7.5	7.5	7.1	8.2	8.1	8.6	9.8	10.1
Hubei	25.3	24.5	25.5	26.0	25.7	24.0	23.7	25.3	24.3	20.3	17.2
Hunan	11.2	11.0	10.7	10.9	10.9	9.8	9.5	10.2	10.5	9.8	9.6
Guangdong	48.2	54.7	57.1	56.8	64.9	63.9	68.1	81.6	95.4	69.5	109.3
Guangxi	1.4	1.4	1.6	1.2	1.4	1.5	1.8	1.9	2.1	2.9	2.5
Hainan	–	–	–	0.1	0.1	0.1	0.1	0.1	0.1	–	1.6
South-west	2.2	2.3	2.0	2.3	2.3	2.5	2.6	3.0	4.2	5.4	12.5
Sichuan	1.3	1.4	1.0	1.3	1.2	1.3	1.2	1.4	1.7	2.7	1.1
Guizhou	0.5	0.6	0.7	0.7	0.7	0.9	1.1	1.2	2.0	1.9	0.5
Yunnan	0.3	0.4	0.4	0.4	0.4	0.3	0.4	0.4	0.4	0.7	10.9
Tibet	–	–	–	–	–	–	–	–	–	–	–
West	19.5	19.1	21.3	20.8	32.4	37.5	40.2	44.8	45.3	40.9	50.7
Shaanxi	4.3	4.4	4.4	4.3	4.1	4.0	3.6	4.1	3.7	3.9	15.2
Gansu	8.3	8.5	9.2	7.3	9.5	11.3	10.8	11.4	11.4	9.8	1.6
Qinghai	2.1	1.8	1.7	1.6	1.5	1.7	2.2	2.4	1.4	0.9	0.6
Ningxia	0.3	0.3	0.2	1.1	3.0	4.2	4.3	5.6	6.0	4.9	4.4
Xinjiang	4.6	4.1	5.8	6.4	14.3	16.3	19.3	21.4	22.9	21.3	29.0

Note: Consumption is estimated based on national total production, imports/exports, and partial regional consumption data, and is notional consumption excluding inventory buildup or withdraw.
Sources: *China Statistical Yearbook, 1994, China Energy Statistical Yearbook, 1991*, and State Statistical Bureau.

Table A3.27 Crude Oil Balance by Region (000 b/d)

	1985	1986	1987	1988	1989	1990	1991	1992	1993	1994	1995
National Total	780.1	761.8	727.8	567.8	484.0	445.0	386.9	328.7	246.8	289.0	85.6
North	68.9	56.9	8.6	(54.8)	(70.6)	(69.0)	(53.1)	(68.8)	(96.2)	(95.2)	(108.9)
Beijing	(110.6)	(118.9)	(122.6)	(135.5)	(136.7)	(136.5)	(133.9)	(132.7)	(132.8)	(127.8)	(130.2)
Tianjin	3.3	16.5	20.0	13.7	12.3	14.7	17.1	22.0	16.2	32.4	20.3
Hebei	176.5	159.6	111.4	68.2	56.3	54.5	43.8	38.2	16.6	17.2	16.7
Shanxi	–	–	–	(1.1)	(2.3)	(1.7)	(0.0)	–	–	–	(0.2)
Inner Mongolia	(0.2)	(0.3)	(0.2)	(0.1)	(0.1)	(0.1)	19.9	3.8	3.8	(17.0)	(15.5)
North-east	708.9	707.1	721.4	698.6	698.2	690.2	677.9	698.1	685.3	730.3	667.3
Liaoning	(206.3)	(203.0)	(191.1)	(206.1)	(202.2)	(196.2)	(199.2)	(210.3)	(218.0)	(191.9)	(193.5)
Jilin	(25.9)	(28.4)	(28.0)	(22.7)	(18.8)	(21.1)	(22.0)	(16.0)	(15.4)	(13.9)	(25.3)
Heilongjiang	941.1	938.5	940.5	927.4	919.2	907.6	899.1	924.3	918.7	936.0	886.1
East	90.7	99.5	89.1	65.4	19.5	9.2	(39.8)	(108.7)	(157.8)	(196.1)	(284.5)
Shanghai	(131.7)	(137.0)	(143.5)	(152.7)	(156.9)	(160.4)	(171.7)	(179.5)	(184.9)	(184.0)	(192.8)
Jiangsu	(82.2)	(89.9)	(114.3)	(140.0)	(140.6)	(146.7)	(155.9)	(161.9)	(165.6)	(165.3)	(178.5)
Zhejiang	(37.0)	(43.7)	(45.1)	(48.6)	(56.1)	(59.1)	(68.8)	(81.8)	(93.2)	(105.0)	(116.1)
Anhui	(40.1)	(42.4)	(48.6)	(51.6)	(50.6)	(51.1)	(50.5)	(53.3)	(52.9)	(47.8)	(55.5)
Fujian	–	–	–	–	–	–	–	(32.8)	(32.8)	(31.7)	(44.4)
Jiangxi	(15.3)	(22.5)	(25.1)	(28.7)	(30.2)	(31.0)	(36.2)	(41.8)	(46.1)	(40.4)	(46.5)
Shandong	396.9	434.9	465.7	487.1	453.9	457.5	443.2	442.3	417.7	378.1	349.4
South-central	(102.8)	(118.4)	(106.6)	(145.9)	(162.9)	(184.6)	(198.3)	(201.9)	(205.3)	(187.1)	(225.8)
Henan	139.1	150.8	160.9	155.9	152.6	134.3	121.9	114.9	100.5	79.7	45.6
Hubei	(62.1)	(69.8)	(64.9)	(79.0)	(83.7)	(84.1)	(81.9)	(83.5)	(80.8)	(80.2)	(83.4)
Hunan	(52.8)	(56.1)	(57.3)	(63.4)	(63.5)	(64.1)	(67.1)	(72.9)	(69.5)	(68.0)	(70.0)
Guangdong	(127.0)	(143.4)	(145.9)	(160.3)	(169.7)	(172.4)	(172.8)	(159.6)	(154.3)	(117.2)	(112.1)
Guangxi	0.1	0.1	0.6	0.9	1.4	1.7	1.6	(0.8)	(1.2)	(1.4)	(5.9)
Hainan	–	–	–	(0.0)	(0.0)	(0.0)	(0.0)	–	–	–	–
South-west	0.1	0.3	(0.1)	0.2	0.2	0.0	(0.6)	1.9	2.2	1.6	(5.8)
Sichuan	0.1	0.3	(0.1)	0.2	0.2	0.0	(0.6)	(1.0)	(0.6)	(0.8)	(1.3)
Guizhou	–	–	–	–	–	–	–	–	–	–	–
Yunnan	–	–	–	–	–	–	–	2.8	2.8	2.4	(4.5)
Tibet	–	–	–	–	–	–	–	–	–	–	–
West	14.3	16.3	15.3	4.2	(0.4)	(0.8)	0.8	8.1	18.7	35.5	43.3
Shaanxi	1.6	1.8	2.8	2.2	4.4	2.6	(0.2)	4.9	6.5	8.8	10.1
Gansu	(41.0)	(45.7)	(49.0)	(56.9)	(61.8)	(65.5)	(68.9)	(60.9)	(77.7)	(67.9)	(81.4)
Qinghai	0.6	3.8	8.0	8.6	9.4	11.3	16.3	17.6	18.4	5.4	7.3
Ningxia	8.0	7.0	6.2	6.4	5.7	5.2	2.0	(4.0)	(6.9)	(5.2)	(5.1)
Xinjiang	45.1	49.4	47.4	43.9	41.9	45.6	51.4	50.5	78.3	94.4	112.4

Note: Notional consumption is used excluding inventory buildup or withdraw.
Sources: *China Statistical Yearbook, 1994, 1997, China Energy Statistical Yearbook, 1991*, and State Statistical Bureau.

Table A3.28 Petroleum Product Balance by Region (000 b/d)

	1985	1986	1987	1988	1989	1990	1991	1992	1993	1994	1995
National Total	37.4	69.1	41.6	(4.4)	(12.1)	50.3	10.2	(17.3)	(170.8)	(86.2)	(173.1)
North	(57.4)	(76.2)	(83.0)	(122.4)	(100.7)	(98.9)	(111.3)	(110.2)	(156.0)	(90.0)	(144.0)
Beijing	(19.0)	(22.1)	(17.1)	(54.5)	(31.9)	(29.4)	(36.1)	(43.5)	(86.0)	(54.0)	(121.8)
Tianjin	(0.9)	(8.4)	(11.8)	(14.0)	(6.7)	(6.9)	(3.6)	(4.4)	(5.2)	12.5	24.8
Hebei	0.8	(2.5)	(4.1)	0.5	(3.7)	(2.6)	(5.3)	(3.0)	2.4	5.0	3.0
Shanxi	(19.1)	(21.2)	(24.7)	(25.8)	(29.2)	(29.8)	(32.6)	(33.5)	(35.6)	(35.5)	(35.6)
Inner Mongolia	(19.2)	(22.1)	(25.3)	(28.6)	(29.1)	(30.2)	(33.7)	(25.8)	(31.6)	(18.1)	(14.4)
North-east	255.9	239.8	230.6	224.5	230.8	240.3	241.7	234.4	234.9	162.8	242.2
Liaoning	206.9	184.9	186.5	191.9	193.4	196.3	197.0	194.9	204.9	181.8	174.2
Jilin	16.7	19.6	24.5	19.6	18.7	26.5	27.0	24.7	10.1	16.2	21.9
Heilongjiang	32.3	35.4	19.7	12.9	18.8	17.5	17.7	14.8	19.9	(23.9)	46.2
East	(22.3)	2.9	11.2	14.7	8.7	19.8	28.9	35.5	3.7	(16.3)	(15.6)
Shanghai	(27.6)	(44.9)	(42.7)	(58.0)	(42.9)	(38.3)	(36.0)	(46.7)	(40.1)	(50.5)	(55.6)
Jiangsu	21.7	19.5	28.7	42.2	23.8	27.0	21.0	16.9	(1.6)	(10.8)	(3.4)
Zhejiang	2.8	1.7	1.1	6.2	7.4	(1.5)	13.4	20.3	17.8	21.3	23.3
Anhui	14.0	9.6	13.5	13.2	4.5	11.9	10.6	10.8	8.8	11.6	12.6
Fujian	(15.7)	(16.4)	(19.6)	(20.8)	(19.6)	(18.8)	(21.0)	(0.0)	(3.1)	(1.4)	1.2
Jiangxi	(7.3)	2.4	4.9	3.6	5.2	6.5	10.7	11.8	9.2	3.0	(1.5)
Shandong	(10.1)	31.1	25.3	28.3	30.3	33.2	30.2	22.4	12.8	31.4	7.9
South-central	(66.7)	(21.5)	(31.5)	(26.7)	(53.6)	(22.0)	(48.7)	(58.5)	(127.4)	(130.2)	(151.5)
Henan	(18.3)	(25.2)	(22.4)	(19.7)	(24.0)	(16.4)	(20.1)	(21.8)	(24.2)	(17.9)	(12.0)
Hubei	(11.4)	7.7	3.2	9.8	8.6	13.3	7.0	2.9	(6.3)	(6.7)	(9.9)
Hunan	2.9	9.2	7.1	6.9	6.6	11.1	11.0	13.0	4.5	10.3	9.5
Guangdong	(40.5)	2.9	1.0	3.1	(15.1)	(1.9)	(10.5)	(17.6)	(60.0)	(58.2)	(90.9)
Guangxi	0.7	(16.2)	(20.2)	(23.3)	(24.3)	(22.9)	(28.9)	(28.5)	(32.3)	(27.5)	(29.2)
Hainan	–	–	–	(3.4)	(5.3)	(5.2)	(7.3)	(6.6)	(9.3)	(5.7)	(19.0)
South-west	(85.6)	(88.8)	(96.7)	(109.4)	(107.6)	(104.2)	(114.7)	(121.7)	(149.4)	(130.0)	(152.7)
Sichuan	(47.3)	(56.7)	(54.4)	(60.1)	(61.9)	(59.8)	(66.2)	(72.4)	(90.7)	(80.1)	(86.3)
Guizhou	(10.6)	(10.9)	(15.6)	(17.1)	(15.8)	(15.0)	(15.6)	(14.5)	(22.1)	(15.3)	(26.3)
Yunnan	(22.2)	(21.2)	(26.7)	(32.2)	(29.9)	(29.4)	(32.9)	(34.8)	(36.6)	(29.6)	(40.0)
Tibet	(5.6)										
West	13.5	12.8	10.9	14.9	10.3	15.2	14.3	3.3	23.4	49.9	48.5
Shaanxi	(15.9)	(18.4)	(19.4)	(19.5)	(17.0)	(13.4)	(13.7)	(16.9)	(17.9)	(10.6)	(23.2)
Gansu	26.6	24.5	26.3	24.6	25.6	25.0	30.2	28.1	33.6	33.0	59.3
Qinghai	(4.3)	(4.1)	(4.6)	(5.6)	(5.6)	(7.3)	(6.8)	(8.5)	(5.6)	4.4	4.2
Ningxia	(2.0)	(2.4)	(2.1)	(3.5)	(4.9)	(6.6)	(4.8)	(10.7)	(5.0)	(3.8)	(1.7)
Xinjiang	9.1	13.4	10.8	18.9	12.1	17.6	9.3	11.3	18.2	33.0	9.9

Note: Notional consumption is used excluding inventory buildup or withdraw.
Sources: *China Statistical Yearbook, 1994, 1997, China Energy Statistical Yearbook, 1991*, and State Statistical Bureau.

Table A3.29 Gasoline Balance by Region (000 b/d)

	1985	1986	1987	1988	1989	1990	1991	1992	1993	1994	1995
National Total	16.8	43.0	26.3	23.8	26.9	67.4	45.2	58.2	42.7	39.2	39.6
North	(7.2)	(0.0)	0.2	(1.7)	(2.8)	2.4	(9.5)	(6.1)	(3.7)	(3.9)	1.0
Beijing	11.5	16.7	18.2	14.3	18.3	20.7	18.0	17.4	15.3	14.0	10.8
Tianjin	5.6	6.0	6.0	6.1	4.8	7.2	6.0	5.9	8.5	6.4	13.1
Hebei	(3.5)	(0.2)	0.3	3.8	1.2	2.2	(1.3)	(0.5)	5.4	2.9	3.5
Shanxi	(10.9)	(11.2)	(12.2)	(12.4)	(14.0)	(14.0)	(15.6)	(16.7)	(17.1)	(17.7)	(17.2)
Inner Mongolia	(9.8)	(11.3)	(12.1)	(13.6)	(13.0)	(13.7)	(16.5)	(12.2)	(15.8)	(9.6)	(9.3)
North-east	54.5	66.7	58.8	68.0	70.6	86.5	86.2	90.0	87.6	78.0	83.4
Liaoning	45.6	52.5	47.5	58.7	58.7	66.9	69.9	73.9	82.2	68.9	54.5
Jilin	4.0	7.7	7.8	9.8	9.5	13.5	14.9	14.5	3.7	3.9	13.3
Heilongjiang	4.9	6.4	3.5	(0.5)	2.4	6.1	1.4	1.6	1.7	5.3	15.7
East	5.0	15.7	11.7	16.6	14.0	15.5	17.4	30.2	23.8	3.2	8.2
Shanghai	4.2	8.0	5.2	4.7	8.2	11.2	10.1	7.1	7.3	(4.6)	(1.0)
Jiangsu	0.5	1.5	0.6	5.1	(2.4)	(1.2)	(4.1)	(3.5)	(5.2)	(12.0)	(12.0)
Zhejiang	0.8	1.9	0.4	1.9	4.5	0.0	11.1	16.0	14.9	11.7	8.4
Anhui	(0.4)	0.6	0.9	1.6	1.5	1.9	(0.5)	0.1	(1.7)	1.7	2.8
Fujian	(7.1)	(6.5)	(7.4)	(8.9)	(7.8)	(7.7)	(9.1)	4.3	4.4	4.1	11.8
Jiangxi	(2.9)	(0.5)	0.2	(0.8)	0.1	1.2	2.4	4.6	3.0	(0.9)	0.3
Shandong	9.9	10.6	11.8	12.9	9.9	!0.0	7.5	1.7	1.1	3.3	(2.2)
South-central	11.0	8.8	6.4	0.9	6.1	21.3	17.6	17.6	14.3	8.1	10.7
Henan	(6.9)	(11.0)	(4.7)	(2.1)	(2.0)	(0.0)	(2.3)	(4.8)	(3.7)	(3.4)	(0.8)
Hubei	11.0	9.7	4.3	5.5	4.8	6.7	3.1	1.3	4.3	(1.2)	0.1
Hunan	3.7	3.2	3.3	2.1	1.5	4.3	3.1	3.2	0.1	2.8	2.6
Guangdong	9.4	14.1	11.4	7.4	13.2	21.6	27.1	31.2	27.2	19.5	25.0
Guangxi	(6.2)	(7.2)	(7.9)	(9.2)	(8.8)	(9.0)	(9.7)	(10.3)	)0.0)	(6.9)	(7.6)
Hainan	–	–	–	(2.6)	(2.6)	(2.3)	(3.6)	(2.9)	(3.5)	(2.7)	(8.5)
South-west	(49.1)	(48.6)	(53.2)	(62.5)	(59.0)	(60.4)	(69.7)	(74.5)	(93.3)	(69.7)	(82.5)
Sichuan	(24.2)	(26.0)	(28.9)	(32.7)	(32.3)	(33.1)	(38.2)	(41.7)	(54.3)	(41.9)	(45.7)
Guizhou	(7.0)	(7.3)	(7.4)	(8.6)	(7.7)	(7.7)	(8.7)	(8.5)	(15.2)	(8.8)	(15.9)
Yunnan	(14.7)	(15.2)	(17.0)	(21.2)	(19.0)	(19.6)	(22.8)	(24.3)	(23.8)	(19.0)	(20.9)
Tibet	(3.2)										
West	2.6	0.5	2.3	2.3	(1.9)	2.1	3.1	1.0	13.9	23.5	18.8
Shaanxi	(8.2)	(8.7)	(8.9)	(9.7)	(8.4)	(7.4)	(9.1)	(8.9)	(10.3)	(4.7)	(7.6)
Gansu	9.8	5.8	8.3	5.9	5.1	4.6	8.7	8.7	12.9	9.2	14.0
Qinghai	(3.3)	(2.9)	(3.3)	(3.7)	(3.6)	(3.6)	(4.2)	(4.2)	(3.2)	2.7	0.8
Ningxia	(1.8)	(1.8)	(1.8)	(2.2)	(1.7)	(2.0)	(1.3)	(3.7)	(0.5)	(0.1)	1.0
Xinjiang	6.2	8.2	8.0	11.9	6.7	10.6	8.9	9.1	15.0	16.4	10.5

Note: Notional consumption is used excluding inventory buildup or withdraw.

Table A3.30 Diesel Oil Balance by Region (000 b/d)

	1985	1986	1987	1988	1989	1990	1991	1992	1993	1994	1995
National Total	16.2	24.4	5.0	(25.3)	(39.3)	(16.4)	(29.1)	(33.4)	(103.7)	(51.2)	(98.4)
North	(44.4)	(35.1)	(43.4)	(83.4)	(68.0)	(71.4)	(76.8)	(79.7)	(102.3)	(82.3)	(120.4)
Beijing	(18.4)	(14.5)	(15.3)	(47.8)	(36.4)	(39.0)	(42.2)	(48.3)	(63.0)	(46.8)	(101.6)
Tianjin	(12.2)	(7.4)	(4.8)	(10.6)	(3.3)	(2.8)	(2.9)	(2.4)	(6.8)	(8.9)	1.8
Hebei	(2.4)	0.2	(4.4)	(3.2)	(3.5)	(4.3)	(4.7)	(5.0)	(5.9)	(2.8)	0.1
Shanxi	(5.7)	(7.5)	(10.1)	(10.8)	(12.7)	(13.5)	(14.4)	(14.0)	(15.4)	(15.0)	(15.5)
Inner Mongolia	(5.7)	(6.0)	(8.7)	(11.1)	(12.1)	(11.8)	(12.6)	(10.1)	(11.1)	(8.8)	(5.1)
North-east	97.2	105.2	107.2	107.2	104.4	103.3	108.4	104.0	118.5	105.9	117.3
Liaoning	81.1	84.0	84.9	84.9	85.2	85.9	87.8	85.3	93.9	86.5	87.0
Jilin	3.7	5.0	7.1	8.5	8.0	9.9	9.4	10.8	5.6	9.6	6.8
Heilongjiang	12.4	16.1	15.2	13.9	11.2	7.6	11.2	7.8	19.0	21.2	23.6
East	4.4	0.7	(0.9)	8.2	0.1	0.4	10.8	15.3	0.9	(10.5)	6.8
Shanghai	(13.3)	(17.9)	(16.8)	(16.2)	(16.5)	(18.7)	(17.2)	(23.1)	(22.4)	(25.2)	(26.5)
Jiangsu	4.9	7.4	5.5	8.4	0.7	3.1	4.4	3.2	(6.7)	(7.8)	(0.8)
Zhejiang	(0.0)	0.3	1.3	4.4	3.5	2.6	10.1	13.0	16.4	19.6	18.8
Anhui	4.7	3.5	4.7	7.5	6.6	5.4	5.6	7.2	6.6	6.2	8.7
Fujian	(6.0)	(6.7)	(8.3)	(8.0)	(7.9)	(7.4)	(7.9)	0.2	(2.1)	1.6	1.9
Jiangxi	(1.0)	0.5	0.8	0.9	1.7	1.7	4.5	4.3	3.3	0.6	(3.9)
Shandong	15.0	13.6	12.0	11.2	12.0	13.6	11.1	10.5	5.7	15.3	8.6
South-central	(18.4)	(23.8)	(28.7)	(23.7)	(45.3)	(25.6)	(43.0)	(41.8)	(92.7)	(111.6)	(94.7)
Henan	(10.7)	(9.8)	(13.3)	(13.5)	(16.9)	(12.2)	(12.6)	(11.9)	(16.2)	(9.3)	(7.6)
Hubei	(0.0)	(6.2)	(2.9)	2.6	(0.4)	3.3	1.7	0.1	(9.4)	(7.0)	(10.1)
Hunan	2.1	3.1	2.4	4.4	3.6	4.8	5.2	8.0	4.3	6.2	5.5
Guangdong	(2.5)	(3.1)	(5.4)	(5.3)	(16.0)	(6.9)	(17.1)	(18.1)	(45.2)	(55.9)	(57.4)
Guangxi	(7.2)	(7.8)	(9.6)	(11.2)	(13.1)	(11.8)	(16.6)	(16.5)	(20.6)	(18.2)	(19.2)
Hainan	–	–	–	(0.7)	(2.6)	(2.8)	(3.6)	(3.5)	(5.6)	(3.0)	(5.9)
South-west	(27.4)	(31.3)	(36.4)	(39.0)	(41.0)	(36.9)	(38.5)	(39.3)	(47.1)	(49.5)	(48.6)
Sichuan	(17.2)	(24.9)	(21.2)	(22.2)	(24.8)	(22.3)	(24.5)	(25.9)	(31.6)	(32.4)	(34.3)
Guizhou	(2.0)	(2.2)	(6.9)	(7.2)	(6.5)	(5.9)	(5.3)	(4.3)	(4.7)	(4.5)	(9.7)
Yunnan	(6.0)	(4.1)	(8.2)	(9.5)	(9.7)	(8.7)	(8.7)	(9.1)	(10.8)	(7.6)	(4.7)
Tibet	(2.2)	n.a.	n.a.	n.a.	n.a.	n.a.	n.a.	n.a.	n.a.	n.a.	n.a.
West	4.7	8.6	7.2	5.4	10.6	13.8	10.1	8.3	19.0	29.1	41.3
Shaanxi	(3.7)	(3.8)	(4.4)	(3.4)	(3.4)	(2.3)	(2.2)	(2.9)	(1.9)	(0.3)	0.6
Gansu	8.1	9.7	9.4	9.0	10.8	11.6	11.5	11.3	13.2	15.1	25.6
Qinghai	(1.4)	(1.2)	(1.7)	(1.9)	(1.9)	(2.0)	(1.8)	(2.0)	(0.9)	2.7	1.5
Ningxia	(0.4)	(0.7)	(0.5)	(0.5)	(0.6)	(0.6)	(0.1)	(1.7)	1.3	0.9	(0.2)
Xinjiang	2.2	4.7	4.4	2.3	5.6	7.1	2.7	3.5	7.2	16.8	13.8

Note: Notional consumption is used excluding inventory buildup or withdraw.
Sources: China Statistical Yearbook, 1994, 1997, *China Energy Statistical Yearbook, 1991*, and State Statistical Bureau.

Table A3.31 Kerosene Balance by Region (000 b/d)

	1985	1986	1987	1988	1989	1990	1991	1992	1993	1994	1995
National Total	6.6	6.6	8.5	3.6	5.4	9.0	5.0	4.0	(12.4)	(7.9)	(8.3)
North	(21.4)	(23.2)	(18.8)	(22.6)	(16.4)	(12.2)	(11.4)	(10.3)	(16.0)	(9.6)	(20.1)
Beijing	(21.6)	(23.5)	(20.0)	(23.3)	(17.3)	(13.5)	(12.8)	(11.8)	(17.2)	(12.9)	(21.3)
Tianjin	1.3	1.5	1.5	1.4	1.6	1.5	1.4	1.2	1.0	3.2	1.8
Hebei	(0.3)	(0.5)	0.2	0.1	(0.1)	0.3	0.4	0.5	0.4	0.4	0.2
Shanxi	(0.7)	(0.6)	(0.5)	(0.7)	(0.5)	(0.3)	(0.3)	(0.2)	(0.2)	(0.2)	(0.7)
Inner Mongolia	(0.1)	(0.2)	(0.1)	(0.1)	(0.1)	(0.1)	(0.0)	(0.0)	(0.1)	(0.1)	(0.1)
North-east	27.1	26.2	23.3	17.3	19.6	18.9	15.8	16.2	9.5	9.2	13.7
Liaoning	24.1	23.8	21.5	15.4	17.4	16.5	13.8	13.9	8.5	8.3	12.3
Jilin	(0.1)	(0.4)	(0.5)	(0.6)	(0.6)	(0.5)	(0.4)	(0.3)	(0.5)	(0.7)	(1.0)
Heilongjiang	3.1	2.8	2.3	2.5	2.9	2.9	2.4	2.5	1.5	1.6	2.3
East	2.6	4.5	5.9	4.6	3.6	0.7	(1.1)	(1.7)	(4.2)	(4.5)	(3.0)
Shanghai	(2.7)	(0.8)	(1.4)	(2.7)	(3.5)	(6.6)	(7.1)	(6.6)	(5.3)	(6.0)	(11.6)
Jiangsu	3.2	4.6	7.2	7.4	7.0	6.8	5.9	5.0	3.1	2.8	8.3
Zhejiang	(0.5)	(0.5)	(0.7)	(1.0)	(0.6)	(0.2)	(0.7)	(0.5)	(1.0)	1.5	5.1
Anhui	(0.3)	(0.7)	(0.6)	(0.5)	(0.5)	(0.3)	(0.3)	(0.4)	(0.6)	(0.1)	(0.2)
Fujian	(0.5)	(0.7)	(0.8)	(1.0)	(0.9)	(1.0)	(0.7)	(1.1)	(1.9)	(3.5)	(7.5)
Jiangxi	(0.4)	(0.6)	(0.5)	(0.3)	(0.3)	(0.0)	0.4	0.5	0.2	(0.3)	(0.5)
Shandong	3.7	3.1	2.6	2.7	2.4	2.0	1.6	1.5	1.3	1.1	3.5
South-central	3.4	3.5	1.4	9.6	2.9	4.7	4.5	4.1	6.6	5.6	12.7
Henan	(0.6)	(0.6)	(0.4)	(0.4)	(0.5)	0.4	0.6	0.6	1.3	0.9	3.3
Hubei	2.8	2.2	2.2	2.5	2.2	2.4	2.0	1.6	0.7	0.4	0.0
Hunan	2.8	2.7	1.1	0.7	1.0	0.7	1.2	1.2	0.6	0.4	0.9
Guangdong	(1.3)	(0.6)	0.0	8.8	1.3	2.2	1.7	0.8	4.1	4.0	11.8
Guangxi	(0.3)	(0.1)	(1.6)	(2.0)	(1.1)	(1.0)	(1.0)	(0.2)	(0.2)	(0.1)	(0.4)
Hainan	–	–	–	(0.0)	(0.0)	(0.0)	(0.0)	(0.0)	(0.0)	(0.0)	(2.9)
South-west	(7.4)	(7.0)	(5.4)	(5.9)	(5.7)	(4.8)	(4.3)	(5.4)	(5.4)	(6.1)	(9.8)
Sichuan	(5.0)	(4.8)	(3.6)	(4.2)	(3.9)	(3.5)	(2.8)	(3.9)	(3.7)	(3.7)	(6.0)
Guizhou	(1.1)	(0.8)	(0.7)	(0.6)	(0.9)	(0.6)	(0.5)	(0.5)	(0.2)	(0.1)	(0.2)
Yunnan	(1.2)	(1.4)	(1.2)	(1.1)	(0.9)	(0.8)	(1.0)	(1.0)	(1.6)	(2.3)	(3.6)
Tibet	(0.2)	n.a.	n.a.	n.a.	n.a.	n.a.	n.a.	n.a.	n.a.	n.a.	n.a.
West	2.4	2.7	2.3	0.6	1.4	1.8	1.4	1.1	(2.8)	(2.5)	(1.8)
Shaanxi	(1.2)	(3.0)	(2.9)	(3.7)	(2.6)	(2.4)	(2.3)	(2.4)	(3.2)	(3.3)	(4.3)
Gansu	3.9	5.9	5.5	5.1	4.4	4.1	4.2	4.1	2.8	2.7	7.3
Qinghai	(0.1)	(0.0)	(0.0)	(0.0)	(0.0)	(0.0)	(0.0)	(0.0)	(0.0)	(0.0)	(0.0)
Ningxia	(0.0)	(0.1)	(0.1)	(0.1)	(0.0)	(0.0)	(0.0)	(0.0)	(0.0)	(0.0)	–
Xinjiang	(0.3)	(0.1)	(0.3)	(0.8)	(0.4)	0.1	(0.5)	(0.7)	(2.4)	(1.9)	(4.7)

Note: Notional consumption is used excluding inventory buildup or withdraw.
Sources: *China Statistical Yearbook, 1994, 1997, China Energy Statistical Yearbook, 1991*, and State Statistical Bureau.

Table A3.32 Fuel Oil Balance by Region (000 b/d)

	1985	1986	1987	1988	1989	1990	1991	1992	1993	1994	1995
National Total	(2.2)	(4.9)	1.8	(6.5)	(5.1)	(9.6)	(10.9)	(46.1)	(97.5)	(66.3)	(106.0)
North	15.6	(17.8)	(21.0)	(14.7)	(13.5)	(17.7)	(13.6)	(14.2)	(34.0)	5.9	(4.5)
Beijing	9.6	(0.8)	(0.0)	2.2	3.4	2.4	0.9	(0.9)	(21.1)	(8.2)	(9.7)
Tianjin	4.4	(8.5)	(14.5)	(10.9)	(9.8)	(12.8)	(8.1)	(9.1)	(7.9)	11.7	8.1
Hebei	7.0	(1.9)	(0.3)	(0.3)	(1.3)	(0.7)	0.4	1.9	2.6	4.5	(0.8)
Shanxi	(1.8)	(1.9)	(1.9)	(2.0)	(2.0)	(2.0)	(2.3)	(2.6)	(2.9)	(2.6)	(2.2)
Inner Mongolia	(3.6)	(4.7)	(4.3)	(3.8)	(3.9)	(4.6)	(4.4)	(3.5)	(4.7)	0.4	0.1
North-east	77.1	41.8	41.3	31.9	36.3	31.5	31.2	24.3	19.2	(30.4)	27.8
Liaoning	56.0	24.5	32.6	32.9	32.2	27.0	25.5	21.8	20.2	18.1	20.3
Jilin	9.2	7.3	!0.0	2.0	1.7	3.5	3.0	(0.4)	1.3	3.4	2.8
Heilongjiang	11.9	10.0	(1.3)	(3.0)	2.3	1.0	2.7	2.8	(2.3)	(52.0)	4.7
East	(34.2)	(18.0)	(5.5)	(14.8)	(8.9)	3.3	1.8	(8.3)	(16.8)	(4.4)	(27.6)
Shanghai	(15.8)	(34.2)	(29.8)	(43.9)	(31.1)	(24.3)	(21.7)	(24.0)	(19.7)	(14.7)	(16.4)
Jiangsu	13.1	5.9	15.4	21.3	18.5	18.2	14.7	12.2	7.1	6.2	1.1
Zhejiang	2.4	(0.0)	0.1	0.8	0.1	(4.0)	(7.2)	(8.2)	(12.5)	(11.6)	(9.1)
Anhui	!0.0	6.2	8.6	4.6	(3.1)	5.0	5.8	3.9	4.5	3.9	1.3
Fujian	(2.1)	(2.5)	(3.0)	(2.8)	(3.0)	(2.8)	(3.4)	(3.3)	(3.6)	(3.5)	(5.0)
Jiangxi	(3.1)	3.0	4.4	3.8	3.7	3.5	3.4	2.5	2.7	3.6	2.5
Shandong	(38.7)	3.7	(1.1)	1.5	6.0	7.6	10.0	8.7	4.8	11.7	(2.0)
South-central	(62.8)	(10.1)	(10.6)	(13.6)	(17.2)	(22.3)	(27.8)	(38.4)	(55.7)	(32.3)	(80.2)
Henan	(0.2)	(3.7)	(4.1)	(3.7)	(4.6)	(4.6)	(5.7)	(5.8)	(5.6)	(6.2)	(6.9)
Hubei	(25.2)	2.1	(0.5)	(0.9)	1.9	1.0	0.2	(0.0)	(1.9)	1.1	0.1
Hunan	(5.7)	0.2	0.3	(0.3)	0.5	1.3	1.6	0.5	(0.4)	0.8	0.4
Guangdong	(46.1)	(7.6)	(5.1)	(7.8)	(13.7)	(18.7)	(22.2)	(31.5)	(46.2)	(25.8)	(70.2)
Guangxi	14.3	(1.1)	(1.1)	(0.8)	(1.2)	(1.2)	(1.6)	(1.5)	(1.6)	(2.3)	(1.9)
Hainan	–	–	–	(0.1)	(0.1)	(0.1)	(0.1)	(0.1)	(0.1)	–	(1.6)
South-west	(1.7)	(1.9)	(1.7)	(2.0)	(1.9)	(2.0)	(2.2)	(2.5)	(3.6)	(4.8)	(11.7)
Sichuan	(0.9)	(1.0)	(0.7)	(1.0)	(0.8)	(0.8)	(0.7)	(0.9)	(1.2)	(2.1)	(0.3)
Guizhou	(0.5)	(0.6)	(0.7)	(0.7)	(0.7)	(0.9)	(1.1)	(1.2)	(2.0)	(1.9)	(0.5)
Yunnan	(0.3)	(0.4)	(0.4)	(0.4)	(0.4)	(0.3)	(0.4)	(0.4)	(0.4)	(0.7)	(10.9)
Tibet	–	–	–	–	–	–	–	–	–	–	–
West	3.8	1.0	(0.8)	6.7	0.3	(2.4)	(0.2)	(7.0)	(6.7)	(0.2)	(9.8)
Shaanxi	(2.8)	(2.9)	(3.2)	(2.7)	(2.6)	(1.3)	(0.1)	(2.8)	(2.6)	(2.3)	(11.8)
Gansu	4.8	3.1	3.1	4.6	5.3	4.7	5.8	4.0	4.8	5.9	12.3
Qinghai	0.5	0.0	0.4	0.1	0.0	(1.7)	(0.8)	(2.4)	(1.4)	(0.9)	1.9
Ningxia	0.2	0.2	0.2	(0.7)	(2.6)	(3.9)	(3.3)	(5.3)	(5.8)	(4.6)	(2.6)
Xinjiang	1.0	0.6	(1.3)	5.5	0.2	(0.2)	(1.8)	(0.6)	(1.7)	1.8	(9.6)

Note: Notional consumption is used excluding inventory buildup or withdraw.
Sources: *China Statistical Yearbook, 1994, 1997, China Energy Statistical Yearbook, 1991,* and State Statistical Bureau.

Table A3.33 Refinery Runs and Product Output by Source (000 b/d)

	1984	1985	1986	1987	1988	1989	1990	1991	1992	1993	1994	1995	1996
Refinery Runs													
Sinopec	1533	1581	1713	1800	1866	1924	1938	2050	2163	2258	2199	2264	2361
National total	1636	1718	1852	1955	2173	2269	2321	2433	2529	2651	2639	2881	3000
Sinopec's share	94%	92%	92%	92%	86%	85%	83%	84%	86%	85%	83%	79%	79%
Gasoline Production													
Sinopec	291	314	360	370	400	430	443	492	546	615	531	530	567
National total	311	343	392	405	442	459	506	560	628	736	665	709	753
Sinopec's share	93%	92%	92%	91%	90%	94%	87%	88%	87%	84%	80%	75%	75%
Diesel Oil Production													
Sinopec	376	386	428	448	458	473	469	511	558	612	59	632	706
National total	395	414	456	484	503	528	533	583	622	724	711	794	883
Sinopec's share	95%	93%	94%	93%	91%	90%	88%	88%	90%	85%	8%	80%	80%
Kerosene Production													
Sinopec	85	85	88	88	80	83	81	86	83	79	85	88	105
National total	87	87	89	90	82	85	83	88	85	81	86	92	111
Sinopec's share	98%	98%	98%	98%	98%	98%	98%	98%	98%	98%	98%	95%	95%
Fuel Oil Production													
Sinopec	489	482	486	526	535	533	530	522	520	498	461	420	346
National total	523	517	545	571	587	604	596	592	604	583	556	509	425
Sinopec's share	93%	93%	89%	92%	91%	88%	89%	88%	86%	85%	83%	83%	82%
Lubricant Production													
Sinopec	25	27	29	31	31	31	29	30	33	33	30	29	35
National total	28	30	32	34	35	35	32	33	36	36	33	36	43
Sinopec's share	88%	90%	91%	89%	89%	90%	91%	90%	91%	90%	89%	81%	81%
Hydrocarbon solvents Production													
Sinopec	5	5	7	7	7	7	7	7	8	7	9	10	13
National total	8	8	9	10	9	9	10	10	11	9	11	14	17
Sinopec's share	65%	62%	73%	70%	74%	79%	76%	74%	74%	78%	76%	76%	76%
Petroleum wax Production													
Sinopec	9	10	11	12	12	13	13	12	13	13	14	16	18
National total	10	11	11	12	13	13	13	13	13	14	15	16	18
Sinopec's share	96%	96%	97%	97%	97%	97%	98%	97%	97%	96%	97%	97%	97%
Petroleum coke Production													
Sinopec	18	17	19	22	23	24	25	29	31	33	38	43	53
National total	18	18	20	22	23	25	26	30	32	33	39	44	54
Sinopec's share	96%	96%	97%	97%	97%	97%	97%	97%	98%	97%	97%	97%	97%
Petroleum asphalt Production													
Sinopec	37	41	43	43	43	41	43	47	49	47	47	46	46
National total	40	46	49	51	52	50	52	60	66	64	61	61	61
Sinopec's share	92%	90%	87%	86%	82%	82%	83%	78%	74%	73%	77%	76%	75%
Light oil for chemical Ind. Production													
Sinopec	86	87	80	119	125	135	141	144	149	146	178	209	237
National total	94	94	89	128	135	145	150	154	160	156	190	224	253
Sinopec's share	92%	92%	90%	93%	92%	93%	94%	94%	93%	94%	94%	94%	94%

Notes:

(1) Data presented in this table may differ from those presented in the previous chapters, which were published by the State Statistical Bureau.

(2) Light oil for chemical Industry includes naphtha.

Sources: *China Petrochemical Yearbook, 1991, 1994, 1997*, China Petrochemical Press.

Table A3.34 Refinery Throughput (000 b/d) by Major Refineries

	Location	1983	1984	1985	1986	1987	1988	1989	1990	1991	1992	1993	1994	1995	1996
Yanshan Petrochemical	Beijing	112.8	117.8	118.0	123.6	127.6	128.6	130.4	130.8	130.6	132.6	132.8	127.9	128.9	126.7
Tianjin Petrochemical	Tianjin	54.1	54.3	56.0	56.2	59.9	59.6	62.6	59.5	60.6	61.2	68.7	64.2	68.4	71.8
Fushun Petrochemical	Liaoning	158.2	158.0	157.4	160.6	162.2	163.6	162.0	160.2	161.6	163.6	164.8	152.2	141.6	149.2
Jinzhou Petrochemical	Liaoning	40.2	41.6	42.4	45.8	48.0	48.0	48.2	49.0	48.0	53.8	58.5	67.5	63.3	67.1
Dalian Petrochemical	Liaoning	86.8	85.0	86.4	89.0	90.8	93.2	95.0	95.2	99.0	106.0	108.2	104.4	105.7	103.4
Liaoyang Petrochemical Fibre	Liaoning	19.6	19.6	20.2	21.8	22.0	21.0	24.4	24.4	24.6	24.0	24.1	23.1	32.0	40.0
Anshan Refomery	Liaoning	–	–	–	–	–	–	29.0	29.2	28.8	28.4	28.0	25.6	24.4	24.2
Gaoqiao Petrochemical	Shanghai	81.2	80.2	81.0	85.0	85.0	88.8	91.4	91.6	95.0	99.0	92.8	90.0	96.4	102.0
Jinling Petrochemical	Jiangsu	93.4	94.0	94.4	102.4	110.0	112.0	109.2	106.4	110.2	116.8	117.8	114.1	115.7	112.9
Yangzi Petrochemical	Jiangsu	–	–	–	–	16.6	32.0	36.2	46.0	53.1	52.5	53.5	56.0	64.3	72.0
Qilu Petrochemical	Shandong	106.0	107.2	109.6	113.6	120.4	128.6	138.4	134.4	145.2	146.8	143.8	142.1	141.2	142.2
Baling Petrochemical	Hubei	55.5	57.2	57.7	60.6	61.6	63.4	63.5	64.1	67.0	70.4	67.3	68.1	69.3	69.0
Maoming Petrochemical	Guangdong	90.4	92.4	93.2	105.2	109.6	110.8	114.2	115.2	126.9	133.3	141.0	131.0	133.7	142.5
Lanzhou Chemical Industry	Gansu	–	–	3.6	8.6	11.6	12.6	13.2	12.6	13.2	13.7	14.5	16.9	21.2	23.1
Daqing General Petrochemical Works	Heilongjiang	86.8	88.2	92.2	96.4	100.2	102.2	106.2	106.4	110.6	114.0	115.0	97.4	107.0	108.6
Shanghai Jinshan Industrial	Shanghai	34.8	35.2	41.0	45.2	47.4	48.6	53.0	67.8	76.6	80.4	91.9	94.1	94.5	97.1
Lanzhou Petro. Processing & Chem. Complex	Gansu	47.0	48.4	50.2	53.2	54.6	50.4	52.0	55.2	60.0	62.0	62.4	64.1	69.3	73.2
Jinxi Petro. Processing & Chem. Complex	Liaoning	66.8	68.2	68.6	72.4	74.6	73.6	75.6	76.0	76.3	78.4	83.5	80.1	80.2	77.8
Zhenhai Refining & Chem.	Zhejiang	29.4	34.0	34.8	42.4	45.4	45.2	52.8	55.6	65.4	78.2	88.0	99.6	110.1	118.8
Anqing General Petrochemical Works	Anhui	39.6	42.4	44.0	45.6	52.6	52.2	51.0	51.4	51.0	53.3	52.9	47.8	55.0	60.3
Guangzhou General Petrochemical Works	Guangdong	45.0	47.4	44.8	51.4	54.6	52.6	58.0	61.0	73.4	84.6	84.0	77.4	86.6	84.9
Urumqi General Petrochemical Works	Xinjiang	15.8	17.0	22.4	26.0	29.4	31.4	32.6	34.2	36.2	41.4	46.4	44.8	46.6	53.6
Shijiazhuang Refinery	Hebei	6.0	13.0	14.0	20.0	23.8	27.6	27.8	27.6	29.0	31.0	34.0	33.4	37.2	38.1
Cangzhou Refinery	Hebei	9.8	9.8	10.6	13.4	14.4	15.2	15.2	15.4	16.3	18.1	20.0	19.5	20.0	20.5
Qianguo Refinery	Heilongjiang	12.2	12.0	12.6	15.2	15.2	16.6	16.8	18.0	16.3	17.3	16.4	16.1	18.4	20.5
Harbin Refinery	Heilongjiang	12.8	11.6	11.8	15.6	15.0	17.6	17.0	16.8	16.8	17.6	20.4	22.0	25.0	25.2
Lingyuan Refinery	Heilongjiang	22.4	25.6	27.8	26.8	28.6	28.2	28.4	27.2	27.2	27.4	28.0	27.0	26.0	26.8
Fujian Refinery	Fujian	–	–	–	–	–	–	–	–	–	9.7	32.6	31.7	45.0	50.0
Jiujiang General Petrochemical Works	Jiangxi	13.6	14.0	16.6	24.2	27.0	28.6	30.0	31.0	36.2	41.8	46.1	40.5	46.1	46.3
Jinan Refinery	Shandong	15.3	15.9	17.0	18.4	19.8	22.5	26.3	27.2	29.0	31.4	37.0	38.2	37.8	37.9
Luoyang General Petrochemical Works	Henan	–	1.4	12.6	19.4	20.2	30.6	28.0	31.4	30.4	32.6	38.0	39.8	52.2	50.9
Wuhan Petrochemical Works	Hubei	30.0	30.4	30.8	36.4	31.2	39.0	40.0	40.4	40.5	44.0	45.6	46.5	45.7	50.3
Jingmen General Petrochemical Works	Hubei	54.6	55.8	55.8	57.6	58.0	58.4	58.0	56.0	52.5	51.7	48.1	47.7	50.0	52.7
Dushanzi Refinery	Xinjiang	24.2	25.4	26.0	28.0	30.8	32.0	34.2	36.8	35.8	42.2	46.6	n.a.	n.a.	n.a.
Dalia West Pacific Petrochemical Corp.	Liaoning	–	–	–	–	–	–	–	–	–	–	–	–	–	17.2
Total		1464.4	1503.0	1553.5	1680.1	1768.0	1834.6	1920.6	1953.9	2043.3	2158.9	2252.6	2150.8	2259.0	2356.6

Sources: *China Petrochemical Yearbook, 1991, 1994, 1997,* China's Petrochemical Press.

Table A3.35 Gasoline Production (000 b/d) by Major Refineries

	Location	1983	1984	1985	1986	1987	1988	1989	1990	1991	1992	1993	1994	1995	1996
Yanshan Petrochemical	Beijing	17.7	19.8	20.4	24.8	28.9	28.1	30.3	32.8	31.4	33.2	32.7	27.0	23.0	28.3
Tianjin Petrochemical	Tianjin	10.2	10.5	11.5	12.2	13.3	13.1	13.1	14.4	15.7	15.0	17.6	16.1	16.9	18.7
Fushun Petrochemical	Liaoning	24.9	25.9	26.9	28.8	30.7	39.4	42.4	42.7	45.4	48.5	50.1	42.0	34.6	36.8
Jinzhou Petrochemical	Liaoning	9.6	9.6	9.7	9.9	9.3	8.8	9.0	9.9	12.3	13.9	15.3	15.7	14.1	14.5
Dalian Petrochemical	Liaoning	12.1	12.0	14.2	17.1	17.4	17.7	21.2	21.8	24.0	27.7	31.2	24.1	21.3	21.6
Liaoyang Petrochemical Fibre	Liaoning	–	–	–	–	–	–	–	–	–	–	–	–	–	–
Gaoqiao Petrochemical	Shanghai	12.1	13.3	13.8	15.8	15.1	15.8	19.2	19.0	21.1	22.5	22.5	19.9	23.2	23.2
Jinling Petrochemical	Jiangsu	13.7	15.8	15.0	16.6	17.4	19.2	18.7	19.0	21.9	26.7	32.8	23.6	21.3	22.8
Yangzi Petrochemical	Jiangsu	–	–	–	–	–	–	–	–	–	–	–	–	–	–
Qilu Petrochemical	Shandong	17.5	19.3	20.8	21.1	24.8	24.5	25.7	22.5	24.3	23.7	24.4	24.8	24.5	25.3
Baling Petrochemical	Hubei	13.7	15.2	15.1	16.7	17.1	17.2	17.6	18.7	20.2	22.7	24.2	23.0	21.0	21.0
Maoming Petrochemical	Guangdong	15.2	15.7	17.2	21.5	20.0	20.2	21.8	24.5	28.7	32.1	36.1	28.9	33.0	31.3
Lanzhou Chemical Industry	Gansu	–	–	–	–	–	–	–	–	–	–	–	–	–	–
Daqing General Petrochemical Works	Heilongjiang	16.5	16.7	17.0	17.4	15.9	15.5	15.8	14.9	16.0	16.4	20.9	16.9	21.0	19.8
Shanghai Jinshan Industrial	Shanghai	1.8	1.8	2.6	14.7	12.9	12.2	10.8	13.4	13.9	14.7	15.3	15.9	17.5	19.3
Lanzhou Petro. Processing & Chem. Complex	Gansu	12.1	12.7	13.6	14.7	12.9	12.2	10.8	13.4	13.9	14.7	15.3	15.9	17.5	19.3
Jinxi Petro. Processing & Chem. Complex	Liaoning	12.6	12.7	13.1	14.0	11.7	14.3	15.3	15.6	15.3	15.3	18.6	16.0	16.3	16.1
Zhenhai Refining & Chem.	Zhejiang	7.4	8.1	9.1	10.6	10.6	11.0	16.2	20.2	25.2	31.9	34.6	28.6	27.5	30.8
Anqing General Petrochemical Works	Anhui	6.0	6.8	6.1	10.1	10.9	12.6	13.5	13.5	13.6	16.1	16.9	14.5	17.3	18.3
Guangzhou General Petrochemical Works	Guangdong	8.7	9.0	9.3	10.4	10.3	10.9	13.0	15.6	21.1	25.2	27.0	23.2	25.8	26.6
Urumqi General Petrochemical Works	Xinjiang	5.2	6.0	8.8	9.3	10.3	10.4	10.4	10.2	11.0	13.3	17.2	14.9	13.1	17.4
Shijiazhuang Refinery	Hebei	2.1	5.6	5.9	9.0	10.2	13.0	13.2	12.3	12.7	13.9	16.3	13.6	14.9	14.8
Cangzhou Refinery	Hebei	2.3	2.5	2.8	3.8	4.0	4.2	4.8	5.9	5.9	6.2	7.5	6.9	6.8	7.2
Qianguo Refinery	Heilongjiang	3.6	3.4	3.8	4.7	4.9	5.5	5.9	6.6	8.3	9.1	8.3	6.8	7.8	7.3
Harbin Refinery	Heilongjiang	2.6	2.5	2.8	3.6	3.3	4.7	5.3	5.3	5.0	5.2	7.6	9.3	10.0	10.0
Lingyuan Refinery	Heilongjiang	2.2	2.6	3.1	4.3	4.9	5.0	5.6	7.3	7.6	8.7	9.0	8.5	9.5	10.6
Fujian Refinery	Fujian	–	–	–	–	–	–	–	–	–	4.0	14.3	14.0	19.0	19.5
Jiujiang General Petrochemical Works	Jiangxi	4.0	4.4	5.7	8.5	9.5	10.1	10.4	11.1	13.1	16.8	18.5	13.8	14.8	12.6
Jinan Refinery	Shandong	3.2	3.4	3.3	3.7	4.3	4.9	7.0	7.6	7.9	9.5	12.0	10.4	10.5	11.3
Luoyang General Petrochemical Works	Henan	–	0.6	5.7	9.0	9.6	15.0	15.5	15.7	16.5	17.5	19.8	20.4	25.3	21.6
Wuhan Petrochemical Works	Hubei	7.1	7.9	7.9	9.5	7.2	10.8	11.7	11.9	13.1	14.7	16.8	15.4	15.4	20.1
Jingmen General Petrochemical Works	Hubei	14.2	14.9	15.5	16.1	15.7	16.1	15.7	15.4	14.1	13.9	14.4	12.8	12.9	14.4
Dushanzi Refinery	Xinjiang	9.9	10.4	10.7	11.3	12.2	13.0	13.6	15.0	16.5	17.4	18.1	n.a.	n.a.	n.a.
Dalia West Pacific Petrochemical Corporation	Liaoning	–	–	–	–	–	–	–	–	–	–	–	–	–	5.8
Total		268.4	288.9	311.7	369.2	375.3	405.6	433.8	456.4	495.8	550.1	615.5	523.1	535.8	560.5

Sources: *China Petrochemical Yearbook, 1991, 1994, 1997*, China's Petrochemical Press.

Table A3.36 Naphtha Production (000 b/d) by Major Refineries

	Location	1983	1984	1985	1986	1987	1988	1989	1990	1991	1992	1993	1994	1995	1996
Yanshan Petrochemical	Beijing	–	–	–	–	–	–	–	–	–	–	–	–	–	–
Tianjin Petrochemical	Tianjin	–	–	–	–	–	–	–	–	–	–	–	–	–	–
Fushun Petrochemical	Liaoning	9.2	9.1	8.4	7.0	6.7	5.6	4.2	0.5	–	–	–	–	–	–
Jinzhou Petrochemical	Liaoning	–	–	–	–	–	–	–	–	–	–	–	–	–	–
Dalian Petrochemical	Liaoning	–	–	–	–	–	–	–	–	–	–	–	–	–	–
Liaoyang Petrochemical Fibre	Liaoning	–	–	–	–	–	–	–	–	–	–	–	–	–	–
Gaoqiao Petrochemical	Shanghai	–	–	–	–	–	–	–	–	–	–	–	–	–	–
Jinling Petrochemical	Jiangsu	9.2	7.9	8.2	7.8	8.2	8.2	7.8	8.1	7.4	6.2	3.6	4.2	4.4	5.7
Yangzi Petrochemical	Jiangsu	–	–	–	–	–	–	–	–	–	–	–	–	–	–
Qilu Petrochemical	Shandong	–	–	–	–	–	–	–	–	–	–	–	–	–	–
Baling Petrochemical	Hubei	–	–	–	–	–	–	–	–	–	–	–	–	–	–
Maoming Petrochemical	Guangdong	–	–	–	–	–	–	–	–	–	–	–	–	–	–
Lanzhou Chemical Industry	Gansu	–	–	–	–	–	–	–	–	–	–	–	–	–	–
Daqing General Petrochemical Works	Heilongjiang	2.2	3.4	3.3	4.5	6.2	7.0	6.9	8.6	11.7	12.3	12.7	9.6	10.1	11.0
Shanghai Jinshan Industrial	Shanghai	–	0.4	9.3	11.7	13.0	12.8	12.5	14.2	14.1	13.5	16.5	17.4	15.8	15.3
Lanzhou Petro. Processing & Chem. Complex	Gansu	–	–	–	–	–	–	–	–	–	–	–	–	–	–
Jinxi Petro. Processing & Chem. Complex	Liaoning	–	–	–	–	–	–	–	–	–	–	–	–	–	–
Zhenhai Refining & Chem.	Zhejiang	1.3	1.9	1.6	1.6	0.9	0.9	0.9	0.9	0.9	1.4	1.6	4.4	3.6	3.2
Anqing General Petrochemical Works	Anhui	–	0.6	1.0	0.7	0.7	1.0	0.9	1.0	0.8	0.5	0.4	–	–	–
Guangzhou General Petrochemical Works	Guangdong	–	0.9	1.0	1.2	0.8	0.9	0.8	–	–	–	–	–	–	–
Urumqi General Petrochemical Works	Xinjiang	–	–	–	–	–	–	–	–	–	–	–	–	–	–
Shijiazhuang Refinery	Hebei	–	–	–	–	–	–	–	–	–	–	–	–	–	–
Cangzhou Refinery	Hebei	–	–	–	–	–	–	–	–	–	–	–	–	–	–
Qianguo Refinery	Heilongjiang	–	–	–	–	–	–	–	–	–	–	–	–	–	–
Harbin Refinery	Heilongjiang	1.2	1.0	0.8	0.7	0.7	0.7	0.7	0.5	0.7	0.9	1.0	1.1	0.9	0.8
Lingyuan Refinery	Heilongjiang	2.5	2.3	2.1	1.7	2.0	2.0	2.0	1.7	2.0	2.7	1.9	3.5	2.9	2.9
Fujian Refinery	Fujian	–	–	–	–	–	–	–	–	–	–	–	–	–	–
Jiujiang General Petrochemical Works	Jiangxi	0.5	0.5	0.5	–	–	–	–	–	–	–	–	–	–	–
Jinan Refinery	Shandong	–	–	–	–	–	–	–	–	–	–	–	–	–	–
Luoyang General Petrochemical Works	Henan	–	–	–	–	–	–	–	–	–	–	–	–	–	–
Wuhan Petrochemical Works	Hubei	–	–	–	–	–	–	–	–	–	–	–	–	–	–
Jingmen General Petrochemical Works	Hubei	–	–	–	–	–	–	–	–	–	–	–	–	–	–
Dushanzi Refinery	Xinjiang	–	–	–	–	–	–	–	–	–	–	–	–	–	–
Dalia West Pacific Petrochemical Corporation	Liaoning	–	–	–	–	–	–	–	–	–	–	–	–	–	–
Total		26.1	28.1	36.2	36.8	39.2	39.2	36.9	35.6	37.5	37.5	37.6	40.2	37.7	38.9

Note: Production does not include naphtha as feedstock in petrochemical plant.
Sources: *China Petrochemical Yearbook, 1991, 1994, 1997*, China's Petrochemical Press.

Table A3.37 LPG Production (000 b/d) by Major Refineries

	Location	1983	1984	1985	1986	1987	1988	1989	1990	1991	1992	1993	1994	1995	1996
Yanshan Petrochemical	Beijing	6.2	7.3	7.2	7.5	7.3	7.3	7.4	7.6	7.2	7.9	8.1	7.8	7.9	8.6
Tianjin Petrochemical	Tianjin	1.7	1.8	1.9	2.3	2.5	2.2	4.0	3.5	2.8	2.9	2.9	2.8	3.4	3.9
Fushun Petrochemical	Liaoning	–	–	–	1.9	2.3	5.1	5.6	5.2	6.6	6.9	6.8	6.8	6.1	7.2
Jinzhou Petrochemical	Liaoning	–	–	–	–	–	–	–	–	–	–	–	–	–	–
Dalian Petrochemical	Liaoning	–	–	–	–	–	–	–	–	–	–	–	–	–	–
Liaoyang Petrochemical Fibre	Liaoning	–	–	–	–	–	–	–	–	–	–	–	–	–	–
Gaoqiao Petrochemical	Shanghai	–	–	–	–	–	–	–	–	–	–	–	–	–	–
Jinling Petrochemical	Jiangsu	1.5	1.7	1.7	1.8	2.2	2.9	2.8	2.4	2.9	4.5	4.7	5.2	5.9	5.4
Yangzi Petrochemical	Jiangsu	–	–	–	–	–	–	–	–	–	–	–	–	–	–
Qilu Petrochemical	Shandong	1.7	2.0	2.4	2.5	3.1	3.2	3.2	3.2	3.0	3.3	3.5	n.a.	n.a.	n.a.
Baling Petrochemical	Hubei	–	–	–	–	–	–	–	–	–	–	–	–	–	–
Maoming Petrochemical	Guangdong	–	–	–	–	–	–	–	–	–	–	–	–	–	–
Lanzhou Chemical Industry	Gansu	–	–	–	–	–	–	–	–	–	–	–	–	–	–
Daqing General Petrochemical Works	Heilongjiang	–	–	–	–	–	–	–	–	–	–	–	–	–	–
Shanghai Jinshan Industrial	Shanghai	–	–	–	–	–	–	–	–	–	–	–	–	–	–
Lanzhou Petro. Processing & Chem. Complex	Gansu	–	–	–	–	–	–	–	–	–	–	–	–	–	–
Jinxi Petro. Processing & Chem. Complex	Liaoning	–	–	–	–	–	–	–	–	–	–	–	–	–	–
Zhenhai Refining & Chem.	Zhejiang	0.7	1.1	0.9	1.1	1.2	0.9	1.1	1.1	1.4	1.8	2.7	3.7	5.1	5.3
Anqing General Petrochemical Works	Anhui	–	–	–	–	–	–	–	–	–	–	–	–	–	–
Guangzhou General Petrochemical Works	Guangdong	1.5	1.5	1.7	2.1	2.2	2.0	2.6	3.5	4.9	5.1	5.0	4.4	4.0	4.3
Urumqi General Petrochemical Works	Xinjiang	–	–	–	–	–	–	–	–	–	–	–	–	–	–
Shijiazhuang Refinery	Hebei	0.4	0.8	1.1	1.8	2.1	2.5	1.9	1.5	1.5	1.9	2.0	1.6	2.7	2.5
Cangzhou Refinery	Hebei	0.3	0.3	0.3	0.5	0.5	0.5	0.5	0.9	0.9	0.9	1.1	1.2	1.2	1.3
Qianguo Refinery	Heilongjiang	0.3	0.3	0.4	0.5	0.5	0.6	0.6	0.7	1.1	1.5	1.5	1.5	1.3	1.5
Harbin Refinery	Heilongjiang	0.4	0.4	0.4	0.6	0.5	0.6	0.7	0.6	0.7	0.7	0.7	1.0	1.2	1.6
Lingyuan Refinery	Heilongjiang	0.2	0.3	0.4	0.5	0.5	0.5	0.7	1.3	1.4	1.3	1.7	1.5	1.9	2.3
Fujian Refinery	Fujian	–	–	–	–	–	–	–	–	–	1.0	2.9	4.3	5.5	4.1
Jiujiang General Petrochemical Works	Jiangxi	0.4	0.4	0.7	0.8	1.2	1.4	1.4	1.4	1.7	1.5	1.8	2.1	2.2	2.6
Jinan Refinery	Shandong	0.3	0.3	0.3	0.4	0.4	0.5	1.6	1.7	1.5	1.7	2.2	2.7	2.1	2.5
Luoyang General Petrochemical Works	Henan	–	0.1	0.6	1.0	1.4	2.6	2.3	2.7	3.0	2.9	2.8	3.2	4.2	3.3
Wuhan Petrochemical Works	Hubei	–	–	–	–	1.6	2.3	1.2	1.1	0.9	1.9	2.6	2.9	3.5	4.5
Jingmen General Petrochemical Works	Hubei	–	–	–	–	–	–	–	–	–	–	–	–	–	–
Dushanzi Refinery	Xinjiang	–	–	–	–	–	–	–	–	–	–	–	–	–	–
Dalia West Pacific Petrochemical Corporation	Liaoning	–	–	–	–	–	–	–	–	–	–	–	–	–	–
Total		15.6	18.3	20.1	25.3	29.4	35.2	37.7	38.4	41.7	48.0	53.1	52.6	58.4	60.7

Sources: *China Petrochemical Yearbook, 1991, 1994, 1997*, China's Petrochemical Press.

Table A3.38 Diesel Oil Production (000 b/d) by Major Refinery

	Location	1983	1984	1985	1986	1987	1988	1989	1990	1991	1992	1993	1994	1995	1996
Yanshan Petrochemical	Beijing	15.7	17.6	15.1	17.6	19.3	18.0	19.3	19.3	18.1	17.6	19.6	21.4	16.1	19.6
Tianjin Petrochemical	Tianjin	13.0	12.8	12.8	12.9	13.9	13.3	14.3	13.5	13.7	14.8	17.3	17.9	16.1	21.2
Fushun Petrochemical	Liaoning	47.1	46.4	45.6	48.9	51.8	51.4	50.0	49.1	47.7	50.1	52.1	46.1	38.6	41.5
Jinzhou Petrochemical	Liaoning	13.3	14.0	14.4	14.8	14.9	13.3	12.9	13.0	17.5	18.7	22.0	25.7	24.0	27.3
Dalian Petrochemical	Liaoning	19.7	18.8	19.7	20.8	21.0	21.1	21.0	20.5	22.9	27.4	29.0	26.9	28.3	25.8
Liaoyang Petrochemical Fibre	Liaoning	–	–	–	–	–	–	–	–	–	–	–	–	–	–
Gaoqiao Petrochemical	Shanghai	19.5	19.1	19.0	19.7	19.5	19.7	20.8	20.2	21.7	23.5	24.0	24.7	24.3	26.3
Jinling Petrochemical	Jiangsu	26.2	26.9	27.4	30.1	32.0	32.3	32.4	30.1	32.2	35.1	35.4	33.5	35.0	37.3
Yangzi Petrochemical	Jiangsu	–	–	–	–	–	–	–	–	–	–	–	–	–	–
Qilu Petrochemical	Shandong	31.4	32.0	32.3	32.7	34.0	31.4	32.5	30.4	31.5	30.5	32.9	33.9	31.9	37.5
Baling Petrochemical	Hubei	17.2	15.2	15.0	15.8	17.4	18.9	18.7	18.1	18.6	21.1	21.5	20.2	21.3	22.1
Maoming Petrochemical	Guangdong	21.8	20.7	20.4	24.0	25.4	24.8	26.2	26.9	30.8	36.5	41.4	36.9	40.5	44.8
Lanzhou Chemical Industry	Gansu	–	–	–	–	–	–	–	–	–	–	–	–	–	–
Daqing General Petrochemical Works	Heilongjiang	29.2	28.5	29.1	31.1	32.2	33.2	34.4	34.8	36.0	37.8	36.0	31.0	36.5	39.9
Shanghai Jinshan Industrial	Shanghai	–	0.1	1.1	1.8	2.1	1.6	1.0	0.9	1.3	0.2	0.2	–	1.7	4.5
Lanzhou Petro. Processing & Chem. Complex	Gansu	8.5	6.7	9.1	10.7	10.9	10.4	11.4	12.2	13.3	14.6	15.3	17.9	19.6	20.8
Jinxi Petro. Processing & Chem. Complex	Liaoning	15.9	16.7	16.0	17.2	17.8	17.4	18.9	19.2	19.8	18.9	20.7	18.1	26.1	27.2
Zhenhai Refining & Chem.	Zhejiang	7.1	8.6	9.2	10.8	11.5	12.0	15.2	18.7	22.7	28.1	34.3	36.3	40.1	40.6
Anqing General Petrochemical Works	Anhui	11.3	11.0	11.0	12.0	13.5	14.8	15.5	15.4	16.4	17.9	17.8	15.3	19.8	24.5
Guangzhou General Petrochemical Works	Guangdong	9.3	9.8	9.7	10.6	10.1	10.7	11.8	14.4	20.2	25.5	24.9	24.4	30.2	30.9
Urumqi General Petrochemical Works	Xinjiang	6.1	6.4	7.5	8.8	9.4	10.2	10.3	10.2	11.9	13.6	13.9	16.2	16.8	22.2
Shijiazhuang Refinery	Hebei	1.6	3.9	4.0	5.8	6.8	8.8	8.2	7.4	8.1	9.0	10.6	11.3	15.3	15.1
Cangzhou Refinery	Hebei	2.5		3.0	4.2	4.8	4.8	5.5	5.3	5.4	6.2	6.9	7.4	8.1	9.3
Qianguo Refinery	Heilongjiang	4.1	3.9	3.2	4.3	4.4	4.8	5.4	5.3	5.2	6.5	6.2	5.9	6.7	8.2
Harbin Refinery	Heilongjiang	3.2	3.2	3.3	4.5	4.2	5.3	5.3	5.3	5.4	5.8	7.1	8.7	10.0	8.1
Lingyuan Refinery	Heilongjiang	3.0	4.1	5.0	6.8	8.0	7.8	7.6	8.4	9.2	8.9	8.5	8.9	9.3	11.8
Fujian Refinery	Fujian	–	–	–	–	–	–	–	–	–	2.6	11.3	11.1	17.4	19.7
Jiujiang General Petrochemical Works	Jiangxi	3.7	3.6	4.3	6.2	7.6	8.2	9.4	9.4	12.2	14.6	16.0	12.9	16.9	18.1
Jinan Refinery	Shandong	4.3	4.1	4.0	4.7	4.9	5.4	7.1	7.3	7.9	8.7	9.6	10.8	12.3	12.8
Luoyang General Petrochemical Works	Henan	–	0.5	4.6	7.3	7.7	9.4	9.6	11.2	11.9	12.0	11.9	13.5	17.7	19.0
Wuhan Petrochemical Works	Hubei	3.9	5.4	6.2	6.9	5.6	8.3	8.5	8.5	9.4	10.4	11.6	11.7	12.8	17.1
Jingmen General Petrochemical Works	Hubei	15.0	16.0	14.8	15.8	15.6	16.3	15.5	16.0	14.8	14.6	15.7	14.7	16.9	16.9
Dushanzi Refinery	Xinjiang	7.8	8.1	8.4	9.5	10.4	10.7	12.0	13.0	13.8	14.7	16.7	n.a.	n.a.	n.a.
Dalia West Pacific Petrochemical Corporation	Liaoning	–	–	–	–	–	–	–	–	–	–	–	–	–	5.8
Total		361.6	364.3	375.2	416.1	436.8	444.6	460.9	464.1	499.2	546.1	590.6	563.4	610.7	670.2

Sources: *China Petrochemical Yearbook, 1991, 1994, 1997*, China's Petrochemical Press.

Table A3.39 Kerosene Production (000 b/d) by Major Refineries

	Location	1983	1984	1985	1986	1987	1988	1989	1990	1991	1992	1993	1994	1995	1996
Yanshan Petrochemical	Beijing	0.2	0.2	0.2	0.3	0.6	0.6	0.6	0.6	0.6	0.6	0.4	0.4	0.0	0.0
Tianjin Petrochemical	Tianjin	1.4	1.4	1.8	2.1	2.0	1.9	2.1	2.2	2.2	1.9	2.0	1.9	1.9	1.7
Fushun Petrochemical	Liaoning	16.8	16.7	17.4	16.2	16.3	12.2	14.4	14.1	14.3	13.2	11.6	10.8	8.3	7.6
Jinzhou Petrochemical	Liaoning	1.7	2.0	2.3	2.7	2.4	1.5	1.4	1.3	0.9	1.8	2.1	2.3	2.0	2.7
Dalian Petrochemical	Liaoning	5.3	4.8	5.5	5.3	5.1	4.9	5.8	6.1	6.7	7.1	6.3	6.6	5.1	7.1
Liaoyang Petrochemical Fibre	Liaoning	–	–	–	–	–	–	–	–	–	–	–	–	–	–
Gaoqiao Petrochemical	Shanghai	7.4	7.6	7.6	7.7	7.8	8.0	7.8	7.6	7.2	7.2	6.6	6.3	6.6	7.0
Jinling Petrochemical	Jiangsu	4.8	4.9	4.5	6.0	9.2	9.4	9.0	8.6	9.3	7.5	6.2	8.2	8.5	7.6
Yangzi Petrochemical	Jiangsu	–	–	–	–	–	–	–	–	–	–	–	–	–	–
Qilu Petrochemical	Shandong	4.7	4.7	4.9	4.4	4.2	3.6	3.4	2.6	2.3	2.1	2.5	3.0	3.2	4.2
Baling Petrochemical	Hubei	4.0	3.5	4.0	4.0	2.0	1.8	2.0	1.5	2.3	2.0	1.7	1.8	1.6	1.9
Maoming Petrochemical	Guangdong	7.7	9.9	9.5	10.4	11.8	12.3	12.3	12.4	12.7	11.9	11.2	10.6	11.4	15.1
Lanzhou Chemical Industry	Gansu	–	–	–	–	–	–	–	–	–	–	–	–	–	–
Daqing General Petrochemical Works	Heilongjiang	5.9	5.4	5.5	5.1	4.5	5.1	5.1	5.1	5.1	5.3	4.8	5.0	4.8	4.6
Shanghai Jinshan Industrial	Shanghai	–	–	–	–	–	–	–	–	–	–	–	–	–	–
Lanzhou Petro. Processing & Chem. Complex	Gansu	6.5	6.6	6.9	7.0	6.9	6.3	5.3	4.9	5.8	5.9	5.5	6.3	7.3	7.8
Jinxi Petro. Processing & Chem. Complex	Liaoning	5.4	5.1	5.8	5.8	5.4	3.5	3.2	3.0	2.5	2.1	2.4	2.4	2.6	2.1
Zhenhai Refining & Chem.	Zhejiang	0.7	0.4	0.3	0.5	0.2	0.2	0.3	0.2	0.3	0.3	0.8	4.9	6.7	9.4
Anqing General Petrochemical Works	Anhui	–	–	–	–	–	–	–	–	–	–	–	–	–	–
Guangzhou General Petrochemical Works	Guangdong	1.4	1.3	1.0	1.2	0.9	0.8	2.3	2.1	2.3	2.8	2.8	2.0	4.2	6.1
Urumqi General Petrochemical Works	Xinjiang	1.2	1.0	0.8	1.5	1.2	1.3	1.2	1.4	1.4	1.6	1.7	1.7	1.7	1.7
Shijiazhuang Refinery	Hebei	0.3	0.5	0.4	0.5	0.9	0.9	0.9	0.8	1.0	1.1	1.2	1.3	0.8	1.5
Cangzhou Refinery	Hebei	0.2	0.1	0.2	0.2	0.3	0.2	0.0	–	–	–	–	–	–	–
Qianguo Refinery	Heilongjiang	0.2	0.1	0.1	0.1	0.1	0.1	0.1	0.2	0.0	0.1	–	–	–	0.0
Harbin Refinery	Heilongjiang	–	–	–	–	–	–	–	–	–	–	–	–	–	–
Lingyuan Refinery	Heilongjiang	–	–	–	–	–	–	–	–	–	–	–	–	–	–
Fujian Refinery	Fujian	–	–	–	–	–	–	–	–	–	–	–	0.1	0.8	1.3
Jiujiang General Petrochemical Works	Jiangxi	0.5	0.6	0.4	0.3	0.9	1.1	0.8	0.2	1.0	1.2	0.9	0.3	0.5	0.9
Jinan Refinery	Shandong	0.8	0.8	0.8	0.8	0.8	0.9	0.9	1.0	1.0	1.0	0.9	0.8	0.7	0.6
Luoyang General Petrochemical Works	Henan	–	–	–	–	–	–	–	0.9	1.2	1.2	2.9	2.9	4.3	4.6
Wuhan Petrochemical Works	Hubei	1.2	1.1	0.6	0.9	0.5	0.6	0.4	0.5	0.4	0.6	0.5	0.3	0.1	0.6
Jingmen General Petrochemical Works	Hubei	3.8	3.8	3.6	3.5	3.5	3.5	3.6	3.6	3.5	2.9	2.1	2.3	2.8	3.5
Dushanzi Refinery	Xinjiang	1.0	1.0	0.9	0.9	0.9	0.9	1.0	0.9	1.0	1.0	1.3	n.a.	n.a.	n.a.
Dalia West Pacific Petrochemical Corporation	Liaoning	–	–	–	–	–	–	–	–	–	–	–	–	–	–
Total		83.0	83.7	85.0	87.4	88.6	81.5	83.7	81.9	85.0	82.2	78.3	82.0	86.1	99.6

Note: Kerosene production includes kero–type jetfuel.
Sources: *China Petrochemical Yearbook, 1991, 1994, 1997*, China's Petrochemical Press.

Table A3.40 Light Oil (000 b/d) for Chemical Industry Production by Major Refineries

	Location	1983	1984	1985	1986	1987	1988	1989	1990	1991	1992	1993	1994	1995	1996
Yanshan Petrochemical	Beijing	–	–	–	–	–	–	–	–	–	–	–	–	–	–
Tianjin Petrochemical	Tianjin	3.8	3.8	3.7	3.1	3.5	3.5	3.4	3.3	3.3	3.7	3.8	4.1	4.4	6.6
Fushun Petrochemical	Liaoning	–	–	–	–	–	–	–	–	–	–	–	–	–	–
Jinzhou Petrochemical	Liaoning	1.8	2.4	1.9	1.5	2.1	2.3	1.5	0.6	0.5	0.5	0.3	0.1	0.1	0.9
Dalian Petrochemical	Liaoning	–	–	–	–	–	–	–	–	–	–	–	–	–	–
Liaoyang Petrochemical Fibre	Liaoning	–	–	–	–	–	–	–	–	–	–	–	–	–	–
Gaoqiao Petrochemical	Shanghai	–	–	–	–	–	–	–	–	–	–	–	–	–	–
Jinling Petrochemical	Jiangsu	–	–	–	–	–	–	–	–	–	–	–	–	–	–
Yangzi Petrochemical	Jiangsu	–	–	–	–	–	–	–	–	–	–	–	–	–	–
Qilu Petrochemical	Shandong	–	–	–	–	–	–	–	–	–	–	–	–	–	–
Baling Petrochemical	Hubei	–	–	–	–	–	–	–	–	–	–	–	–	–	–
Maoming Petrochemical	Guangdong	–	–	–	–	–	–	–	–	–	–	–	–	–	–
Lanzhou Chemical Industry	Gansu	–	–	–	–	–	–	–	–	–	–	–	–	–	–
Daqing General Petrochemical Works	Heilongjiang	–	–	–	–	–	–	–	–	–	–	–	–	–	–
Shanghai Jinshan Industrial	Shanghai	–	–	–	–	–	–	–	–	–	–	–	–	–	–
Lanzhou Petro. Processing & Chem. Complex	Gansu	–	–	–	–	–	–	–	–	–	–	–	–	–	–
Jinxi Petro. Processing & Chem. Complex	Liaoning	–	–	–	–	–	–	–	–	1.4	1.1	1.0	1.0	1.0	1.7
Zhenhai Refining & Chem.	Zhejiang	–	–	–	–	–	–	–	–	–	–	–	–	–	–
Anqing General Petrochemical Works	Anhui	–	–	–	–	–	–	–	–	–	–	–	–	–	–
Guangzhou General Petrochemical Works	Guangdong	–	–	–	–	–	–	–	–	–	–	–	–	–	–
Urumqi General Petrochemical Works	Xinjiang	–	–	–	–	–	–	–	–	–	–	–	–	–	–
Shijiazhuang Refinery	Hebei	–	–	–	–	–	–	–	–	–	–	–	–	–	–
Cangzhou Refinery	Hebei	–	–	–	–	–	–	–	–	–	–	–	–	–	–
Qianguo Refinery	Heilongjiang	–	–	–	–	–	–	–	–	–	–	–	0.4	0.9	2.1
Harbin Refinery	Heilongjiang	–	–	–	–	–	–	–	–	–	–	–	–	–	–
Lingyuan Refinery	Heilongjiang	–	–	–	–	–	–	–	–	–	–	–	–	–	–
Fujian Refinery	Fujian	–	–	–	–	–	–	–	–	–	–	–	–	–	–
Jiujiang General Petrochemical Works	Jiangxi	–	–	–	–	–	–	–	–	–	–	–	–	–	–
Jinan Refinery	Shandong	–	–	–	–	–	–	–	–	–	–	–	–	–	–
Luoyang General Petrochemical Works	Henan	–	–	–	–	–	–	–	–	–	–	–	–	–	1.2
Wuhan Petrochemical Works	Hubei	4.1	2.6	2.9	3.2	3.3	3.1	3.1	3.1	3.1	3.1	3.2	3.6	3.8	3.5
Jingmen General Petrochemical Works	Hubei	–	–	–	–	–	–	–	–	–	–	–	–	–	–
Dushanzi Refinery	Xinjiang	–	–	–	–	–	–	–	–	–	–	–	–	–	–
Dalia West Pacific Petrochemical Corporation	Liaoning	–	–	–	–	–	–	–	–	–	–	–	–	–	1.9
Total		9.7	8.7	8.5	7.8	8.9	8.9	8.0	7.1	8.3	8.4	8.4	9.2	10.3	15.9

Sources: *China Petrochemical Yearbook, 1991, 1994, 1997*, China's Petrochemical Press.

Table A3.41 Fuel Oil Production (000 b/d) by Major Refineries

	Location	1983	1984	1985	1986	1987	1988	1989	1990	1991	1992	1993	1994	1995	1996
Yanshan Petrochemical	Beijing	40.7	42.8	43.2	42.4	41.6	42.8	41.9	42.8	41.3	44.3	40.7	40.1	40.0	29.4
Tianjin Petrochemical	Tianjin	20.6	21.0	21.4	20.9	21.7	22.8	24.1	22.3	22.6	22.5	22.7	22.5	22.4	19.3
Fushun Petrochemical	Liaoning	37.4	36.6	35.0	33.7	28.6	24.0	19.9	19.9	18.3	18.6	18.3	12.5	12.3	6.5
Jinzhou Petrochemical	Liaoning	10.2	10.3	10.3	12.5	3.5	17.2	17.9	16.3	4.7	5.2	4.1	6.1	4.4	2.3
Dalian Petrochemical	Liaoning	33.3	32.2	31.3	31.4	31.3	31.8	29.2	28.5	27.9	26.3	24.6	23.9	23.8	23.1
Liaoyang Petrochemical Fibre	Liaoning	–	–	–	–	–	–	–	–	–	–	–	–	–	–
Gaoqiao Petrochemical	Shanghai	26.5	25.0	24.6	24.6	24.4	25.4	23.6	24.5	23.9	23.8	18.2	19.0	18.5	15.3
Jinling Petrochemical	Jiangsu	26.8	24.9	24.1	25.0	26.6	26.5	24.4	23.1	22.0	22.9	20.0	16.5	11.7	7.2
Yangzi Petrochemical	Jiangsu	–	–	–	–	–	–	–	–	–	–	–	–	–	–
Qilu Petrochemical	Shandong	30.9	29.1	28.9	30.9	29.6	32.9	39.4	38.0	41.8	43.3	38.7	n.a.	n.a.	n.a.
Baling Petrochemical	Hubei	10.4	10.7	11.1	11.1	11.0	10.6	11.0	11.0	11.0	10.2	9.6	9.7	10.0	9.3
Maoming Petrochemical	Guangdong	27.1	23.7	23.9	23.7	25.5	26.5	27.3	24.0	27.5	25.4	25.0	23.1	21.7	17.2
Lanzhou Chemical Industry	Gansu	–	–	–	–	–	–	–	–	–	–	–	–	–	–
Daqing General Petrochemical Works	Heilongjiang	23.9	24.7	27.0	26.5	27.1	26.2	27.4	25.3	22.7	22.9	18.6	17.5	16.4	14.5
Shanghai Jinshan Industrial	Shanghai	–	5.5	15.3	16.8	18.0	18.3	20.2	26.9	30.0	31.5	35.7	36.2	32.5	27.1
Lanzhou Petro. Processing & Chem. Complex	Gansu	15.5	15.8	14.7	15.8	19.1	18.0	22.8	26.3	27.9	24.3	24.6	21.8	19.8	19.4
Jinxi Petro. Processing & Chem. Complex	Liaoning	–	–	–	–	–	–	–	–	–	–	–	–	–	–
Zhenhai Refining & Chem.	Zhejiang	9.8	11.4	8.4	8.8	11.6	11.2	11.0	8.8	8.4	6.2	2.6	4.1	4.0	3.2
Anqing General Petrochemical Works	Anhui	13.6	14.6	15.6	15.2	18.1	15.2	11.9	12.0	11.3	9.0	9.2	7.9	8.3	5.2
Guangzhou General Petrochemical Works	Guangdong	18.8	20.0	18.4	21.3	24.8	22.5	22.3	20.6	20.3	22.8	21.9	16.9	15.3	6.5
Urumqi General Petrochemical Works	Xinjiang	–	–	–	–	–	–	1.5	1.7	8.9	10.1	8.8	8.7	8.5	8.5
Shijiazhuang Refinery	Hebei	0.6	2.0	2.1	2.6	3.5	3.1	2.7	4.2	4.5	3.9	2.8	1.2	0.2	0.8
Cangzhou Refinery	Hebei	3.8	3.6	3.7	3.9	4.0	4.3	3.1	2.2	2.7	3.2	2.8	1.9	1.9	0.9
Qianguo Refinery	Heilongjiang	3.6	3.8	4.4	4.8	4.5	4.7	4.7	4.4	2.1	1.2	0.9	0.8	0.7	0.4
Harbin Refinery	Heilongjiang	5.1	4.4	4.3	5.6	5.4	5.3	4.1	4.2	4.3	4.5	3.1	–	–	–
Lingyuan Refinery	Heilongjiang	14.5	15.0	15.9	12.6	12.1	12.4	11.8	8.3	8.0	7.4	8.2	5.5	3.1	2.0
Fujian Refinery	Fujian	–	–	–	–	–	–	–	–	–	0.9	2.6	2.8	1.9	2.3
Jiujiang General Petrochemical Works	Jiangxi	4.1	3.8	3.4	6.9	8.1	8.3	7.9	7.7	7.4	7.1	8.1	8.6	7.9	6.3
Jinan Refinery	Shandong	4.9	5.1	5.6	6.0	5.4	6.6	5.3	5.3	5.3	5.3	5.9	4.8	3.0	3.7
Luoyang General Petrochemical Works	Henan	–	–	0.3	0.9	0.6	2.0	0.9	0.9	0.9	0.5	0.4	0.4	0.5	1.4
Wuhan Petrochemical Works	Hubei	1.1	11.7	11.7	13.7	12.4	13.5	13.7	13.4	12.8	12.6	13.3	10.8	10.0	6.5
Jingmen General Petrochemical Works	Hubei	13.8	13.0	11.4	11.4	11.7	10.9	12.5	9.9	9.2	8.9	5.7	8.6	6.6	5.4
Dushanzi Refinery	Xinjiang	1.1	1.0	1.2	1.1	1.1	1.1	1.4	1.6	1.6	1.9	1.6	n.a.	n.a.	n.a.
Dalia West Pacific Petrochemical Corporation	Liaoning	–	–	–	–	–	–	–	–	–	–	–	–	–	0.6
Total		398.3	411.7	417.1	429.9	431.3	444.2	443.9	434.2	429.3	426.8	398.9	331.7	305.4	243.7

Sources: *China Petrochemical Yearbook, 1991, 1994, 1997*, China's Petrochemical Press.

Table A3.42 Lubricant Oil Production (000 b/d) by Major Refineries

	Location	1983	1984	1985	1986	1987	1988	1989	1990	1991	1992	1993	1994	1995	1996
Yanshan Petrochemical	Beijing	3.0	3.0	2.9	3.0	2.9	3.5	3.3	3.1	3.4	3.5	3.0	2.1	2.1	3.4
Tianjin Petrochemical	Tianjin	–	–	–	–	–	–	–	–	–	–	–	–	–	–
Fushun Petrochemical	Liaoning	0.9	0.8	1.0	0.8	1.1	0.8	0.8	0.7	0.7	0.6	0.6	0.4	0.4	0.3
Jinzhou Petrochemical	Liaoning	–	–	–	–	–	–	–	–	–	–	–	–	–	–
Dalian Petrochemical	Liaoning	3.7	3.5	4.2	3.6	3.6	3.8	3.3	2.7	2.7	2.3	2.5	2.3	2.1	3.4
Liaoyang Petrochemical Fibre	Liaoning	–	–	–	–	–	–	–	–	–	–	–	–	–	–
Gaoqiao Petrochemical	Shanghai	3.5	3.5	3.5	3.6	3.9	4.0	3.8	3.8	3.4	3.5	3.8	4.0	4.1	6.6
Jinling Petrochemical	Jiangsu	0.3	0.3	0.3	0.3	0.4	0.3	0.4	0.3	0.2	0.2	0.1	0.0	0.0	–
Yangzi Petrochemical	Jiangsu	–	–	–	–	–	–	–	–	–	–	–	–	–	–
Qilu Petrochemical	Shandong	–	–	–	–	–	–	–	–	–	–	–	–	–	–
Baling Petrochemical	Hubei	–	–	–	–	–	–	–	–	–	–	–	–	–	–
Maoming Petrochemical	Guangdong	2.2	2.4	2.6	2.7	2.9	2.9	3.1	2.8	2.7	3.0	3.4	3.4	3.5	3.7
Lanzhou Chemical Industry	Gansu	–	–	–	–	–	–	–	–	–	–	–	–	–	–
Daqing General Petrochemical Works	Heilongjiang	0.0	0.9	1.5	2.2	2.7	2.6	3.0	3.4	3.4	4.3	4.2	2.4	3.1	3.3
Shanghai Jinshan Industrial	Shanghai	–	–	–	–	–	–	–	–	–	–	–	–	–	–
Lanzhou Petro. Processing & Chem. Complex	Gansu	4.6	5.2	5.7	6.0	6.0	5.9	5.8	5.3	5.5	5.7	5.6	5.4	5.9	6.0
Jinxi Petro. Processing & Chem. Complex	Liaoning	2.4	2.5	2.4	3.0	2.9	2.6	2.8	2.3	2.0	2.9	2.7	3.0	2.4	2.6
Zhenhai Refining & Chem.	Zhejiang	–	–	–	–	–	–	–	–	–	–	–	–	–	–
Anqing General Petrochemical Works	Anhui	–	–	–	–	–	–	–	–	–	–	–	–	–	–
Guangzhou General Petrochemical Works	Guangdong	–	–	–	–	–	–	–	–	–	–	–	–	–	–
Urumqi General Petrochemical Works	Xinjiang	–	–	–	–	–	–	–	–	–	–	–	–	–	–
Shijiazhuang Refinery	Hebei	–	–	–	–	–	–	–	–	–	–	–	–	–	–
Cangzhou Refinery	Hebei	–	–	–	–	–	–	–	–	–	–	–	–	–	–
Qianguo Refinery	Heilongjiang	–	–	–	–	–	–	–	–	–	–	–	–	–	–
Harbin Refinery	Heilongjiang	–	–	–	–	–	–	–	–	–	–	–	–	–	–
Lingyuan Refinery	Heilongjiang	–	–	–	–	–	–	–	–	–	–	–	–	–	–
Fujian Refinery	Fujian	–	–	–	–	–	–	–	–	–	–	–	–	–	–
Jiujiang General Petrochemical Works	Jiangxi	–	–	–	–	–	–	–	–	–	–	–	–	–	–
Jinan Refinery	Shandong	0.3	0.4	0.6	0.5	0.9	0.9	1.1	1.3	1.5	1.6	1.8	1.5	2.2	1.6
Luoyang General Petrochemical Works	Henan	–	–	–	–	–	–	–	–	–	–	–	–	–	–
Wuhan Petrochemical Works	Hubei	–	–	–	–	–	–	–	–	–	–	–	–	–	–
Jingmen General Petrochemical Works	Hubei	–	0.4	0.9	1.0	1.2	1.5	1.5	1.3	1.3	1.5	1.6	1.9	2.0	2.9
Dushanzi Refinery	Xinjiang	1.3	1.5	1.7	1.7	1.8	2.0	1.8	1.8	2.0	2.3	2.1	n.a.	n.a.	n.a.
Yiping Chemical Works	Sichuan	–	–	–	–	0.1	0.2	0.2	0.5	0.6	1.3	0.9	n.a.	n.a.	n.a.
Changcheng Premium–grade Lube Oil Company	Beijing	0.2	0.2	0.4	0.5	0.6	0.7	0.8	0.8	1.0	1.2	1.2	n.a.	n.a.	n.a.
Total		22.4	24.7	27.7	28.8	30.9	31.5	31.6	30.3	30.4	33.9	33.6	26.5	27.9	33.8

Sources: *China Petrochemical Yearbook, 1991, 1994, 1997*, China's Petrochemical Press.

Table A3.43 Petroleum Wax Production (000 b/d) by Major Refineries

	Location	1983	1984	1985	1986	1987	1988	1989	1990	1991	1992	1993	1994	1995	1996
Yanshan Petrochemical	Beijing	1.0	0.9	0.9	1.0	0.9	1.2	1.1	1.3	1.3	1.3	1.4	1.3	1.4	1.4
Tianjin Petrochemical	Tianjin	–	–	–	–	–	–	–	–	–	–	–	–	–	–
Fushun Petrochemical	Liaoning	3.1	3.2	3.2	3.3	3.4	3.2	3.3	3.7	3.0	3.4	3.7	4.3	4.1	4.8
Jinzhou Petrochemical	Liaoning	–	–	–	–	–	–	–	–	–	–	–	–	–	–
Dalian Petrochemical	Liaoning	1.7	1.8	1.9	2.0	2.2	2.4	2.3	2.3	2.4	2.4	2.4	2.0	3.5	3.6
Liaoyang Petrochemical Fibre	Liaoning	–	–	–	–	–	–	–	–	–	–	–	–	–	–
Gaoqiao Petrochemical	Shanghai	1.1	1.2	1.1	1.1	1.3	1.4	1.5	1.5	1.4	1.4	1.2	1.5	1.5	2.3
Jinling Petrochemical	Jiangsu	–	–	–	–	–	–	–	–	–	–	–	–	–	–
Yangzi Petrochemical	Jiangsu	–	–	–	–	–	–	–	–	–	–	–	–	–	–
Qilu Petrochemical	Shandong	–	–	–	–	–	–	–	–	–	–	–	–	–	–
Baling Petrochemical	Hubei	–	–	–	–	–	–	–	–	–	–	–	–	–	–
Maoming Petrochemical	Guangdong	1.0	1.1	1.3	1.4	1.7	1.6	1.7	1.4	1.5	1.5	1.9	1.8	1.7	1.8
Lanzhou Chemical Industry	Gansu	–	–	–	–	–	–	–	–	–	–	–	–	–	–
Daqing General Petrochemical Works	Heilongjiang	0.9	1.1	1.2	1.2	1.4	1.7	2.0	1.9	2.1	2.0	2.3	1.8	2.4	2.8
Shanghai Jinshan Industrial	Shanghai	–	–	–	–	–	–	–	–	–	–	–	–	–	–
Lanzhou Petro. Processing & Chem. Complex	Gansu	0.2	0.3	0.3	0.3	0.3	0.3	0.4	0.4	0.5	0.4	0.5	0.6	0.6	0.6
Jinxi Petro. Processing & Chem. Complex	Liaoning	0.7	0.8	0.9	0.9	0.9	0.8	0.8	0.7	0.6	0.5	0.3	0.5	0.6	0.7
Zhenhai Refining & Chem.	Zhejiang	–	–	–	–	–	–	–	–	–	–	–	–	–	–
Anqing General Petrochemical Works	Anhui	–	–	–	–	–	–	–	–	–	–	–	–	–	–
Guangzhou General Petrochemical Works	Guangdong	–	–	–	–	–	–	–	–	–	–	–	–	–	–
Urumqi General Petrochemical Works	Xinjiang	–	–	–	–	–	–	–	–	–	–	–	–	–	–
Shijiazhuang Refinery	Hebei	–	–	–	–	–	–	–	–	–	–	–	–	–	–
Cangzhou Refinery	Hebei	–	–	–	–	–	–	–	–	–	–	–	–	–	–
Qianguo Refinery	Heilongjiang	–	–	–	–	–	–	–	–	–	–	–	–	–	–
Harbin Refinery	Heilongjiang	–	–	–	–	–	–	–	–	–	–	–	–	–	–
Lingyuan Refinery	Heilongjiang	–	–	–	–	–	–	–	–	–	–	–	–	–	–
Fujian Refinery	Fujian	–	–	–	–	–	–	–	–	–	–	–	–	–	–
Jiujiang General Petrochemical Works	Jiangxi	–	–	–	–	–	–	–	–	–	–	–	–	–	–
Jinan Refinery	Shandong	–	–	–	–	–	–	0.1	0.1	0.1	0.1	0.1	0.1	0.2	0.2
Luoyang General Petrochemical Works	Henan	–	–	–	–	–	–	–	–	–	–	–	–	–	–
Wuhan Petrochemical Works	Hubei	–	–	–	–	–	–	–	–	–	–	–	–	–	–
Jingmen General Petrochemical Works	Hubei	–	0.1	1.0	1.1	1.0	1.2	1.4	1.2	1.4	1.4	1.5	1.8	1.4	1.6
Dushanzi Refinery	Xinjiang	–	–	–	–	–	–	–	–	–	–	–	–	–	–
Dalia West Pacific Petrochemical Corporation	Liaoning	–	–	–	–	–	–	–	–	–	–	–	–	–	–
Total		9.7	10.4	11.6	12.2	13.0	13.9	14.6	14.4	14.2	14.5	15.2	15.8	17.4	19.9

Sources: *China Petrochemical Yearbook, 1991, 1994, 1997*, China's Petrochemical Press.

Table A3.44 Asphalt Production (000 b/d) by Major Refineries

	Location	1983	1984	1985	1986	1987	1988	1989	1990	1991	1992	1993	1994	1995	1996
Yanshan Petrochemical	Beijing	–	–	–	–	–	–	–	–	–	–	–	–	–	–
Tianjin Petrochemical	Tianjin	0.9	0.9	1.0	1.2	1.0	0.8	0.9	0.8	0.9	1.1	1.3	0.1	0.1	0.2
Fushun Petrochemical	Liaoning	–	–	–	–	–	–	–	–	–	–	–	–	–	–
Jinzhou Petrochemical	Liaoning	–	–	–	–	–	–	–	–	–	–	–	–	–	–
Dalian Petrochemical	Liaoning	–	–	–	–	–	–	–	–	–	–	–	–	–	–
Liaoyang Petrochemical Fibre	Liaoning	–	–	–	–	–	–	–	–	–	–	–	–	–	–
Gaoqiao Petrochemical	Shanghai	0.6	0.7	0.7	0.7	0.7	0.7	0.6	0.6	0.5	0.7	0.7	0.6	0.5	0.6
Jinling Petrochemical	Jiangsu	0.8	1.0	1.0	1.1	1.2	1.1	1.0	1.0	0.9	1.0	1.1	0.8	1.0	0.7
Yangzi Petrochemical	Jiangsu	–	–	–	–	–	–	–	–	–	3.4	–	–	–	–
Qilu Petrochemical	Shandong	8.0	8.1	9.8	8.7	9.5	!0.0	10.5	10.9	11.8	11.6	12.1	10.9	10.4	9.9
Baling Petrochemical	Hubei	–	–	–	–	–	–	–	–	–	–	–	–	–	–
Maoming Petrochemical	Guangdong	5.4	8.6	8.3	9.8	9.2	9.5	8.4	8.4	8.5	8.2	8.5	7.4	4.7	3.8
Lanzhou Chemical Industry	Gansu	–	–	–	–	–	–	–	–	–	–	–	–	–	–
Daqing General Petrochemical Works	Heilongjiang	0.4	0.4	0.5	0.5	0.5	0.4	0.5	0.4	0.5	0.5	0.3	0.3	0.1	0.0
Shanghai Jinshan Industrial	Shanghai	–	–	–	–	–	–	–	–	–	–	–	–	–	–
Lanzhou Petro. Processing & Chem. Complex	Gansu	2.7	2.6	2.8	2.9	2.8	2.6	2.8	3.1	3.5	3.4	2.9	3.5	4.0	3.6
Jinxi Petro. Processing & Chem. Complex	Liaoning	1.2	1.4	1.7	2.6	3.2	3.1	2.1	2.2	2.6	2.8	3.2	3.9	3.5	3.7
Zhenhai Refining & Chem.	Zhejiang	0.3	0.5	0.5	0.5	0.5	0.4	0.4	0.3	0.4	0.4	0.3	0.4	0.5	0.1
Anqing General Petrochemical Works	Anhui	0.5	0.8	0.8	0.7	0.6	0.5	0.6	0.4	0.3	0.4	0.3	0.5	0.2	0.0
Guangzhou General Petrochemical Works	Guangdong	–	–	–	–	–	–	–	–	–	–	–	–	–	–
Urumqi General Petrochemical Works	Xinjiang	0.3	0.1	0.1	0.2	0.3	0.2	0.2	0.2	0.2	0.4	0.4	0.7	0.5	0.3
Shijiazhuang Refinery	Hebei	–	–	–	–	0.0	–	–	–	–	–	–	–	–	–
Cangzhou Refinery	Hebei	0.3	0.4	0.3	0.4	0.4	0.6	0.4	0.5	0.5	0.5	0.4	1.0	1.0	1.0
Qianguo Refinery	Heilongjiang	0.0	0.1	0.1	0.3	0.3	0.1	0.1	0.1	0.2	0.4	0.2	n.a.	n.a.	n.a.
Harbin Refinery	Heilongjiang	–	–	–	–	–	–	–	–	–	–	–	–	–	–
Lingyuan Refinery	Heilongjiang	–	–	–	–	–	–	–	–	–	–	–	–	–	–
Fujian Refinery	Fujian	–	–	–	–	–	–	–	–	–	–	–	–	–	–
Jiujiang General Petrochemical Works	Jiangxi	–	0.2	0.2	0.3	0.1	0.2	0.1	0.2	0.2	0.2	0.3	0.3	0.8	0.9
Jinan Refinery	Shandong	–	–	–	–	–	–	–	–	1.7	1.7	1.7	1.7	3.5	1.8
Luoyang General Petrochemical Works	Henan	–	–	–	–	–	–	–	–	–	–	–	–	–	–
Wuhan Petrochemical Works	Hubei	–	–	–	–	–	–	–	–	–	–	–	–	–	–
Jingmen General Petrochemical Works	Hubei	1.2	1.3	1.9	2.0	1.8	1.6	1.2	1.3	1.0	1.0	0.4	0.5	0.8	1.5
Dushanzi Refinery	Xinjiang	0.9	0.9	0.9	0.9	0.9	0.9	0.9	0.8	0.9	1.0	0.8	n.a.	n.a.	n.a.
Sinopec Total		23.6	27.9	30.9	32.6	33.1	32.7	30.7	31.2	34.5	38.7	34.8	32.6	31.8	28.2

Sources: *China Petrochemical Yearbook, 1991, 1994, 1997*, China's Petrochemical Press.

Table A3.45 Refinery Gas Production by Major Refineries (000 b/d)

	Location	1983	1984	1985	1986	1987	1988	1989	1990	1991	1992	1993	1994	1995	1996
Yanshan Petrochemical	Beijing	–	–	–	–	–	–	–	–	–	–	–	–	–	–
Tianjin Petrochemical	Tianjin	–	–	–	–	–	–	–	–	–	–	–	–	–	–
Fushun Petrochemical	Liaoning	–	–	–	–	–	–	–	–	–	–	–	–	–	–
Jinzhou Petrochemical	Liaoning	–	–	–	–	–	–	–	–	–	–	–	–	–	–
Dalian Petrochemical	Liaoning	–	–	–	–	–	–	–	–	–	–	–	–	–	–
Liaoyang Petrochemical Fibre	Liaoning	–	–	–	–	–	–	–	–	–	–	–	–	–	–
Gaoqiao Petrochemical	Shanghai	–	–	–	–	–	–	–	–	–	–	–	–	–	–
Jinling Petrochemical	Jiangsu	–	–	–	–	–	–	–	–	–	–	–	–	–	–
Yangzi Petrochemical	Jiangsu	–	–	–	–	–	–	–	–	–	–	–	–	–	–
Qilu Petrochemical	Shandong	–	–	–	–	–	–	–	–	–	–	–	–	–	–
Baling Petrochemical	Hubei	–	–	–	–	–	–	–	–	–	–	–	–	–	–
Maoming Petrochemical	Guangdong	–	–	–	–	–	–	–	–	–	–	–	–	–	–
Lanzhou Chemical Industry	Gansu	–	–	–	–	–	–	–	–	–	–	–	–	–	–
Daqing General Petrochemical Works	Heilongjiang	–	–	–	–	–	–	–	–	–	–	–	–	–	–
Shanghai Jinshan Industrial	Shanghai	–	–	–	–	–	–	–	–	–	–	–	–	–	–
Lanzhou Petro. Processing & Chem. Complex	Gansu	–	–	–	–	–	–	–	–	–	–	–	–	–	–
Jinxi Petro. Processing & Chem. Complex	Liaoning	–	–	–	–	–	–	–	–	–	–	–	–	–	–
Zhenhai Refining & Chem.	Zhejiang	–	–	–	–	–	–	–	–	–	–	–	–	–	–
Anqing General Petrochemical Works	Anhui	–	–	–	–	–	–	–	–	–	–	–	–	–	–
Guangzhou General Petrochemical Works	Guangdong	–	–	–	–	–	–	–	–	–	–	–	–	–	–
Urumqi General Petrochemical Works	Xinjiang	–	–	–	–	–	–	–	–	–	–	–	–	–	–
Shijiazhuang Refinery	Hebei	–	–	–	–	–	–	–	–	–	–	–	–	–	–
Cangzhou Refinery	Hebei	0.3	0.2	0.3	0.4	0.5	0.5	0.5	0.9	0.9	0.7	0.9	n.a.	n.a.	n.a.
Qianguo Refinery	Heilongjiang	0.4	0.5	0.5	0.6	0.5	0.6	0.6	0.2	0.4	0.4	0.8	n.a.	n.a.	n.a.
Harbin Refinery	Heilongjiang	–	–	–	–	–	–	–	–	–	–	–	–	–	–
Lingyuan Refinery	Heilongjiang	–	–	–	–	–	–	–	–	–	–	–	–	–	–
Fujian Refinery	Fujian	–	–	–	–	–	–	–	–	–	–	–	–	–	–
Jiujiang General Petrochemical Works	Jiangxi	–	–	–	–	–	–	–	–	–	–	–	–	–	–
Jinan Refinery	Shandong	–	–	–	–	–	–	–	–	–	–	–	–	–	–
Luoyang General Petrochemical Works	Henan	–	–	–	–	–	–	–	–	–	–	–	–	–	–
Wuhan Petrochemical Works	Hubei	2.0	2.2	2.2	2.5	2.0	3.2	3.2	1.9	–	–	–	–	–	–
Jingmen General Petrochemical Works	Hubei	–	–	–	–	–	–	–	–	–	–	–	–	–	–
Dushanzi Refinery	Xinjiang	2.0	2.1	1.9	2.2	2.4	2.6	2.6	2.4	1.9	2.2	3.5	n.a.	n.a.	n.a.
Total		5	5	5	6	5	7	7	5	3	3	5	n.a.	n.a.	n.a.

Sources: *China Petrochemical Yearbook, 1991, 1994, 1997*, China's Petrochemical Press.

Table A3.46 Hydrocarbon Solvents Production (000 b/d) by Major Refineries

	Location	1983	1984	1985	1986	1987	1988	1989	1990	1991	1992	1993	1994	1995	1996
Yanshan Petrochemical	Beijing	–	–	–	–	–	–	–	–	–	–	–	–	–	–
Tianjin Petrochemical	Tianjin	0.2	0.3	0.3	0.3	0.5	0.5	0.6	0.5	0.6	0.7	0.7	0.7	0.8	0.7
Fushun Petrochemical	Liaoning	0.2	0.3	0.3	0.1	0.2	0.2	0.2	0.2	–	–	–	–	–	–
Jinzhou Petrochemical	Liaoning	0.7	0.6	0.6	0.5	0.4	0.4	0.4	0.5	0.7	1.0	0.7	0.8	0.8	0.9
Dalian Petrochemical	Liaoning	–	–	–	–	–	–	–	–	–	–	–	–	–	–
Liaoyang Petrochemical Fibre	Liaoning	–	–	–	–	–	–	–	–	–	–	–	–	–	–
Gaoqiao Petrochemical	Shanghai	–	–	–	–	–	–	–	–	–	–	–	–	–	–
Jinling Petrochemical	Jiangsu	–	–	–	–	–	–	0.7	0.6	0.7	0.8	0.8	0.8	1.0	1.1
Yangzi Petrochemical	Jiangsu	–	–	–	–	–	–	–	–	–	–	–	–	–	–
Qilu Petrochemical	Shandong	–	–	–	–	–	–	–	–	–	–	–	–	–	–
Baling Petrochemical	Hubei	–	–	–	–	–	–	–	–	–	–	–	–	–	–
Maoming Petrochemical	Guangdong	–	–	–	–	–	–	–	–	–	–	–	–	–	–
Lanzhou Chemical Industry	Gansu	–	–	–	–	–	–	–	–	–	–	–	–	–	–
Daqing General Petrochemical Works	Heilongjiang	–	–	–	–	–	–	–	–	–	–	–	–	–	–
Shanghai Jinshan Industrial	Shanghai	–	–	–	–	–	–	–	–	–	–	–	–	–	–
Lanzhou Petro. Processing & Chem. Complex	Gansu	–	–	–	–	–	–	–	–	–	–	–	–	–	–
Jinxi Petro. Processing & Chem. Complex	Liaoning	–	–	–	–	–	–	–	–	–	–	–	–	–	–
Zhenhai Refining & Chem.	Zhejiang	–	–	–	–	–	–	–	–	–	–	–	–	–	–
Anqing General Petrochemical Works	Anhui	–	–	–	–	–	–	–	–	–	–	–	–	–	–
Guangzhou General Petrochemical Works	Guangdong	–	–	–	–	–	–	–	–	–	–	–	–	–	–
Urumqi General Petrochemical Works	Xinjiang	–	–	–	–	–	–	–	–	–	–	–	–	–	–
Shijiazhuang Refinery	Hebei	–	–	–	–	–	–	–	–	–	–	–	–	–	–
Cangzhou Refinery	Hebei	–	–	–	0.1	0.1	0.1	0.1	0.0	0.1	0.1	0.0	0.1	0.2	0.4
Qianguo Refinery	Heilongjiang	–	–	–	–	–	–	–	–	–	–	–	–	–	–
Harbin Refinery	Heilongjiang	–	–	–	0.2	0.2	0.2	0.2	0.3	0.3	0.3	0.1	0.0	0.5	0.2
Lingyuan Refinery	Heilongjiang	–	–	–	–	–	–	–	–	–	–	–	–	–	–
Fujian Refinery	Fujian	–	–	–	–	–	–	–	–	–	–	–	0.0	0.1	0.4
Jiujiang General Petrochemical Works	Jiangxi	–	–	–	–	–	–	–	–	0.1	0.3	0.3	0.3	0.3	1.9
Jinan Refinery	Shandong	–	–	–	–	–	–	–	–	0.3	0.6	0.4	0.4	0.5	0.4
Luoyang General Petrochemical Works	Henan	–	–	–	–	–	–	–	–	–	–	–	0.3	0.4	0.4
Wuhan Petrochemical Works	Hubei	–	–	–	–	–	–	–	–	–	–	–	–	–	–
Jingmen General Petrochemical Works	Hubei	–	–	–	–	–	–	–	–	–	–	–	–	–	–
Dushanzi Refinery	Xinjiang	–	–	–	0.0	0.0	0.1	0.0	0.1	0.1	0.0	0.0	n.a.	n.a.	n.a.
Total		1.1	1.2	1.1	1.3	1.4	1.4	2.2	2.2	2.8	3.7	3.0	3.6	4.5	6.5

Sources: *China Petrochemical Yearbook, 1991, 1994, 1997*, China's Petrochemical Press.

Table A3.47 Petroleum Coke Production (000 b/d) by Major Refineries

	Location	1983	1984	1985	1986	1987	1988	1989	1990	1991	1992	1993	1994	1995	1996
Yanshan Petrochemical	Beijing	–	–	–	–	–	–	–	–	–	–	–	–	–	–
Tianjin Petrochemical	Tianjin	–	–	–	–	–	–	–	–	–	–	–	–	–	–
Fushun Petrochemical	Liaoning	3.2	3.4	3.7	4.1	4.3	4.0	4.1	4.3	4.0	4.0	4.3	4.3	3.6	4.0
Jinzhou Petrochemical	Liaoning	–	–	–	–	–	–	–	–	–	–	–	–	–	–
Dalian Petrochemical	Liaoning	–	–	–	–	–	–	–	–	–	–	–	–	–	–
Liaoyang Petrochemical Fibre	Liaoning	–	–	–	–	–	–	–	–	–	–	–	–	–	–
Gaoqiao Petrochemical	Shanghai	0.2	0.3	0.2	0.2	0.2	0.2	0.3	0.3	–	–	–	–	–	0.4
Jinling Petrochemical	Jiangsu	1.5	1.8	1.8	2.0	2.3	2.3	2.5	2.6	2.6	2.5	2.3	3.2	3.6	4.3
Yangzi Petrochemical	Jiangsu	–	–	–	–	–	–	–	–	–	–	–	–	–	–
Qilu Petrochemical	Shandong	–	–	–	–	–	–	–	–	–	–	–	–	–	–
Baling Petrochemical	Hubei	–	–	–	–	–	–	–	–	–	–	–	–	–	–
Maoming Petrochemical	Guangdong	1.0	0.9	0.8	1.4	2.1	1.8	1.7	1.7	1.6	1.6	1.8	1.5	1.9	2.0
Lanzhou Chemical Industry	Gansu	–	–	–	–	–	–	–	–	–	–	–	–	–	–
Daqing General Petrochemical Works	Heilongjiang	1.5	1.5	1.4	1.8	1.7	1.8	2.2	2.6	2.4	2.6	2.3	2.0	2.3	2.5
Shanghai Jinshan Industrial	Shanghai	–	–	–	–	–	–	–	–	–	–	–	–	–	–
Lanzhou Petro. Processing & Chem. Complex	Gansu	–	–	–	–	–	–	–	–	–	–	–	–	–	–
Jinxi Petro. Processing & Chem. Complex	Liaoning	–	–	–	–	–	–	–	–	–	–	–	–	–	–
Zhenhai Refining & Chem.	Zhejiang	–	–	–	–	–	–	–	–	–	–	–	–	–	–
Anqing General Petrochemical Works	Anhui	–	–	–	–	0.0	0.7	1.1	1.3	1.5	1.7	1.7	1.6	1.8	2.6
Guangzhou General Petrochemical Works	Guangdong	–	–	–	–	–	–	–	–	–	–	–	–	–	–
Urumqi General Petrochemical Works	Xinjiang	0.3	0.3	0.2	0.1	0.1	–	0.0	0.1	0.4	0.4	0.6	0.4	0.4	0.4
Shijiazhuang Refinery	Hebei	–	–	–	–	–	–	–	–	–	0.0	0.0	0.7	1.6	1.6
Cangzhou Refinery	Hebei	–	–	–	–	–	–	–	–	–	–	–	–	–	–
Qianguo Refinery	Heilongjiang	–	–	–	–	–	–	–	–	–	–	–	–	–	–
Harbin Refinery	Heilongjiang	–	–	–	–	–	–	–	–	–	–	–	–	–	–
Lingyuan Refinery	Heilongjiang	–	–	–	–	–	–	–	–	–	–	–	–	–	–
Fujian Refinery	Fujian	–	–	–	–	–	–	–	–	–	–	0.0	0.3	1.1	1.2
Jiujiang General Petrochemical Works	Jiangxi	–	–	–	–	–	–	–	–	–	–	–	–	–	–
Jinan Refinery	Shandong	–	–	–	–	–	–	–	–	–	–	–	–	–	–
Luoyang General Petrochemical Works	Henan	–	–	–	–	–	–	–	–	–	–	–	–	–	–
Wuhan Petrochemical Works	Hubei	–	–	–	–	–	–	–	–	–	–	–	–	–	–
Jingmen General Petrochemical Works	Hubei	0.5	0.8	0.6	0.8	0.8	0.9	0.9	0.9	0.9	1.0	0.9	1.1	1.1	1.3
Dushanzi Refinery	Xinjiang	0.7	0.8	0.7	0.9	0.8	0.9	0.9	1.0	1.2	1.2	1.3	n.a.	n.a.	n.a.
Total		9.0	9.7	9.5	11.3	12.4	12.6	13.9	14.7	14.6	15.0	15.2	15.1	17.3	20.5

Sources: *China Petrochemical Yearbook, 1991, 1994, 1997*, China's Petrochemical Press.

Table A3.48 Length and Traffic of Petroleum and Gas Pipelines (End of 1996)

	Number of Pipelines	Length of Pipelines (km)	Capacity of Pipeline Traffic (million tons)	Pipeline Traffic (million tons or billion cu.m) kilometer	Ton-kilometer or Cubic Meter (million ton-km or billion cu.m-km)
Lanzhou Refinery	9.00	36.20	5.00	0.18	0.65
Qilu Petrochemical Co.	1.00	3.00	0.60	0.88	2.65
Jihua Refinery	1.00	11.00	0.30	0.02	0.23
Natural Gas Pipelines	204.00	8909.90	10.83	10.51	1999.84
Pipeline Bureau of Ministry of Petroleum	1.00	361.89	0.40	0.31	113.82
Sichuan Petroleum Administration Bureau	125.00	5686.76	1.96	7.02	1730.39
Daqing Petroleum Administration Bureau	7.00	208.70	0.60	0.68	16.54
Shengli Petroleum Administration Bureau	1.00	142.60	0.16	0.26	36.85
North China Petroleum Administration Bureau	6.00	102.93	2.83	0.18	10.90
Liaohe Petroleum Prospecting Bureau	21.00	816.22	1.00	0.70	16.79
Dagang Petroleum Administration Bureau	2.00	132.07	0.93	0.29	18.37
Zhongyuan Petroleum Prospecting Bureau		416.61	0.12	0.38	23.76
Jilin Provincial Oil Fields Administration Bureau	2.00	56.34	0.78	0.03	1.00
Yunna-Guizhou-Guangxi Petroleum Prospecting Bureau	5.00	18.63	0.11	0.12	0.06
Jianghan Petroleum Administration Bureau	3.00	12.50	0.60	0.05	0.22
Xinjiang Petroleum Administration Bureau	6.00	441.90	0.05	0.36	27.33
Qinghai Petroleum Administration Bureau	3.00	16.46	0.40		
Changqing Petroleum Prospecting Bureau	2.00	112.06		0.05	2.59
Eastern Hebei Petroleum Prospecting and Exploiting Co.	1.00	15.10			
Turpan-Hami Petroleum Prospecting Headquarters	6.00	369.13	0.90	0.08	1.23
Other Gas Pipelines	27.00	261.30	3.31	0.62	6.19
Yanshan Petrochemical Co.	1.00	3.00	0.21	0.14	0.43
Tianjin Petrochemical Co.	2.00	35.00	0.67	0.05	0.72
Cangzhou Refinery	1.00	0.50	0.10	0.01	0.00
Dalian West Pacific Ocean Ltd.	1.00	6.00	0.12	0.02	0.14
Anshan Refinery	1.00	2.70	0.05	0.00	0.01
Jinzhou Petrochemical Co.	1.00	3.00	0.03	0.02	0.05
Jinxi Refinery & Petrochemical Corporation	1.00	56.50	0.06		
Jinling Petrochemical Corporation	1.00	16.00	0.06	0.04	0.68
Juijiang Petrochemical Corporation		4.00	0.50	0.02	0.09
Zhenhai Petrochemical Corporation	4.00	26.60	0.59	0.05	0.36
Wuhan Petroleum Factory	2.00	22.60	0.16	0.03	0.47
Lanzhou Refinery	1.00	2.00	0.01	0.00	0.01
Jingmen Refinery	1.00	3.00	0.01	0.01	0.02
Shijiazhuang Refinery	2.00	36.50	0.34	0.03	0.94
Guangzhou Petrochemical corporation	1.00	0.50	0.01	0.00	0.00
Maoming Petrochemical Co.	5.00	21.40	0.38	0.17	1.84
Qilu Petrochemical Co.	1.00	22.00	0.01	0.02	0.46

Table A3.48 (Continued)

	Number of Pipelines	Length of Pipelines (km)	Capacity of Pipeline Traffic (million tons)	Pipeline Traffic (million tons or billion cu.m) kilometer	Ton-kilometer or Cubic Meter (million ton-km or billion cu.m-km)
Total	419.00	19,261.36	310.06	159.92	58,465.31
Petroleum Pipelines	188.00	10,090.16	295.93	148.79	56,459.27
Gas Pipelines	231.00	9,171.20	14.14	11.12	20,006.03
Crude Oil Pipelines	84.00	8698.27	245.94	135.01	56,110.58
China National Petroleum Pipeline Bureau	12.00	4485.49	112.30	65.16	50,123.75
Daqing Petroleum Administration Bureau	8.00	111.65	7.97	8.35	80.61
Xinjiang Petroleum Administration Bureau	5.00	928.35	16.45	12.18	1599.98
Jianghan Petroleum Administration Bureau	3.00	118.30	2.80	0.51	38.15
Shengli Petroleum Administration Bureau	6.00	421.59	20.60	9.70	717.18
Jilin Provincial Oil Fields Administration Bureau	4.00	24.46	2.40	2.40	19.16
North China Petroleum Administration Bureau	7.00	733.62	15.28	2.70	562.98
Dagang Petroleum Administration Bureau	4.00	80.80	12.00	3.94	91.92
Yumen Petroleum Prospecting Bureau	4.00	31.75	4.20	1.40	7.13
Liaohe Petroleum Prospecting Bureau	8.00	466.30	13.90	10.79	1096.60
Qinghai Petroleum Adiminstration Bureau	4.00	448.62	2.62	2.58	547.65
Henan Petroleum Prospecing Bureau	3.00	92.70	4.00	1.74	56.78
Maoming Petroleum Industrial Co.	2.00	165.16	17.00	7.17	602.71
Changqing Petroleum Prospecging Bueau	3.00	150.85	1.75	0.30	13.73
Zhongyuan Petroleum Prospecing Bureau	1.00	3.64	0.40	0.39	1.05
Turpan-Hami Petroleum Prospecing Headquarters	8.00	81.67	3.28	2.79	26.75
Tarim Prospecting Exploration Headquarters	2.00	353.32	9.00	2.92	524.46
Refined Oil Pipelines	104.00	1391.89	49.98	13.78	348.69
Yanshan Petrochemical Co.	7.00	25.80	4.69	0.64	3.10
Tianjin Petrochemical Co.	6.00	28.27	3.36	0.79	2.47
Cangzhou Refinery	4.00	12.90	0.71	0.10	0.43
Fushun Petrochemical Co.	9.00	826.70	4.63	1.30	247.99
Jinzhou Petrochemical Co.	5.00	93.91	2.83	1.54	33.91
Jinxi Refinery & Petrochemical Corporation	2.00	3.80	1.31	0.02	0.08
Anshan Refinery	2.00	6.12	4.81	0.56	2.38
Dalian Petrochemical Co.	2.00	10.50	0.80	0.37	1.88
Dalian West Picific Ocean Ltd.	6.00	37.60	4.22	0.47	2.80
Daqing Petrochemical Corporation	7.00	51.54	1.89	0.75	0.85
Zhenhai Petrochemical Corporation	5.00	46.50	1.13	0.39	3.29
Juijiang Refinery	4.00	18.73	2.69	0.04	0.18
Jingmen Refinery		11.90	0.56	0.19	0.37
Wuhan Petrochemical Factory	4.00	15.50	3.49	0.40	1.57
Anqing Petrochemical Corporation	3.00	2.10	0.11	0.28	0.23
Guangzhou Petrochemical corporation	13.00	50.31	3.23	3.07	10.49
Maoming Petroleum Industrial Co.	9.00	99.51	3.63	1.80	33.16

Source: China Statistical Yearbook, 1997.

Appendix 4
Oil Regulations and Official Contracts

I. Regulations of the People's Republic of China Concerning the Exploitation of Onshore Petroleum Resources in Cooperation with Foreign Enterprises

(Promulgated by the State Council on, and effective from, October 7, 1993)

Chapter 1 General provisions

Article 1. These Regulations are formulated in order to safeguard the development of the petroleum industry and to promote international economic cooperation and technical cooperation.

Article 2. These Regulations must be observed in all activities involving Chinese–foreign cooperative exploitation of onshore petroleum resources in the People's Republic of China.

Article 3. Ownership of the petroleum resources within the territory of the People's Republic of China shall vest in the State of the People's Republic of China.

Article 4. The government of China protects according to law the co-operative exploitation activities, investments, profits and other lawful rights and interest of foreign enterprises participating in the cooperative exploitation of onshore petroleum resources in the People's Republic of China.

Activities in the People's Republic of China involving the cooperative exploitation of onshore petroleum resources must comply with the relevant laws, rules and regulations of the People's Republic of China and shall be subject to supervision and control by the relevant authorities of the Chinese government.

Article 5. The State will not requisition the investments and gains of foreign enterprises participating in the cooperative exploitation of onshore petroleum resources. In special circumstances, when the public interest requires, part or all of the petroleum due to the foreign enterprise in connection with cooperative exploitation may be requisitioned according to legal procedures. In case of such requisitioning appropriate compensation shall be paid.

Article 6. A department or unit authorized by the State Council shall be responsible for demarcating cooperation blocks, determining the forms of cooperation, arranging for the formulating of relevant plans and policies, and examining and approving overall development programs for oil/gasfields in cooperation with foreign enterprises, for the cooperation areas approved by the State Council.

Article 7. China National Petroleum Corporation (CNPC) shall be responsible for business matters in respect of the exploitation of onshore petroleum resources in cooperation with foreign enterprises, and for negotiating, entering into and implementing contracts for the cooperative exploitation of onshore petroleum resources with foreign enterprises. CNPC shall have the exclusive right to engage in petroleum exploration, development and production in cooperation with foreign enterprises in areas approved by the State Council for exploitation of onshore petroleum resources in cooperation with foreign enterprises.

Article 8. Following invitations to bid or negotiations, CNPC shall enter into contracts with foreign enterprises for the cooperative exploitation of onshore petroleum resources in the cooperation blocks demarcated in the areas approved by the State Council for exploitation of onshore petroleum resources in cooperation with foreign enterprises. Such contracts shall be formed only after they have been approved by the Ministry of Foreign Trade and Economic Cooperation of the People's Republic of China.

In addition to the contracts referred in the preceding paragraph, CNPC may also enter into other cooperation contracts with foreign enterprises in respect of areas approved by the State Council for cooperative exploitation of onshore petroleum resources. Such contracts must be submitted to the Ministry of Foreign Trade and Economic Cooperation of the People's Republic of China for the record.

Article 9. After a block has been declared open to foreign cooperation, CNPC may engage in cooperative exploitation of onshore petroleum resources with a foreign enterprises and, other than that, no enterprise may enter such block to prospect for petroleum or may enter into an economic or technical cooperation agreement with a foreign enterprise for petroleum exploitation in such block.

Any enterprise that has entered and is prospecting for petroleum (in the area appraisal prospecting stage) in a block declared open to foreign cooperation prior to the time such declaration has been made shall withdraw from such block when CNPC and the foreign enterprise have entered into a contract. CNPC shall be responsible for selling the prospecting data obtained by such withdrawing enterprise, so as to compensate the enterprise for its investment in an appropriate manner. After an oil/gasfield with commercial exploitation value is discovered in such block, the enterprise that withdraws from the block may participate in development by investing.

A department or unit authorized by the State Council shall periodically adjust the blocks, which have been determined, as open blocks based on the circumstances of entry into and implementation of the contracts.

Article 10. In the exploitation of onshore petroleum resources in cooperation with foreign enterprises, the principle of taking both central and local interests into consideration shall be observed. Appropriate consideration shall be given to local interests by methods such as attracting funds from the place where an oil/gasfield with commercial exploitation value is located for investment in the development of such field.

The relevant local People's governments shall protect normal production and business activities carried out in cooperation areas according to law and shall render effective assistance in respect of land use, road access, living amenities, etc.

Article 11. In the cooperative exploitation of onshore petroleum resources, taxes shall be paid according to law; in addition, royalties shall be paid.

Employees of enterprises engaged in the cooperative exploitation of onshore petroleum resources shall pay tax on their income according to law.

Article 12. Duties and taxes on equipment and materials imported for the purpose of implementing contracts shall be reduced or exempted, or other preferential treatment shall be granted in respect of such duties and taxes, in accordance with relevant regulations of the State. The measures therefore shall be formulated by the Ministry of Finance in conjunction with the General Administration of Customs.

Chapter 2 Rights and obligations of foreign contractors

Article 13. To engage in cooperative exploitation of onshore petroleum resources, CNPC and foreign enterprises must conclude contracts. Unless laws, regulations or the contract provide otherwise, the foreign enterprise entering to the contract (hereafter referred to as "Foreign contractor") shall be solely provide the investment to carry out exploration, be responsible for the exploration operations and bear all exploration risks. After the discovery of an oil/gasfield with commercial exploitation value, the Foreign contractor and CNPC shall jointly invest in cooperative development. The Foreign contractor shall undertake the development and production operations, until production operations are taken over by CNPC as stipulated in the contract.

Article 14. Foreign contractors may, in accordance with the provisions of the contract, recover their investment and expenses and obtain remuneration out of the petroleum produced.

Article 15. Foreign contractors may transport the petroleum due to them and the petroleum purchased by them out of the country, and they may also remit the investment recovered by them, their profits and their other lawful gains out of the country, in accordance with the relevant regulations of the State and the provisions of the contractors.

If a foreign contractor sells the petroleum due to it within the territory of the People's Republic of China, such petroleum shall be purchased exclusively by CNPC.

Article 16. Foreign contractors shall open accounts with banks within the territory of the People's Republic of China that have obtained approval to engage in foreign exchange business, and shall abide by State regulations on exchange control.

Article 17. Foreign contractors shall establish branches or representative offices within the territory of the People's Republic of China according to law.

Article 18. In implementing the contracts, foreign contractors shall employ advanced technology and management experience. In addition, they shall transfer technology and pass on experience to Chinese personnel and train such personnel.

In their petroleum operations, foreign contractors shall gradually increase the ratio of Chinese personnel used.

Article 19. In the course of implementing the contracts, foreign contractors shall promptly and accurately report on the petroleum operations to CNPC, obtain complete and accurate data, records, samples, vouchers and other original information in respect of technical, economic, financial, accounting and administrative aspects to CNPC in accordance with regulations.

Article 20. After a foreign contractor has been compensated for its investment in accordance with the contract or after the production period of the oil/gasfield in question has expired, title to all assets purchased and manufactured by the foreign contractor according to the plan and the budget for implementation of the contracts, except for equipment leased from third parties, shall vest in CNPC. During the term of the contract, the foreign contractor may use such assets in accordance with the contract.

Chapter 3 Petroleum operations

Article 21. The operations must formulate overall development programs for oil/gasfields in accordance with the regulations of the State concerning the exploitation of petroleum resources, and shall carry out development and production operations after such programs have been approved by a department or unit authorized by the State Council.

Article 22. Given equal conditions in and outside China, the operators shall to the extent possible give preference to purchase in China of the machinery, equipment, raw materials, accessories, means of transportation and office articles required for the petroleum operations.

With respect to the personnel and subcontractors required for the petroleum operations, the operators shall give preference to the employment of Chinese citizens and subcontractors, as stipulated in the contracts.

Article 23. In carrying out petroleum operations, operators and subcontractors shall observe State laws, regulations and standards concerning environmental protecting and operational safety, carry out their operations according to international practice, protect farmland, aquatic resources, forest reserves and other natural resources, and prevent pollution of and damage to the atmosphere, oceans, rivers, streams, lakes, groundwater and other land environment.

Article 24. The use of land in the course of petroleum operations shall be handled in accordance with the law of the People's Republic of China on the Administration of Land and other relevant regulations of the State.

Article 25. Title to all data, records, samples, vouchers and other original information in respect of all petroleum operations as referred to in Article 19 hereof shall vest in CNPC.

The use, assignment, donation, exchange, sale, publication, and transport or transmission out of the People's Republic of China, of data, records, samples, vouchers and other original information as mentioned in the preceding paragraph must be in accordance with the relevant regulations of the State.

Chapter 4 Settlement of disputes

Article 26. Any dispute between the parties to a contract for the cooperative exploitation of onshore petroleum resources that arises from the implementation of the contract shall be resolved through consultations or mediation. If the parities are not willing to resolve the dispute through consultations or mediation or if consultation or mediation is unsuccessful, the dispute may be submitted for arbitration by a Chinese arbitration institution or another arbitration institution in accordance with the arbitration Article in the contract or a written arbitration agreement entered into subsequently.

If the parties have neither included an arbitration Article in their contract nor reached a subsequent written arbitration agreement, proceedings may be instituted in a People's Court of China.

Chapter 5 Legal liability

Article 27. Anyone who violates these Regulations in any of the following ways may be warned and ordered to rectify the situation within a prescribed time limit by a department or unit authorized by the State Council and, if the situation is not rectified within the prescribed time limit, be ordered to cease petroleum operations; if the violation constitutes a criminal offense, criminal liability shall be pursued according to law:

Acting in violation of the first paragraph of Article 9 hereof by entering without authorization a block opened to foreign cooperation to prospect for petroleum or by signing without authorization a cooperation agreement with a foreign enterprise for petroleum exploitation in a such block;

Acting in violation of Article 19 hereof by, in the course of implementing the contract, failing promptly and accurately to report on the petroleum operations to CNPC or failing to submit information, samples and various reports in respect of technical, economic, financial, accounting and administrative aspects to CNPC in accordance with regulations;

Acting in violation of Article 21 hereof by carrying out development and production operations on one's own authority without having obtained approval for the overall development program for the oil/gasfield; or

Acting in violation of the second paragraph of Article 25 hereof by making unauthorized use of data, records, samples, vouchers or other original information on petroleum operations, or by assigning, donating, exchanging, selling, publishing the same without authorization, or by transporting or transmitting the same out of the People's Republic of China without authorization.

Article 28. Anyone who violates Article 11, 16, 23 or 24 hereof shall be punished by the relevant department-in-charge of the State in accordance with relevant laws and regulations. If the violation constitutes a criminal offense, criminal liability shall be pursued according to law.

Chapter 6 Supplementary Provisions

Article 29. For the purpose of these Regulations, the following terms shall have the meaning assigned to them below:

The term "petroleum" shall mean underground crude oil and natural gas that is being or has been extracted;

The term "onshore petroleum resources" shall mean underground petroleum resources anywhere within the onshore area (including sea beaches, islands and marine areas extending from the onshore area up to a water depth of 5 meters);

The term "exploitation" shall mean the exploration for and development, production and sale of petroleum as well as activities in connection therewith;

The term "petroleum operations" shall mean exploration, development and production operations carried out in order to implement a contract, as well as activities in connection therewith;

The term "exploration operations" shall mean all work carried out to find oil-bearing traps by various means such as geological, geophysical and geochemical means, including the drilling of exploration wells, as well as all work carried out to determine whether a discovered petroleum trap has commercial value, such as the drilling of appraisal wells, feasibility studies and preparation of overall development programs for the oil/gasfield;

The term "development operations" shall mean all designing, manufacturing, installation and drilling projects, and the corresponding research, carried out as from the date of approval of the overall development program for the oil/gasfield for the purpose of realizing petroleum production, including production activities carried out prior to the commencement of commercial production;

The term "production operations" carried shall means all operations out for the purpose of petroleum production as from the date of commencement of the commercial production of an oil/gasfield, as well as all activities in connection therewith.

Article 30. Articles 4, 11, 12, 15, 16, 17 and 22 hereof shall be applicable to foreign subcontractors.

Article 31. These Regulations shall be implemented as from the date of promulgation.

II. Regulations of the People's Republic of China on the Exploitation of Offshore Petroleum Resources in Cooperation with Foreign Enterprises

(Adopted at the Regular Session of the State Council on January 12, 1982; promulgated by the State Council on January 30, 1982)

Chapter 1 General provisions

Article 1. In the interests of developing the national economy and expanding international economic and technological cooperation, these Regulations are formulated, on the premise of safeguarding national sovereignty and economic interests, to permit foreign enterprises to participate in the cooperative exploitation of offshore petroleum resources of the People's Republic of China.

Article 2. All petroleum resources in the internal water, territorial sea and continental shelf of the People's Republic of China and in all sea areas within the limits of national jurisdiction over the maritime resources of the People's Republic of China are owned by the People's Republic of China.

In the sea areas referred to in the preceding paragraph, all buildings and structures set up and vessels operating to exploit petroleum, as well as the corresponding onshore oil (gas) terminals and bases shall be under the jurisdiction of the People's Republic of China.

Article 3. The government of the People's Republic of China shall protect, in accordance with the law, the investments of foreign enterprises participating in the cooperative exploitation of offshore petroleum resources, the profits due to them and their other legitimate rights and interests, and shall protect, in accordance with the law, the cooperative exploitation activities of foreign enterprises.

All activities for the cooperative exploitation of offshore petroleum resources within the scope of these Regulations shall be subject to the laws and decrees of the People's Republic of China and relevant provisions of the State; all persons and enterprises taking part in petroleum operations shall be subject to the laws of China and shall accept inspection and supervision by the competent authorities concerned of the Chinese government.

Article 4. The Ministry of Petroleum Industry of the People's Republic of China shall be the competent authority in charge of the exploitation of offshore petroleum resources in cooperation with foreign enterprises, and shall determine the forms of cooperation and demarcate areas of cooperation in accordance with the zones and the surface areas of cooperation designated by the State; it shall work out a plan for the exploitation of offshore petroleum resources in cooperation with the foreign enterprises in accordance with long-term State economic plans, formulate operation and management policies for the cooperative exploitation of offshore petroleum resources and examine and approve the overall development program for offshore oil (gas) fields.

Article 5. The China National Offshore Oil Corporation (CNOOC) shall have exclusive and overall responsibility for the work of exploiting offshore petroleum resources in the People's Republic of China in cooperation with foreign enterprises.

CNOOC shall be a State corporation with the qualifications of a juridical person and shall have the exclusive right to explore for, develop, produce and market the petroleum within the zones of cooperation with foreign enterprises.

CNOOC may, as the work requires, establish regional corporations, specialized corporations and overseas representative offices to carry out the tasks delegated by the head office.

Article 6. CNOOC shall, by means of calling for bids and signing petroleum contracts, cooperate with foreign enterprises to exploit petroleum resources in accordance with the zones, surface areas and areas of cooperation with foreign enterprises for the exploitation of petroleum resources.

The petroleum contracts referred to in the preceding paragraph shall come into force after approval by the Foreign Investment Commission of the People's Republic of China.

All documents signed by CNOOC for other forms of cooperative exploitation of petroleum resources utilizing technology and funds provided by foreign enterprises shall also be subject to approval by the Foreign Investment Commission of the People's Republic of China.

Chapter 2 Rights and obligation of the parties to petroleum contracts

Article 7. CNOOC shall cooperate with foreign enterprises to exploit offshore petroleum resources by means of entering into petroleum contracts, and, unless otherwise specified by the Ministry of Petroleum Industry or in a petroleum contract, the foreign enterprise that is one party to the petroleum contract (hereafter "foreign contractor") shall provide the investment to carry out exploration, be responsible for exploration operations and bear all exploration risks; after a commercial oil (gas) field is discovered, both the foreign contractor and CNOOC shall provide the investment for its cooperative development, and the foreign contractor shall be responsible for the development operations and production operations until CNOOC takes over the production operations when conditions permit as provided in the petroleum contract, the foreign contractor, in accordance with the provisions of the petroleum contract, may recover its investment and expenses and receive remuneration out of the petroleum produced.

Article 8. The foreign contractor may export the petroleum due to it and the petroleum it purchases, and may also remit abroad, in accordance with the law, the investment it recovers, its profits and its other legitimate income.

Article 9. All Chinese enterprises and foreign enterprises participating in the cooperative exploitation of offshore petroleum resources shall pay taxes in accordance with the law and shall pay mining royalties.

All employees of the enterprises referred to in the preceding paragraph shall pay individual income tax in accordance with the law.

Article 10. The equipment and material imported for carrying out the petroleum contract shall be subject to tax at a reduced rate, or exempted from tax, or given other preferential tax treatment in accordance with State provisions.

Article 11. The foreign contractor shall open a bank account in accordance with the provisions of the Provisional Regulations on Foreign Exchange Control of the People's Republic of China.

Article 12. In carrying out the petroleum contract, the foreign contractor shall use appropriate and advanced technology and management experience and shall be obliged to transfer the technology and pass on the experience to the personnel of the Chinese side involved in carrying out the petroleum contract (hereafter "Chinese personnel"); in petroleum operations, the foreign contractor must give preference in employment to Chinese personnel, progressively increase the percentage of Chinese personnel and train Chinese personnel in a planned way.

Article 13. In carrying out the petroleum contract, the foreign contractor must promptly and accurately report to CNOOC on the situation of petroleum operations; and it must acquire complete and accurate date, records, samples, vouchers and other original data with respect to various aspects of the petroleum operations, and regularly submit to CNOOC necessary data and samples as well as various technological, economic, financial and accounting, and administrative reports.

Article 14. In carrying out the petroleum contract, the foreign contractor shall establish a branch or subsidiary or representative office within the territory of the People's Republic of China and fulfilf registration formalities in accordance with the law.

The domiciles of the offices referred to in the preceding paragraph shall be determined through consultation with CNOOC.

Article 15. The provisions of Articles 3, 8, 9, 10 and 14 of these Regulations shall apply, by analogy, to foreign subcontractors that render services in connection with the petroleum operations.

Chapter 3 Petroleum operations

Article 16. In order to achieve the highest possible oil recovery factor, the operator must, in accordance with these Regulations and the relevant provisions promulgated by the Ministry of Petroleum Industry on the exploitation of petroleum resources and by taking account of international practice, formulate an overall development program for the oil (gas) field and implement production operations.

Article 17. In carrying out the petroleum contract, the foreign contractor shall use the existing bases within the territory of the People's Republic of China, and, if new bases are needed, they must be established within the territory of the People's Republic of China.

The specific locations of the new bases referred to in the preceding paragraph, and other arrangements that may be necessary in special circumstances, must all be subject to the written approval of CNOOC.

Article 18. CNOOC shall have the right to send personnel to join the foreign operator in making master designs and engineering designs for carrying out the petroleum contract. Design corporations within the territory of the People's Republic of China shall have priority in entering into subcontracts for the master designs and engineering designs mentioned above, provided that their terms are competitive.

Article 19. With respect to all facilities required to be built in carrying out the petroleum contract, including artificial islands, platforms, buildings, and structures, when signing subcontracts, the operator must give preference to manufacturing plants and engineering corporations within the territory of the People's Republic of China, provided that they are competitive in terms of quality, price, term of delivery and services.

Article 20. With respect to the equipment and materials required to carry out the petroleum contract, the operator and subcontractors must give preference to procuring and utilizing equipment and materials manufactured and supplied by the People's Republic of china, provided that these are competitive.

Article 21. With respect to services that are required to carry out the petroleum contract, such as those for geophysical prospecting, well-drilling, diving, aircraft, ships and bases, the operator and subcontractors must enter into subcontracts and service contracts with relevant departments within the territory of the People's Republic of China, provided that they are competitive in terms of price, efficiency and services.

Article 22. The ownership of all assets purchased or built by the foreign contractor to carry out the petroleum contract in accordance with the plan and budget, excluding equipment leased from a third party, shall belong to CNOOC after the foreign contractor's investment has been compensated as provided for, and, within the term of the contract, the foreign contractor may continue to use those assets in accordance with the provisions of the contract.

Article 23. CNOOC shall have the ownership of all of the data, records, samples, vouchers and other original data with respect to the petroleum operations stipulated in Article 13 of these Regulations.

The utilization and transfer, donation, exchange, sale and publication of the previously mentioned data, records, samples, vouchers and other original data and their export and transmission from the People's Republic of China all must be conducted in accordance with the "Provisions for the Control of Data" formulated by the Ministry of Petroleum Industry.

Article 24. In the course of implementing petroleum operations, the operator and subcontractors shall comply with the relevant laws and provisions on environmental protection and safety of the People's Republic of China, and shall, by taking account of international practice when conducting operations, protect fishery resources and other natural resources and prevent the environment, including the air, sea, rivers, lakes and land, from being polluted or damaged.

Article 25. The petroleum produced within the petroleum contract area shall be landed in the People's Republic of China or may be exported from oil (gas) metering points on offshore terminals. In case such petroleum has to be landed at a point outside the People's Republic of China, the approval of the Ministry of Petroleum Industry must be obtained.

Article 26. In case of war, threat of war or other state of emergency, the Chinese government shall have the right of compulsory purchase or requisition with respect to a portion or all of the petroleum earned or purchased by the foreign contractor.

Chapter 4 Supplementary principles

Article 27. Any dispute arising between foreign and Chinese enterprises during the cooperative exploitation of offshore petroleum resources shall be settled through friendly consultations. If it cannot be resolved through consultation, mediation and arbitration may be conducted by an arbitration body of the People's Republic of China, or the parties to the contract may agree upon arbitration by another arbitration body.

Article 28. In case an operator or subcontractor violates the provisions of these Regulations in implementing petroleum operations, the Ministry of Petroleum Industry shall have the right to issue a warning and set a deadline for correction. If no correction can be made prior to the specified deadline, the Ministry of Petroleum Industry shall have the right to adopt necessary measures, even to the extent of suspending implementation of petroleum operation. All economic losses caused as a result of this shall be borne by the responsible party.

A party responsible for serious violation of these Regulations may be fined by and/or even be subject to suit before the judicial authorities filed by the Ministry of Petroleum Industry.

Article 29. The terms used in these Regulations shall have the following definitions:

(1) "Petroleum" means crude oil or natural gas deposited underground, currently being extracted or already extracted.

(2) "Exploitation" means, in general, the exploration for and development, production and marketing of petroleum, as well as other related activities.
(3) "Petroleum contract" means a contract signed, in accordance with the law, between CNOOC and foreign enterprises for the cooperative exploitation of offshore petroleum resources of the People's Republic of China, including the exploration for and development and production of petroleum.
(4) "Contract area" means a surface area designated within a sea area demarcated by geographical coordinates in the petroleum contract for the cooperative exploitation of petroleum resources.
(5) "Petroleum operations" means all exploration, development and production operations and other related activities conducted in carrying out the petroleum contract.
(6) "Exploration operations" means all work done to locate the petroleum-bearing traps by means of geological, geophysical and geochemical methods and including drilling exploratory wells, etc., and all work done to determine the commerciality of discovered petroleum traps, including appraisal drilling, feasibility studies and preparation of the overall development program for an oil (gas) field.
(7) "Development operations" means projects, such as those for design, construction, installation and drilling, and corresponding research work, conducted from the date of the approval of the overall development program for an oil (gas) filed by the Ministry of Petroleum Industry, in order to bring about petroleum production, including production activities carried out before the commencement of commercial production.
(8) "Production operations" means all operations for producing petroleum conducted after the date of commencement of the commercial production of an oil (gas) field and related activities, such as extraction, injection, production stimulation, processing, storage and transportation and lifting of petroleum and other operations.
(9) "Foreign contractor" means a foreign enterprise that signs a petroleum contract with CNOOC. The foreign enterprise may be a corporation or a consortium of corporations.
(10) "Operator" means an entity that is responsible for implementing the operations pursuant to the provisions of the petroleum contract.
(11) "Subcontractor" means an entity that renders services to the operator.

Article 30. Rules for the implementation of these Regulations shall be formulated by the Ministry of Petroleum Industry.

Article 31. These Regulations shall come into force on the day of promulgation.

III. CNPC'S Contract Terms for 3rd Onshore Bidding Round

1. Parties to the contract

The parties to the contract shall be

(I) China National Petroleum Corporation (CNPC), and
(II) Foreign company

2. Type of contract

The contract shall be a production-sharing contract.

3. Contract area

The contract area shall consist of basic blocks, and shall be relinquished stage by stage.

4. Contract term

The contract term shall include three periods: exploration period, development period and production period. The exploration period is 8 years divided into three phases: the 1st phase is 4 years; the 2nd phase is 2 years; and the 3rd phase is 2 years. The foreign company will have the option to terminate the contract at the end of each exploration phase. The production period shall not exceed 15 years. The contract term shall not exceed 30 years.

5. Relinquishment

At the end of the 1st exploration phase, the foreign company shall relinquish 40% of the original contract area. At the end of the 2nd exploration phase, the foreign company shall relinquish 25% of the contract area after deducting each development area and or production area. At the end of the 3rd phase, the foreign company shall relinquish all areas except those converted by an oil and or gasfield.

6. Minimum exploration work commitment

The foreign company shall bid the minimum exploration work program in terms of line km of 2-D seismic surveys and number of wildcats (with stated minimum objective depths).

7. Joint management committee (JMC)

For the purpose of the proper performance of the petroleum operations, CNPC and the foreign company will establish JMC. Any decision of JMC shall be made unanimously through consultation. JMC will have the right to review the work program and budget, determine commerciality of discovery, adopt overall development program and budget, approve or confirm important procurement items, etc.

8. Operator

The foreign company will be the operator for exploration, development and production operations; CNPC has the right to take over the production operations after recovery of the development costs by the foreign company. CNPC may take over the production operations upon an agreement reached through consultation of JMC.

9. Financing

The foreign company shall pay 100% of the exploration costs. Development costs and operation costs shall be provided by CNPC and the foreign company in proportion to their respective participating interests: 51% by CNPC, 49% by the foreign company, unless CNPC decides not to participate or to participate to an extent of less than 51%.

10. Cost recovery

Sixty percent of the annual gross production of crude oil will be used for cost recovery. The exploration costs shall be recovered from the annual gross production of crude oil of the contract area. The development costs with its deemed interest and the operating costs shall be recovered only from the annual gross production of crude oil of the oilfield in which the costs were incurred.

11. Taxes and duties

The foreign company will be required to make a royalty payment on production and is exempt from customs duty on the imports for petroleum operations under the contract. The foreign company will be subject to corporation income tax at 33% of taxable income arising from such contract and value added tax.

12. Remainder oil allocation

After the cost recovery and payments of value added tax and royalty respectively, the remainder of the annual gross production of crude oil (remainder oil) of an oilfield in the contract area will be divided into "share oil" of the Chinese side (1–X) and "allocable remainder oil". The remainder oil times one of the negotiable factors (X) in successive incremental tiers equals the allocable remainder oil which will be shared by both parties in proportion to their respective participating interests.

13. Valuation and adjustment

The valuation of crude oil produced will be made with reference to the prevailing price in arm's length transaction of similar quality crude oil on the main world markets and the adjustment in such price will be made in accordance with such determinants as the quality of the crude oil, the terms of delivery, transportation, payment and other terms.

14. Training of Chinese personnel and technology transfer

The foreign companies have the obligation to train Chinese personnel and transfer advanced technology and managerial experience to CNPC. All costs and expenses incurred shall be charged to the joint account and subject to the cost recovery.

15. Ownership of assets and data

All assets purchased and installed under the work program and budget for an oilfield and/or gasfield shall be owned by CNPC after the foreign company's development costs have been recovered or after the expiration of the production period. Within the term of the contract, the foreign company may continue to use those assets in accordance with the provisions of the contract. All data and samples obtained in the course of the petroleum operations shall vest in CNPC.

16. Associated natural gas

Any associated natural gas will be primarily used for operations. If the associated natural gas has no commercial value, it will be disposed of by the operator. If the associated natural gas has commercial value, CNPC and the foreign company will utilize it.

17. Non-associated natural gas

When CNPC and the foreign company consider a gas reservoir to be commercial, both parties will negotiate and reach an agreement, as a supplementary document to the contract. The price of natural gas will

Appendix 5
Addresses, Phone Numbers, and Fax Numbers of Oil Companies

Crude Producers:

China National Offshore Oil Corporation (CNOOC)
Jia 1 Sidaokou Rd., Haidian
Box 9623
Beijing
Tel: (010)62014906
Telex: 22611; 222301 CNOOC CN Cable: 3113

China National Petroleum Corporation (CNPC)
Liupukang
Beijing
Tel: (010)62095129, Fax: (010)62095126

Dagang Petroleum Administration
Dagang District
Tianjing
Tel: 5922524

Daqing Petroleum Administration
Daqing
Heilongjiang Province, 163003
Tel: 5995196
Telex: 81050 Cable: 2646

Liaohe Petroleum Administration
Puanjin
Liaoning Province
Tel: 7823403

Shengli Petroleum Administration
Dongying
Shendong Province
Tel: 8554436

Sichuan Petroleum Administration
Chendu
Sichuan Province
Tel: 3324911

Xinjiang Petroleum Administration
Karamay
Xinjiang Autonomous
Tel: 888077

Refiners:

China Petrochemical Corp. (Sinopec)
A6, Huixindong Street
Chaoyang District
Beijing 100029
Tel: (010)64225533, Fax: (010)64212429

Anqing General Petrochemical Works
Gaohuating, Anqing
Anhui Province 246001
Tel: (0556)5374133, Fax: (0556)5375852

Baling Petrochemical Company
Nanhu Revenue, Yueyang
Hunan Province 414000
Tel: (0730)8844512, Fax: (0730)8843613

Cangzhou Refinery
Beijiao, Cangzhou
Hebei Province 061000
Tel: (0317)2042221, Fax: (0317)2042688
Telex: 678013 CRCPC CN

Changcheng Premium-Grade Lube Oil Company
Qinghe, Haidian District
Beijing 100085
Tel: (010)62913781, Fax: (010)62913311

Dalian Petrochemical Company
Ganjingzhi District, Dalian
Liaoning Province 116032
Tel: (0411)6671371, Fax: (0411)6671415

Daqing General Petrochemical Works
Longfen District, Daqing
Heilongjiang Province 163714
Tel: (0459)6763028, Fax: (0459)6250208

Fushun Petrochemical Company
Xingfu District, Fushun
Liaoning Province 113008
Tel: (0413)262811, Fax: (0413)2629892

Gaoqiao Petrochemical Company
3000 Pudong Avenue
Shanghai 201208
Tel: (021)58711001, Fax: (021)58712207

Guangzhou General Petrochemical Works
Huangpu District, Guangzhou
Guangdong Province 510726
Tel: (020)82121088, Fax: (020)82399045

Harbin Refinery
Taiping District, Harbin
Heilongjiang Province 150056
Tel: (0451)2402522, Fax: (0451)2423657

Hubei Chemical Fertilizer Plant
Zhijiang
Hubei Province 443200
Tel: (0717)4212012, Fax: (0717)4212660

Jinan Refinery
Jinan
Shandong Province 250101
Tel: (0531)8987330, Fax: (0531)8984709

Jinling Petrochemical Company
Xunwu District, Nanjing
Jiangshu Province 210037
Tel: (025)5410567, Fax: (025)5410220

Jinmen General Petrochemical Works
Jinmen
Hubei Province 448039
Tel: (0724)2211306, Fax: (0724)2211539

Jinxi Petroleum Processing and Chemical Complex
Huludao
Liaoning Province 125001
Tel: (0429)2178015, Fax: (0429)2178199

Jinzhou Petrochemical Company
Chongqing Road, Jinzhou
Liaoning Province 121001
Tel: (0416)4153413, Fax: (0416)4167532

Jiujiang General Petrochemical Works
Juijiang
Jiangxi Province 332004
Tel: (0792)8224911, Fax: (0792)8227005

Lanzhou Petrochemical Company
Xigu District, Lanzhou
Gansu Province 730060
Tel: (0931)7555981, Fax: (0931)7555851

Lanzhou Petroleum Processing and Chemical Complex
Xigu District, Lanzhou
Gansu Province 730060
Tel: (0931)7556160, Fax: (0931)7554386

Liaoyang Petrochemical Fiber Company
Hongwei District, Liaoyang
Liaoning Province 111003
Tel: (0419)5152248, Fax: (0419)5154911

Linyuan Refinery
Daqing
Heilongjiang Province 163813
Tel: (0459)6712345, Fax: (0459)6713185

Luoyang General Petrochemical Works
Jili District, Luoyang
Henan Province 471012
Tel: (0379)6996114, Fax: (0379)6993999

Maoming Petrochemical Company
2 Hongqibeilu, Maoming
Guangdong Province 525011
Tel: (0668)2264248, Fax: (0668)2269317
Telex: 45032 MMPIC CN Cable: 3555

Ningxia Chemical Works
Xingshi District, Yingchuan
Ningxia 750026
Tel: (0951)2072361, Fax: (0951)2021379

Qianguo Refinery
Ningjiang District, Songyuan
Jilin Province 138008
Tel: (0438)2128111, Fax: (0438)2128429

Qilu Petrochemical Company
Zhibo
Shandong Province 255408
Tel: (0533)7180777, Fax: (0533)7180406

Shainghai Jinshan Industrial Company
Jinshan District
Shainghai 200540
Tel: (021)57942942, Fax: (021)57940560

Shijiazhuang Refinery
Dongjiao, Shijiazhuang
Hebei Province 050032
Tel: (0311)5162296, Fax: (0311)5161234

Sichuan Vinylon Plant
Changshou County
Chongqing 631254
Tel: (023)68974716, Fax: (023)68975999

Tianjing Petrochemical Company
Dagang District
Tianjing 300271
Tel: (022)25990393, Fax: (022)25997884

Urumqi General Petrochemical Works
Dongshan District, Urumqi
Xingjiang 830019
Tel: (0991)6891011, Fax: (0991)6896666

Wuhan Petrochemical Works
Qingshan District, Wuhan
Hubei Province 430082
Tel: (027)6515662, Fax: (027)6515962

Yangzi Petrochemical Company
Dachang District, Nanjing
Jiangshu Province 210048
Tel: (025)7791442, Fax: (025)7784389

Yanshan Petrochemical Company
Fangshan District
Beijing 102500
Tel: (010)69342295, Fax: (010)60342736

Yiping Grade Lube Oil Company
Jiulongpo District
Chongqing 630039
Tel: (023)68614966, Fax: (023)68603475

Joint-ventrue and Stock Issuing Oil Companies:

Fujian Petrochemical Co. Ltd.
Xiaocuo, Quanzhou
Fujian Province 362100
Tel: (0595)7789188, Fax: (0595)7789130

Shanghai Petrochemical Co. Ltd.
Jin Shan Wei
Shanghai 200540
Tel: (021)57941941, Fax: (021)57942267
Telex: 33484 SPC CN Cable: 7608

Zhenhai Petrochemical Co. Ltd.
Zhenhai District, Ningbo
Zhejiang Province 315207
Tel: (0574)6444447, Fax: (0574)6456155

Oil Trading Companies:

China International United Petroleum and Chemicals Corporation (Unipec)
Wangfujin Hotel
Beijing
Tel: (010)65223825, Fax: (010)65223817

China National United Oil Corp. (ChinaOil)
No. 135 Xi Zhi Men Wai Street
Beijing 100044
Tel: (010)6316633, Fax: (010)6350989
Telex: 22163 OIL CN

Sinochem International Oil Co.
Sinochem Tower
Fuxingmenwai Dajie
Beijing 100045
Tel: (010)6569684, Fax: (010)6569401

Bibliography and References

Chapter 1

Bornstein, Morris (1989) The comparison of economic systems: an integration, in: Morris Bornstein (ed.) *Comparative Economic Systems: Models and Cases*, Homewood, IL, pp. 3–17.
Chanda, Nayan, Rigoberto Tiglao, and John Mcbeth (1995) Territorial imperative, *Far Eastern Economic Review*, February 23, pp. 14–16.
The China Monitor (1993) The future of Sino–American relations, US China Research Associates, Maryland, January/February, pp. 1–4.
China's OGP (1995) Joint development billed as top priority at spratly, July 15, p. 7.
Gregory, Paul R., and Robert C. Stuart (1992) *Comparative Economic System*, Fourth Edition, Boston: Houghton Mifflin Company.
He, Ping (ed.) (1994) *China after Deng*, Hong Kong: Ming Jing Press.
Schwarz, Adam (1995) Where oil and water mix, *Far Eastern Economic Review*, March 16, pp. 54–58.
Tasker, Rodney (1995) A line in the sand, *Far Eastern Economic Review*, April 6, pp. 14–16.
US General Consul in Beijing (1994) US trade with China, May 16, by electronic mail.
Wang, Haijiang (1993) *Decentralization and Its Impact on Electric Power Sector; An Empirical Study for a Centrally Planned Economy: China*, Ph.D. dissertation, University of Pennsylvania.
Wang, Haijiang (1994) China's oil industry: the economic and political outlook, *Petroleum Politics*, Vol. 5, No. 2, 1994, pp. 55–83.
Wang, Haijiang (1995) Prospects of China's oil market, *Energy*, Summer, pp. 12–14.
Wang, Haijiang (1997) The perplexing dispute over oil, *Resources Policy*, Vol. 23, No. 4, pp. 173–178.
World Bank (1990) *China: Between Plan and Market*, World Bank Country Study.
World Bank (1990) *China: Macroeconomic Stability and Industrial Growth Under Decentralized Socialism*, World Bank Country Study.
Zhao, Hairui (1995) Dispute in the South China Sea and its prospects, (in Chinese), *China Spectrum*, May, pp. 21–23.

Chapter 2

China Oil, Gas & Petrochemicals (1994) Price and distribution reform in force, Vol. 2, No. 10, May 15.
Wang, Haijiang (1995) China's oil policy and its impact, *Energy Policy*, Vol. 23, No. 7, pp. 627–635.
Wang, Haijiang (1996) Evolution of China's oil import policy and its prospects, Chapter 4 of the book, *The China's Road Towards Modernization*, The Edwin Mellon Press, pp. 73–88.
Wang, Haijiang (1997) China's deregulation, expansion not enough to restore market balance, *Oil & Gas Journal*, September 29, pp. 33–39.

Chapter 3

Brogan, Chris (1993) Tantalizing Tarim to see foreign explorers, *Petroleum Economist*, April, p. 12–13.
Canada–China Trade Council (1989) *China's Oil and Gas Industry: Recent Developments and Priorities*, Toronto, Canada. pp. 5–15.
Dorain, James P. and Allen L. Clark (1987) China's energy resources: potential supply, problems, and Implications, *Energy Policy*, February, pp. 73–90.
Energy Information Administration (1987) *The Petroleum Resources of China*, US Department of Energy, Washington, DC.
Fridley, David (1991) *China's Oil Industry: A Decade in Review and Challenges Ahead*, Working paper, East-West Center, Honolulu, Hawaii.
Kim, Woodard (1985) *Development of China's Petroleum Industry: An Overview*, China Energy Ventures, Inc., Washington, DC.
Platt's Oilgram News, various issues between January 1993 and April 1997.
Wang, Haijiang (1995) The prospects of oil supply in China, *Energy Exploration & Exploitation*, Vol. 13, No. 6, pp. 631–648.
World Bank (1985) *China: The Energy Sector*, A World Bank country study, Washington, DC, pp. 109–124.
Zhang, Kang (1994) A forecast of the future production and reserves of Chinese oil and gas, *Guoji Shiyou Jingji (International Petroleum Economics)*, Vol. 2, No. 1, 1994, p. 18–22.

Chapter 4

China National Petrochemical Corporation (1994) *China National Petrochemical Corporation Year Book, 1994*, Sinopec Press, Beijing.
China Oil, Gas & Petrochemicals (1994) China's refiners face massive overhaul, expansion to meet demand growth, new crude slate, May 9, p. 36.
Fridley, David (1991) *China's Oil Industry: A Decade in Review and Challenges Ahead*, A Report of the China Energy Study, East-West Center, Hawaii, April.
Oil & Gas Journal (1994) China's refiners face massive overhaul, expansion to meet demand growth, new crude slate, May 9.
Ou, Canqi (1993) China's refining industry: current situation, future development and opportunity for foreign partners, Proceeding in China's Oil and Gas Investment Conference, September.
Wang, Haijiang (1995) Government will shape China's refining boom, *Oil & Gas Journal*, May 8, pp. 42–49.
Wang, Haijiang (1995) China: lack of refining capacity in booming southern provinces, *Petroleum Economist*, March, with George Beranek, pp. 10–11.
Wang, Haijiang (1997) China's deregulation, expansion not enough to restore market balance, *Oil & Gas Journal*, September 29, pp. 33–39.
Woodard, Kim (1985) Development of China's petroleum industry: an overview, *China's Petroleum Industry*.

Chapter 5

Energy Security Analysis, Inc. (1993) *The Global Gasoline Market to the Year 2003*, Washington DC
Energy Security Analysis, Inc. (1994) *Asia-Pacific Refining: Uncertain Capacity, Surging Product Demand, and Crude Oil Deficits*, Washington, DC

Lu, Yingzhong (1993) *Fueling One Billion: An Insider's Story of Chinese Energy Policy Development*, Washington Institute Press.

Wang, Haijiang (1994) Fuel oil demand, policy impact, and energy substitution, with Edward Krapels, paper presented at IGT conference of The Energy Modeling: Methodology, Models, and Applications, Atlanta, April.

Wang, Haijiang (1996) Petroleum product demand in China, *Energy Exploration & Exploitation*, Vol. 14, No. 3 & 4, pp. 283–293; also presented at *14th CERI International Oil and Gas Markets Conference*, Calgary, Canada, September 25–26, 1995.

Wang, Haijiang (1997) China's deregulation, expansion not enough to restore market balance, *Oil & Gas Journal*, September 29, pp. 34–39.

World Bank (1985) *China: energy sector*, A World Bank Country Economic Report, Washington, DC

Chapter 6

Energy Security Analysis, Inc. (1993) *Global Crude Oil Quality to the Year 2003*, A multi-client study, Washington, DC

Han, Jingcheng, and Gai Hongyu (1996), Auto fuel demand and supply in China, *Guoji Shiyou Jingji, International Petroleum Economics*, in Chinese, Vol. 4, No. 3, pp. 12–18.

Wang, Haijiang (1998) Oil quality and its environmental impact in China, *The Journal of Energy and Development*, Vol. 23, No. 1, pp. 15–28.

Wang, Xi and Robert F. Blomquist (1992) The developing environmental law and policy of the People's Republic of China: an introduction and appraisal, *Georgetown International Environment Law Review*, Vol. 5 (25), pp. 28–75.

Chapter 7

Energy Security Analysis, Inc. (1994) *Asia-Pacific Refining: Uncertain Capacity, Surging Product Demand, and Crude Oil Deficits*, Washington, DC

Hayashi, Taizo (1993) Petroleum product trade centering on Hong Kong and South China, *Energy in Japan*, No. 121, May, pp. 28–36.

Oil & Gas Journal (1994) China's new oil import status underpins world's most dynamic petroleum scene, May 9, pp. 33–35.

Tian, Chunrong (1994) An analysis of Chinese petroleum imports and Exports from 1989 to 1993, *Guoji Shiyou Jingji (International Petroleum Economics)*, Vol. 2, No. 3, 1994, pp. 1–5.

Wang, Haijiang (1995) Refined product import/export in China and their market impact, *Pacific & Asian Journal of Energy*, Vol. 5, No. 1, June, pp. 113–122.

Wang, Haijiang (1996) Evolution of China's oil import policy and its prospects, Chapter 4 of the book, *The China's Road Towards Modernization*, The Edwin Mellon Press, pp. 73–88.

US General Consul in Beijing (1994) US Trade with China, May 16, news from Electronic mail.

Chapter 8

Energy Security Analysis, Inc. (1994) *Asia-Pacific Refining: Uncertain Capacity, Surging Product Demand, and Crude Oil Deficits*, Washington, DC

Norton, Hugh E. (1991) East Asia now important factor in oil world' *Oil & Gas Journal*, October 21, pp. 65–67.

Wang, Haijiang (1995) China's impact on the world crude oil market, *The Journal of Energy and Development*, Vol. 19, No. 1, pp. 81–96.

Wang, Haijiang (1995) Refined product import/Export in China and their market impact, *Pacific & Asian Journal of Energy*, Vol. 5, No. 1, June, pp. 113–122.
Yamada, Kikuo (1993) Future oil demand in Japan and the Pacific Rim to the year 2000 and beyond, a speech in the Third Annual Conference of the Center for Global Energy Studies, London, April.

About the Author

Haijiang Henry Wang is Director of the research sciences group at Quality Strategies, Inc. where he is in charge of survey design, statistical analysis and modeling. Earlier, he was Manager of Pacific Analysis and Senior Analyst at Energy Security Analysis, Inc. where he authored the monthly *China Watch* report and direct data analysis, and econometric forecasting. He has also held the position of Division Chief and Deputy Manager of the Second Credit (Energy Credit) Department of the Bank of China and was affiliated with the World Bank.

He holds a B.A. in economics from the People's University in Beijing, an M.A. in the same discipline from the Chinese Academy of Social Sciences, and M.S. and Ph.D. degrees in energy management and policy from the University of Pennsylvania.

His publications have appeared in international journals, such as, *Energy Policy, Resource Policy, Petroleum Economist, Oil & Gas Journal, Pacific & Asian Journal of Energy, The Journal of Energy and Development, Energy,* and *Energy Exploration & Exploitation.*

Index

Regional Sales Offices

EUROPE, MIDDLE EAST ASIA and AFRICA (not FAR EAST)	Elsevier Science, Regional Sales Office, Customer Support Department, PO Box 211, 1000 AE Amsterdam, The Netherlands Tel: + 31 20 485 3757; Fax: + 31 20 485 3432 E-mail: nlinfo-f@elsevier.nl
AMERICAS (NORTH/ SOUTH/CENTRAL)	Elsevier Science, Regional Sales Office, Customer Support Department, PO Box 945, New York, NY 10159-0945 USA Tel: + 1 212 633 3730; Toll Free number for North-American customers: 1-888-4ES-INFO (437-4636); Fax; + 1 212 633 3680 E-mail: usinfo-f@elsevier.com
JAPAN	Elsevier Science, Regional Sales Office, Customer Support Department, 9-15 Higashi-Azabu, 1-chome, Minato-ku, Tokyo, 106-0044 Japan Tel: + 81 3 5561 5033; Fax: + 81 3 5561 5047 E-mail: info@elsevier.co.jp
FAR EAST and AUSTRALASIA (except JAPAN)	Elsevier Science, Regional Sales Office, Customer Support Department, No 1 Temasek Avenue, #17-01 Millenia Tower, Singapore 039192 Tel: + 65 434 3727; Fax: + 65 337 2230 E-mail: asiainfo@elsevier.com.sg